Karst without Boundaries

Selected papers on hydrogeology

23

Series Editor: Dr. Nick S. Robins
Editor-in-Chief IAH Book Series, British Geological Survey, Wallingford, UK

INTERNATIONAL ASSOCIATION OF HYDROGEOLOGISTS

Karst without Boundaries

Editors

Zoran Stevanović
Faculty of Mining and Geology, University of Belgrade, Belgrade, Serbia

Neven Krešić
Environment and Infrastructure, Amec Foster Wheeler, Kennesaw, GA, USA

Neno Kukurić
IGRAC, Delft, The Netherlands

CRC Press
Taylor & Francis Group
Boca Raton London New York Leiden

CRC Press is an imprint of the
Taylor & Francis Group, an **informa** business

A BALKEMA BOOK

Published by: CRC Press/Balkema
P.O. Box 11320, 2301 EH Leiden, The Netherlands
e-mail: Pub.NL@taylorandfrancis.com
www.crcpress.com – www.taylorandfrancis.com

First issued in paperback 2020

CRC Press/Balkema is an imprint of the Taylor & Francis Group, an informa business

© 2016 Taylor & Francis Group, London, UK

No claim to original U.S. Government works

ISBN 13: 978-0-367-57487-1 (pbk)
ISBN 13: 978-1-138-02968-2 (hbk)

Visit the Taylor & Francis Web site at
http://www.taylorandfrancis.com

and the CRC Press Web site at
http://www.crcpress.com

Typeset by Quick Sort (India) Pvt Ltd., Chennai, India

Library of Congress Cataloging in Publication Data

Table of contents

Preface

Karst covers more than 13% of the continental ice-free surface of our planet, with major coverage in the Middle East and Central Asia where it occupies approximately 23% of the land surface. According to some estimates more than 20% of the world's population is consuming groundwater originated from karst aquifers. At the same time, the management of karst aquifers and their water resources is more problematic compared to any other aquifer type due to uncertainties in defining their boundaries, often irregular drainage regime, and vulnerability to pollution. All these problems are closely related to high heterogeneity of karst aquifers and their locally very high permeability.

During recent decades the first systematic overview and inventory of transboundary aquifers and their internationally shared water resources has been made. This work led by UN agencies (UNESCO, UNECE, IGRAC) and the International Association of Hydrogeologists (IAH) underlined the challenges of international groundwater data availability and highlighted the importance of establishing proper monitoring systems as a prerequisite to the better assessment of shared water resources.

When it comes to karst transboundary aquifer systems there is a rough approximation that not less than 20% of all internationally shared aquifers belong to karst types. This estimate is currently being adjusted upwards as a result of new investigations especially in Central and South Eastern Europe, and the Middle East within the Alpine orogenic belt and its branches, along with the numerous middle- and small-size countries and their dense network of borders.

A typical example of these complexities is the Dinaric mountain system, the region of classical karst, which is composed almost entirely of carbonate rocks (limestones and dolomites), with thickness often exceeding 1000 m. In this part of Europe not only was the term karst born (the German derivation of the local name of the district between Italy and Slovenia called Carso or Kras), but also a new scientific discipline – karstology was established at the end of the ninetenth century. The appearance of several new sovereign states in the 1990s, from what was once Yugoslavia, and complex transboundary inter-linkages have had a considerable impact on water use and water sharing for domestic supply, power generation, and agriculture. For instance, this is also an area where one of the world's largest springs, located in one country, has 95% of its catchment in another country. Similarly, this is an area where one may enter a cave in one country and several kilometers away enter another country. These, among many others, are the reasons why the DIKTAS (Protection and Sustainable Use of the Dinaric Karst Transboundary Aquifer System) project was initiated and included three former Yugoslav countries (Croatia, Bosnia & Herzegovina and Montenegro) and Albania. The project was conducted in the period of 2011–2015, as full-size GEF regional project, implemented by UNDP and executed by UNESCO-IHP.

The task of organising the scientific conference related to karst and transboundary issues was part of the DIKTAS. After four years of work the team of international and national consultants organised the international conference *Karst without Boundaries* which took place on 11–15 June of 2014 in Trebinje, Bosnia & Herzegovina, with a mid-conference field seminar in Dubrovnik, Croatia, and a post-conference excursion in Boka Kotorska Bay, Montenegro. The conference was preceded by an international summer school and a field seminar *Characterization and Engineering of Karst Aquifers* supported by UNESCO and attended by university students from Europe, Asia, and South America. The international summer school is now traditionally organised beyond DIKTAS as one of its capacity building activities.

This book contains 23 selected chapters from the conference *Karst without Boundaries*. These are written by 62 authors coming from 19 countries. The book is divided into four main topics: Management of transboundary karst aquifers, Karst aquifer charaterisation and monitoring, Water flow in karst: from vadose to discharge zone, and Engineering, sustainable use and protection of water in karst.

The chapter *Management of transboundary karst aquifers* contains seven contributions. The first one aims to improve understanding of transboundary groundwater resources of the Dinaric Karst region and to present DIKTAS project goals and achievements. The second opens a general but very sensitive question of relevancy of boundaries in karst. The third presents another important international project for worldwide karst aquifers mapping (WOKAM). The remaining four contributions present concerns and results of transboundary aquifers projects between Syria and Lebanon, Slovakia and Hungary, Bulgaria and Serbia, and Montenegro and Albania.

The chapter *Karst aquifer characterisation and monitoring* is the largest in this book with eight contributions. This is also result of common understanding of hydrogeologists that characterisation and monitoring are essential tools for determining aquifer behaviour and its resources before any instruments of water policy are to be imposed. The introductory paper of this group again deals with Dinaric karst and achievements of implemented large engineering projects which became the basis for socio-economic development of this region. Three papers of this group deal with aquifer recharge processes, spring hydrographs and time series analyses with examples from Italy, Slovenia and Iran. Characterisation of aquifer based on hydrochemistry and several other applied field methods is discussed with case studies from Spain (evaporitic aquifer) and Hungary (shallow aquifer). Complex karst aquifer characterisation including development of physical models is presented with examples from Croatia (Plitvice) and Serbia (Beljanica Mountain).

The *Water flow in karst: from vadose to discharge zone* contains case studies from Italy, south China, eastern Serbia and Croatia. All four are different in content, dealing with aquifers in orogenic belt, tower karst, fresh and thermal waters, and catchment delineation, but all present suggestions how to establish conceptual models applicable to local conditions.

The final chapter *Engineering, sustainable use and protection of water in karst* also contains thematically various subjects. Two examples are related to surface and groundwater damming in karst and related concerns (examples from China and

Bosnia and Herzegovina), one paper discusses consequences of abandoned mine dewatering (Hungary), while the first paper in this group presents possible environmental indicators which may be used to evaluate impact of various human activities in karst environment.

Our hope is that this book will be beneficial to everyone working on the many challenges of sustainable surface water and groundwater use in karst.

The editors would like to express their great appreciation and thank all the authors of this book for their dedication, and all reviewers for their help and suggestions. We also acknowledge guidance and technical support of Nicholas Robins, the IAH book Editor-in-Chief who enthusiastically worked with us to meet high standards of the Publisher. Thanks also go to Janjaap Blom and Germaine Seijger editors in charge of Engineering & Applied Sciences, Taylor & Francis. We very much appreciate our collaboration with DIKTAS project staff, external consultants and advisors since the very beginning of the project, and are grateful for the support of the organisers of the conference in Trebinje in June, 2014.

Zoran Stevanović
Neven Krešić
Neno Kukurić

Part I

Management of transboundary karst aquifers

Chapter 1

Dinaric Karst Aquifer – One of the world's largest transboundary systems and an ideal location for applying innovative and integrated water management

Zoran Stevanović[1], Neno Kukurić[2], Želimir Pekaš[3], Boban Jolović[4], Arben Pambuku[5] & Dragan Radojević[6]
[1]*University of Belgrade – Faculty of Mining and Geology, Centre for Karst Hydrogeology of the Department of Hydrogeology, Belgrade, Serbia*
[2]*IGRAC, Delft, The Netherlands*
[3]*Croatian Waters, Zagreb, Croatia*
[4]*Geological Survey of Republic of Srpska, Zvornik, Bosnia and Herzegovina*
[5]*Albanian Geological Survey, Tirana, Albania*
[6]*Geological Survey of Montenegro, Podgorica, Montenegro*

ABSTRACT

The Dinaric region is a karst holotype with its north west margin in the Crasso area around Trieste in Italy, and south west margin in Albania. There are more than 150 poljes, and in certain areas the density of the dolines can reach 150/km². Not only was the term karst born in the area, but Jovan Cvijić and his followers founded here a new scientific discipline – karstology, at the end of the nineteenth century. The Dinaric region is by far the richest in all of Europe: there are more than 100 springs with a minimum discharge over 500 l/s. However, there are numerous problems for groundwater sustainable utilisation and protection from pollution. The rise of several new sovereign states from what was once Yugoslavia has established complex transboundary inter-linkages that impact water use and water sharing for domestic supply, power generation, and agriculture. DIKTAS (Protection and Sustainable Use of the Dinaric Karst Transboundary Aquifer System) is a GEF project implemented by UNDP and UNESCO's IHP. Its mandate is to improve understanding of shared water resources and to facilitate their equitable and sustainable utilisation, including the protection of dependent ecosystems.

1.1 INTRODUCTION

The Dinaric system (Dinarides) represents a geologically heterogeneous, south European orogenic belt of the Alpine mountain chain (Alpides). The Alpine and Dinaric belts are in contact either directly or through an intermediate zone (Herak, 1972). The main orientation of the system is north west-south east, parallel to the Adriatic Sea. The system extends from the Carso area in Italy in the north over the countries of former Yugoslavia (Slovenia, Croatia, Bosnia and Herzegovina, Montenegro, Serbia, FYR of Macedonia) and ends in the Albanian Alps. The orogenic belt extends further in Albania and into Greece to the southern Alpine branches of Pindes and Hellenides, respectively.

DIKTAS is an acronym of the GEF (Global Environment Facility) regional project Protection and Sustainable Use of the Dinaric Karst Transboundary Aquifer System which commenced in 2011. The Dinaric Karst Aquifer System, shared by several countries and one of the world's largest, has been identified as an ideal location for applying new and integrated management approaches to these unique freshwater resources and ecosystems.

The project aims at focusing the attention of the international community on the huge but vulnerable water resources contained in karst aquifers (carbonate rock formations), which are widespread globally, but often poorly managed. At the regional level the main project objectives are to facilitate the equitable and sustainable utilisation and management of the transboundary water resources, and to protect the unique groundwater dependent ecosystems that characterise the Dinaric karst region from natural and man-made hazards, including climate change. These objectives aim to contribute to the sustainable development of the region.

Partner countries within the framework of the DIKTAS project are Croatia, Bosnia Herzegovina, Montenegro and Albania as GEF-fund recipient countries. The Adriatic coastline and the islands make up the western border of the Dinaric system. A tectonic graben of the Sava River is usually considered as the northern edge of the Dinarides. The fringe of the Dinarides in Croatia and in Bosnia Herzegovina is, therefore, placed some 20–30 km south of the Sava riverbed (Figure 1.1). Montenegro is the only country whose entire territory belongs to the Dinarides; the karstified carbonate rocks cover more than 2/3 of the Montenegrin territory. For the purpose of transboundary diagnostic analysis (TDA) the southern border of project area has been extended to the Vjosa River basin in Albania. As such, the total surface area of the Dinaric system, including non-karstic rocks, is estimated at 140 000 km², out of which some 110 000 km² belongs to the four project countries. Approximately 60% belongs to the Adriatic/Ionian basins, while the rest is a part of the Black Sea basin.

1.2   GEOLOGY AND HYDROGEOLOGY

The Dinaric system is a karst holotype and a classic karst region. Not only was the term "*karst*" born in the area (the German derivation of the local name of the district around Trieste between Italy and Slovenia "*Carso*" or "*Kras*") but also at the end of the nineteenth century a group of researchers including A. Penck, A. Grund, and F. Katzer led by J. Cvijić established a new scientific discipline – karstology. Cvijić undertook most of his work in the Dinaric karst, and in *Das Karstphänomen* (1893), argued that rock dissolution was the key process in the creation of most types of dolines, 'the diagnostic karst landforms'. The Dinaric karst thus becomes the *locus typicus* area for dissolutional landforms (Ford, 2005) and some local terms were accepted, and are still used, in international karst terminology (e.g. ponor, doline, uvala, polje) (Stevanović & Mijatović, 2005). Cvijić stated that 'there is no deeper and more thorough karst development than Herzegovina–Montenegro's karst located between the lower Neretva River, Skadar Lake and the Adriatic Sea'.

Following Cvijić's research, a large number of specialists from former Yugoslavia, Italy and Albania further improved the knowledge of the Dinarides in terms of hydrology, geomorphology, geology, hydrogeology and social/humanistic sciences.

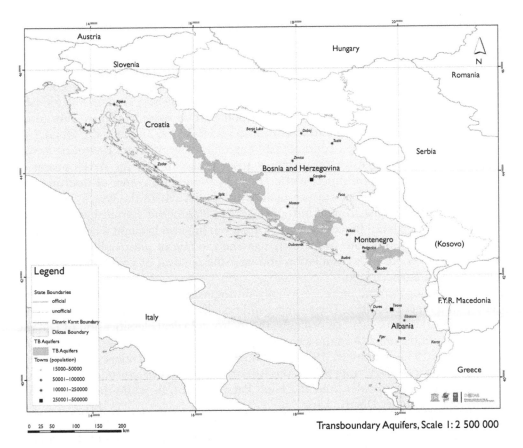

Figure 1.1 Extension of Dinaric karst from Italy to Albania and the boundaries of study area of four DIKTAS project countries. Some selected aquifers of major transboundary concern are also shown.

Preparation of the Basic Geological Map of Yugoslavia (some fifty years ago) on the scale 1:100 000 (with working sheets 1:25 000) substantially improved the geological knowledge on Dinaric karst. During several decades, large scale hydrogeological exploration was carried out for the construction of dams in the Dinaric karst region. Technical applications of control and regulation of karst aquifers through the construction of galleries, batteries of wells, and groundwater reservoirs (storage) represent an important contribution to the international hydrogeological science. Petar Milanović's 'Hydrogeology of karst and methods of investigation' (first edition 1979) became one of the most important references when dealing with karst groundwater distribution and circulation. A monograph 'Hydrogeology of the Dinaric Karst' (Mijatović, 1984) published by IAH, also confirms the wide interest of the hydrogeological community in the Dinaric karst.

The Dinaric karst is almost entirely carbonate (limestones and dolomites), its thickness is often over 1000 m and it is mostly of Mesozoic age (Tethys sedimentary basin). The development of the Dinaric karst was gradual. Herak (1972) stated that at the end of the Triassic or during the Lower Jurassic, the Triassic carbonate rocks were first exposed to the impact of water circulation processes. At the end of the Upper Cretaceous and during the Paleocene, intensive uplifting and folding took place, during which most of the carbonate and flysch rocks were tectonised. After the Laramian tectonic phase, the next intensive movements occurred in the Helvetian phase (Eocene/Oligocene). All the main nappes along the Adriatic/Ionian Sea coastline can be related to this stage and the rising of large land masses, accompanied locally by intensive structural changes providing the potential for more intensified subsurface water circulation and widespread karstification (Herak, 1972). Since the Oligocene, the Dinaric region has been continuously exposed to weathering, providing favorable conditions for intensive subsurface water circulation and the development of karst features. The most distinctive effects can be found in the area of uplifting and subsidence. The areas of subsidence include the poljes where the water was active before and after the diastrophic movements. The Pleistocene started not only with climatic changes (glacial process, lowering sea level) but also brought a new structural and morphologic evolution. Climate change and the rate of diastrophic movements regulated the periods of accumulation and later the removal of young deposits from the poljes (Mijatović, 1983).

Three major tectonic units are usually distinguished in the Dinarides: External, Central and Inner Dinarides. These can be additionally separated into several sub-units (Herak, 1972) (Figure 1.2). Accordingly, the hydrogeological classification of Dinaric karst indicates the units: Adriatic karst belt, High karst belt, Fluviokarst and Isolated karst (Šarin, 1983; Mijatović, 1984). Although no precise equivalency between tectonic and hydrogeology units exists, the Adriatic belt could be considered an equivalent to External Dinarides, and the High karst to Central, while the last two are distributed over Inner Dinaric karst.

The Albanian tectonic classification is different in names and in structures. Nevertheless, the units such as the Adriatic depression (Figure 1.3) and the parts of the Ionian, Krasta-Cukali or Kruja zones, belong to the External Dinarides, while the Mirdita unit could be interpreted as an extension of the Central ophiolitic zone of the Central Dinarides (Meçe & Aliaj, 2000).

Dinaric karst is a mountainous region with a prevalence of highly karstified rocks and large karstic poljes and valleys created in tectonic depressions by perennial or sinking streams. The karstification base is deep; boreholes have locally encountered karstified zones at depths of 2000 m but Milanović (2005) concluded that the average depth does not exceed 350–400 m. The karstification in the near-surface zone (0–10 m) is about 30 times larger than at a depth of 300 m (Milanović, 2000). The Dinaric region contains all types of karst landforms and features including karren (lapies), dolines, jamas (pits), ponors (swallow holes, sinks), dry and blind valleys, caves and caverns as single forms, and uvalas (Figure 1.4), poljes and karst plains as larger complex forms (Cvijić, 1893; Božičević, 1966; Roglić, 1972; Krešić, 1988). As an example, the number of sinkholes (dolines) in certain areas can reach 150 km^2. According to Milanović (2000), in the Dinaric karst region there are approximately 130 poljes. The total area of all these poljes is about 1350 km^2.

Livanjsko Polje, considered the world's largest karst polje, covers an area of 380 km^2, and together with Buško Blato, which morphologically is its integral part, it

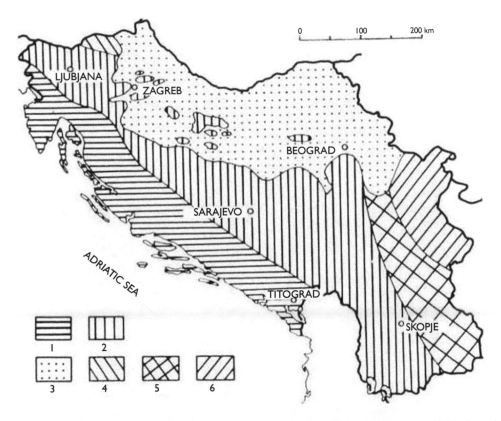

Figure 1.2 Major tectonic units of Yugoslavia (Herak, 1972). 1 = Outer Dinaric units (Adriatic and high karst); 2 = Inner Dinaric and south Alpine units; 3 = Pannonian Basin; 4 = eastern Alps; 5 = Serbian-Macedonian Belt; 6 = Carpathian-Balkanides Belt.

totals 433 km². The poljes are characterised by complicated hydrogeological relations: drainage of surface water is achieved through many ponors. These are frequently located in the polje areas nearest to the prevailing erosion base. In the Nikšićko polje, about 880 ponors and estavelles were identified, 851 of which are located along its southern perimeter. In Popovo Polje there are more than 500 ponors and estavelles. Some of them are lakes or swamps, the others periodically inundated or even dry. In general, the poljes are geologically heterogeneous, but an abundance of impervious rocks implicate tectonic origins.

Herak (1972) stated that more than 12 000 caves have been explored in former Yugoslavia alone, more than 5000 of which are in Croatia. Krešić (1988) listed some 15 potholes (pits, shafts) in former Yugoslavia deeper than 400 m. In the meantime, some much deeper potholes have been discovered: Lukina jama-Trojama (–1392 m) and the Slovačka jama (–1301 m) at Velebit Mountain, Croatia are among the deepest speleological phenomena in the World. At the Kameno more (Stone Sea) and the Orjen Mountain above Kotor Bay (Montenegro), within an area of only 8 km², more

Figure 1.3 The contact zone between large tectonic zones – Adriatic and Ionian (Dhermi, Albanian coast, photo Z. Stevanović).

than 300 vertical shafts were registered (Milanović, 2005). Some of those shafts were speleologicaly investigated to depths of 200–350 m.

As a result of intensive karstification a network of highly permeable underground channels acts as preferential pathways of intensive groundwater circulation. Along with its richness in various karstic features, the Dinaric region is by far the richest in Europe in water resources but these are unequally distributed throughout the year, which results from climate and a high degree of karstification (Bonacci, 1987). Some areas, such as southern Montenegro, are characterised by an intensive water balance where the average specific yield is over 40 l/s/km^2 (Stevanović *et al.*, 2012). In the Dinaric region of former Yugoslavia there are 230 springs with a minimum discharge

Figure 1.4 A small depression – uvala used for crop cultivation (Orjen Mountain, shared between Montenegro and Bosnia & Herzegovina, photo Z. Stevanović).

of over 100 l/s, while about 100 springs have a minimal discharge of over 500 l/s (Komatina, 1983). In the Albanian karst there are roughly 110 springs with an average discharge exceeding 100 l/s (Eftimi, 2010).

There are some distinctive patterns in distribution of karstic features in the main geo-structures of the Dinaric system. In the Adriatic zone, vruljas (submarine springs), (Milanović, 2007), are the most specific hydrogeologic features. They are formed by the sinking of the floor of the northern Adriatic Sea and rising of the sea level, so that karstified land was submerged beneath the sea (Figure 1.5).

The surface and groundwater of the Dinaric karst belongs to two main catchments. The area of External (Outer) Dinarides belongs to the Adriatic catchment

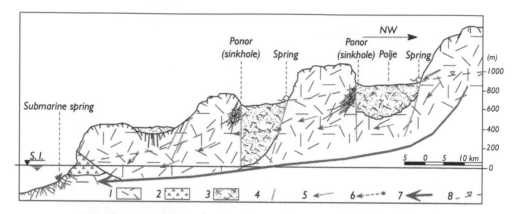

Figure 1.5 Schematic cross-section through the karstic poljes of the Dinarides (after Mijatović 1983, modified by Stevanović). Legend: 1. karstified Mesozoic limestone; 2. flysch barrier; 3. porous aquifer of polje; 4. fault; 5. groundwater flow; 6. direction of flow around the barrier; 7. regional flow; 8. groundwater table.

(a small part to the Ionian Sea), while the Internal (Inner) Dinarides are part of the Sava (i.e. Danube and the Black Sea catchment area). The main river basins in the Adriatic/Ionian catchment area are the Vjosa, Seman, Drini, Buna (Bojana), Zeta, Neretva, Cetina, Krka, Zrmanja and Soča. Karstic groundwater from the river basins of the Tara, Piva, Vrbas, Pliva, Sana, Una, the upper course of the Kupa River and the Krka (in Slovenia) gravitates to the Black Sea catchment area (Stevanović *et al.*, 2012).

Komatina (1983) noted that tracer experiments were conducted at more than 650 sites in the Dinaric karst of former Yugoslavia. In eastern Herzegovina alone, tracers were applied at 281 sites, in the catchment area of the Cetina River at 99 sites and in the Skadarsko Lake catchment area at 77 sites.

According to Milanović (2000) more than several hundred investigations have been performed in the Dinaric karst so far, for the purpose of finding the major routes of underground water circulation (Figure 1.6). Based on experiments, the average flow velocity is estimated to vary between a range of 0.002–55.2 cm/s. Such extreme values are rare, whereas an average velocity is about 5 cm/s. Based on 380 conducted experiments Komatina (1983) concluded that the frequency of groundwater velocities in Dinaric karst is: in 70% of cases from 0 to 5 cm/s; in 20% of cases 5 to 10 cm/s; and in 10% of cases more than 10 cm/s. During the dry season and low aquifer water table, water circulation in the karst system is characterised by slower rates of flow. The water labeled with dye takes two- to five-fold less time to travel the same distance during a season of high hydrologic activity (Milanović, 2000). The same author presents an example: to cover the distance (34 km) from Gatačko Polje to the Trebišnjica Spring (Dinaric karst, Herzegovina), the underground flow takes 35 days when the water table is low and inflow is small. During high water levels and large inflow, the water takes only five days to cover the same distance.

The karst aquifer recharges from precipitation and waters percolating from numerous sinking rivers. Depending on the locality, morphology and karstification properties the average infiltration rate lies between 50% to 80% of the precipitation. In the Cetina

Figure 1.6 Tracing test at the Ponikve Ponor (Dabarsko Polje, East Herzegovina, photo courtesy of Željko Zubac).

River catchment area, more than 80% of the precipitation appears at the terminal water gauge controlled profile, while in the Trebišnjica River catchment area this percentage reaches 90%. Effective infiltration in the Albanian karst is on average made up of 40–55% rainfall (Eftimi, 2010). Considering that the rainfall rate in the region is one of the highest in Europe (over 2000 mm annually in the southern part, and even up to 5000 mm in Boka Kotorska Bay) there are abundant but variable water reserves. However, more than 70% of the precipitation occurred during the wet season (October–March).

Milanović (2005) noted that 'only through three huge springs along the Neretva Valley and Adriatic coast (Buna, Bunica and Ombla) and a few spring zones in the Kotor Bay, more than 150 m³/s is discharged annually into the Adriatic Sea directly or indirectly through the Neretva River'.

Three capital cities in the project countries receive drinking water from the karstic aquifers. Sarajevo obtains a part of its water supply from the Vrelo Bosne springs in

Central Dinarides (1.4–24 m³/s). The Albanian capital Tirana is supplied in part from the spring that discharges the Triassic and Jurassic karstic aquifer of Mali me Gropa plateau (Eftimi, 2010), while downstream from the third important spring, Buvilla, issuing from Dajti Mountain a large reservoir has been constructed. Mareza spring (2.0–10.0 m³/s) in the Skadar basin is the main source of water for the Montenegrin capital city of Podgorica (Radulović, 2000). Along the Adriatic, Ionian and Aegean coast, almost all cities and tourist centres use karstic groundwater (Stevanović, 2010; Stevanović & Eftimi, 2010).

Perhaps the most famous and the largest spring on the northern Italian coast is Timavo, with an average discharge rate of 30 m³/s. Jadro Spring is the main source for the water supply of Split (Figure 1.7). The average minimum discharge of Jadro during the recession period is 3–5 m³/s, while maximum discharge is often over 50 m³/s (Bonacci, 1987).

Figure 1.7 One of the largest springs on the Adriatic coast, Jadro Spring supplies drinking water to Split (Croatia) which is the main reason why the Romans founded the city (photo Z. Stevanović).

Figure 1.8 Ombla spring (Rijeka Dubrovačka source, Croatia, photo Z. Stevanović).

Ombla Spring is the largest permanent karstic spring at the South Adriatic coast (Figure 1.8). It supplies the city of Dubrovnik; at a maximum discharge of about 154 m³/s. Since the completion of the Trebišnjica Hydropower System and the regulation of this longest sinking river in Europe in the catchment of Ombla, the average discharge of Ombla was reduced from 34 m³/s to 24 m³/s. However, the minimum discharge (2.3 m³/s) is not affected (Milanović, 2006).

The main springs along Boka Kotorska Bay in Montenegro are: Gurdić and Škurda spring near Kotor, Ljuta spring at Orahovac, Spila spring at Risan, Morinj springs, Opačica at Herceg Novi and Plavda at Tivat. The Sopot near Risan is a well known submarine spring (Figure 1.9). All these springs are characterised by a high variation in discharges due to a highly karstified catchment and extremely fast propagation of the rainfall. Some of those springs even dry up completely during summer while after intensive rainfall some of them can discharge over 100 m³/s.

In the Skadar Lake basin there is a large number of sublacustrine springs such as Oko Matice, Golač, Kaludjerovo Oko and many other along the edge of Malo Blato (Radulović, 2000). Volač, Karuč, Bolje sestre was recently tapped for the regional water supply of the Montenegrin coastal area (Stevanović, 2010).

It is assumed that 2/3 of the whole groundwater resources in Albania are within karstic aquifers, providing more than 60% of the water consumed in the country (Eftimi, 2010). The average potential yield of coastal karst aquifers in the catchment of the Ionian Sea in Albania is estimated at about 15–20 m³/s (Eftimi, personal communication).

Figure 1.9 Sopot spring near Risan (Montenegro) during peak flow discharge (left, photo Z. Stevanović) and during drought when only submarine flow exists (right, photo S. Milanović).

1.3   ARTIFICIAL CONTROL OF KARST WATERS

A large fluctuation in the water table is common in the region (Pekaš *et al.*, 2012). For instance, the water level can change by 312 m during a period of 183 days at the observation borehole Z-3 in the Nevesinjsko Polje. In the Cetina River basin the maximum recorded water table increase was 3.17 m/h. Along with continual seasonal flooding of cultivated land, this was also the main reason why large projects to regulate river flows were initiated in all the countries in the region after World War Two and many of these were implemented during the 1960s and 1970s. The idea to regulate flows is much older; the Klinje Dam (the Mušnica stream, Boznia Herzegovina) was built in the period 1888–1896, while the hydro-electrical power plant (HPP) at Kraljevac (Cetina, Croatia) was erected in 1912. Today many streams are dammed and their waters are utilised by HPPs. The major dams and reservoirs have been built on the Cetina, Neretva, Trebišnjica, Zeta, and Drini rivers. Dinaric karst becomes, therefore, a reference area for the successful completion of dams in karst, a problematic media for water losses (Milanović, 2000; 2006). About 2/3 of the total existing hydro power facilities of the four countries are located in the DIKTAS karst area, therefore, hydro power generation from the DIKTAS karst system plays significant role in these country's economies.

The Cetina water system is managed by the Croatian water authority although a considerable amount of water originates in the territory of Bosnia & Herzegovina including the catchments of the Kupres, Glamoč, Duvno, and Livno poljes, and Buško Blato (Figure 1.10). Currently there are five HPPs on the Cetina River: the largest

Figure 1.10 Lake Cetina, upstream from the dam (photo Z. Stevanović).

storage reservoir, Buško Blato, has a capacity of 831×10^6 m³. Bonacci (1987) stated that Cetina's surface area is from 3700 to 4300 km² of which the topographic watershed encompasses about 1300 km² and the subsurface watershed about 2700 km². The average flow is 118 m³/s.

The main structures of the Trebišnjica water system are Lake Bilećko (i.e. the Bileća Reservoir) behind the Grančarevo Dam, and the Gorica Dam and reservoir downstream. Active operating HPPs are the Trebinje I (180 MW), Trebinje II (8 MW), Čapljina (420 MW) and Plat (Dubrovnik, 210 MW). Average river flows at the Grančarevo (Figure 1.11) and Gorica dam sites are 74.3 m³/s and 85.6 m³/s, respectively. Losses from the Bileća Reservoir are negligible, while losses from the Gorica Reservoir are approximately 5–7% of the average river flows. These losses appear mostly downstream from the Gorica Dam and represent a guaranteed ecological flow to the regulated Trebišnjica River (Figure 1.12).

The number of artificial reservoirs in Montenegro is small in comparison with the hydropower potential. The total capacity of these reservoirs amounts to slightly more than 1×10^9 m³. With respect to the total amount of surface water (about 14×10^9 m³/year) in the territory of Montenegro, this amounts to about 7% (Hrvačević, 2004). In the Adriatic basin, the reservoirs in the Nikšićko Polje (Krupac, Slano and Vrtac) have been formed on the River Zeta, while the Liverovići Dam controls the flow of the Gračanica River. Water from the reservoirs is utilised by the Perućica HPP (307 MW). All the reservoirs in the Nikšićko Polje have been built in highly karstified rocks. The Slano and Vrtac required intensive and expensive anti-infiltration works. The grout curtain along the southern rim of the Slano is one of the longest in the world. It has a

Figure 1.11 Grančarevo dam (Trebinje, Bosnia & Herzegovina).

length of 7,011 m, depth of 57 m and surface of 396 122 m². The current hydropower capacity of the main Albanian plants is 1750 MW, and most of the dams were built on the Drini catchment.

1.4   KARST AQUIFER MANAGEMENT AND MONITORING – DIKTAS ACHIEVEMENTS

The Transboundary Diagnostic Analysis (TDA) was conducted in the period 2011–2013 by the DIKTAS Project Team in accordance with the GEF guidelines. The TDA is based on a substantial regional analysis that was required in order to fully understand the context of transboundary issues. The regional analysis was particularly important given the complexity of the karst environment and regime and interconnectivity of karst aquifers.

The Project Team was organised into four working groups, reflecting the main issues of the regional analysis:

1) Hydrogeology of the Dinaric karst;
2) Environmental and socio-economic analysis;
3) Legal and institutional framework and policy; and
4) Stakeholder analysis.

Figure 1.12 Channeled Trebišnjica riverbed and highly karstified bedding planes, (photo Z. Stevanović).

One of the main DIKTAS outputs in terms of the regional hydrogeological characterisation is the GIS based digital Hydrogeological map of the Dinaric Karst region (Figure 1.13). Its creation involved harmonisation of data, classifications, methodologies, reference systems, projections and semantics. The map was used as a basis for the development of various thematic maps for the environmental and socio-economical assessment and represents an important tool for various further analyses and karst aquifer management.

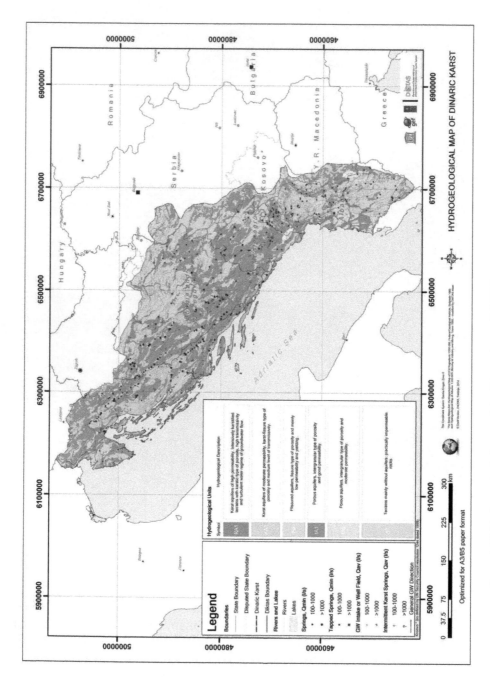

Figure 1.13 Hydrogeological map of the Dinaric karst (DIKTAS database; http://dinaric.iwlearn.org).

Through the TDA the DIKTAS project was focused on selected and prioritised transboundary karstic aquifers (TBAs), examining current and potential issues of concern. Based on five criteria (importance, representativeness, data availability, issues of concern, relevance), eight TBAs were initially selected for detailed analysis: Una, Krka, Cetina, Neretva, Trebišnjica (all shared by Croatia and Bosnia Herzegovina), Bilećko Lake, Piva (Bosnia & Herzegovina and Montenegro) and Cijevna/Cemi (Montenegro and Albania). Six of these TBAs belong to the Adriatic Sea catchment area while two are part of the Black Sea basin (Una and Piva). The TBAs comprise a total surface area of 12 000 km², which is approximately 10% of the entire study area of the four project countries which share Dinaric karst. The surface area of individual TBAs varies from 668 km² (Krka) to 3,455 km² (Cetina). The later diagnostic analysis focussed on six more problematic TBAs, excluding Krka and Piva.

A delineation of the aquifer surface area was the first step in the hydrogeological analysis of each TBA. Further analysis included the characterisation and development of conceptual models. Groundwater budgeting of TBAs created a base for the assessment of groundwater reserves and availability, as well as for proposals and measures aiming to ensure sustainable development of TBAs. Some of the selected TBAs, such as Una, Neretva or Cijevna/Cemi, are of particular importance because they represent parts of designated protected zones, or wetlands, or the habitat for endangered species (Figure 1.14).

Three major shortcomings were identified in the region:

1. *Insufficient knowledge of the distribution and properties of aquifers.* Despite many tracing tests conducted in the region, especially in East Herzegovina and Dalmatia, because the groundwater regime is highly variable many catchments are still not properly delineated and characteristics of deeper aquifer parts not studied in detail. Moreover, systematic monitoring of groundwater quantity and quality is still inadequate despite the existence of many local waterworks and hydro-energy structures.
2. *The minimum discharges of the springs and accordingly reduced minimum river flows during recession periods* (summer and autumn months) result in a water deficit not only for water consumers but also for dependent ecosystems in certain areas. Although the total dynamic reserves in studied Dinaric karst aquifers often surpass the actual exploitation capacities, most of the tapping structures are constructed simply to draw on the natural discharge of the springs and thus depend solely on the natural flow regime.
3. *Unstable water quality and aquifer vulnerability to pollution* are typical characteristics of most of the studied TBAs in the Dinaric karst. The karstic aquifer formed in well-karstified carbonate rocks is recharged mainly from rainfall and from sinking flows.

The TDA indicated that water extraction was still far below the aquifer's replenishment potential, and there is no evidence of significant over-exploitation in the studied TBAs. For instance, in the case of Cetina and Neretva TBAs the average extraction of groundwater is ten times less than the total minimal discharge of the springs (dynamic reserves). However, shortage of water is locally in evidence during summer and early autumn months which coincides with increased demands during the tourist season.

Figure 1.14 The sablacustrian springs in the Skadar basin (Montenegro): the Karuč, previously planned for water supply of the coastal area. Skadar (Shkoder) Lake is included in the Ramsar convention for protected wetlands.

In such circumstances the principles of EU Water Framework Directive regarding ecological flow for downstream consumers have to be fully respected further complicate the water and environmental situation.

Therefore, the Dinaric region has the most dynamic water budget in all of Europe, but there are numerous challenges for sustainable utilisation of groundwater. This includes high annual variation of natural flows and the vulnerability of aquifers to pollution. It is, therefore, important to (1) improve the quality of water by eliminating or mitigating sources of pollution, (2) regulate the minimum spring discharges, (3) ensure ecological flows, and (4) establish proper water monitoring systems. These actions are recommended as priorities during implementation of the DIKTAS Strategic Action Plan.

The karstic water quality is generally assumed to be satisfactory which is mostly due to sparsely populated catchments in mountainous areas and the absence of intensive farming or industrial activities. However, when pollutants are present (for example mines, industrial and domestic waste waters, solid waste dumps, fertilizers), deterioration of water quality in unconfined and vulnerable karstic aquifers is almost certain. Therefore, proven connections between ponors and certain important springs in the territory of the neighbouring countries (for example Plitvice-Klokot, Trebišnjica-Ombla) require strict enforcement of already established sanitary protection zones and preventive measures. The Dinaric karst aquifer system areas require specific solutions and compromises in land use planning, and the protection of nature and water resources. Some of the important springs with catchments shared by the Dinaric countries are included in the Strategic Action Plan as demonstration sites for the application of methodology and design of sanitary protection zones.

The actual monitoring of groundwater in the region and in the studied TBAs is far from satisfactory. Only in Croatia has the characterisation of groundwater bodies been completed and monitored in accordance with EU Water Framework Directive. One of the tasks of the DIKTAS project was to prepare a proposal for the creation of a new monitoring network which will fully respect karst specific behaviour and include local water users (waterworks, dams, irrigation, industry). The Cijevna/Cemi TBA has been identified as the most problematic in terms of available data on water resources, and installation of a modern surface and groundwater monitoring network for observation of climate elements has been proposed.

In all four countries there are on-going efforts for transposition of the fundamental principles, objectives and measures from the EU Water Framework Directive, (2000/60/EC) and the Groundwater Directive (2006/118/EC) in national legislations. Although the 'polluter pays' principle and the principle of 'recovery of the costs' are promoted in national legislative documents, the principle of cost recovery is not fully transposed either in national regulations or in water management practices, with regards to implementation of the environmental and resource costs in water pricing policies. There is no legal or policy document in any of project countries which adequately defines and prescribes the integration of environmental and resource costs into development of pricing policies. It should be noted that the main shortcoming of the legislative framework in all countries is an underdeveloped system of by-laws and insufficient implementation of present legislation due to lack of human resources and financial means for fulfilling legal and policy requirements. Due to the lack of clear development strategies, programmes and plans for water management issues, the Dinaric Karst region cannot be considered as an example of successful implementation of these principles.

Based on TDA a Strategic Action Plan which includes different common activities, proposals for legal and institutional reforms and harmonisation of legislation has been prepared. In order to support sustainable utilisation of groundwater in the Dinaric region, the Strategic Action Plan is focusing on several priority actions: improvement of the quality of water, elimination of sources of pollution, improvement of minimal discharges, assurance of ecological flows, and establishment of proper water monitoring systems (Table 1.1). Conceptual proposals for investment in all of these areas and at designated sites have also been prepared as a part of the Plan.

The DIKTAS Strategic Action Plan adheres to two key environmental management principles. These are:

- The Ecosystem Based Management Approach; and
- Integrated Water Resources Management.

The vision guiding the long-term water resources and ecosystem quality objectives is:

To achieve joint sustainable and equitable use and protection of Dinaric karst aquifer system.

Table 1.1 Proposed actions and suggested TBAs where they can be performed.

Proposed Action	Proposed for TBA(s), not exclusive
1. Establishment of a common groundwater monitoring programme	All TBAs
2. Improvement of wastewater treatment especially in the Bihać region	Una, Neretva (Ljubuški town)
3. Harmonisation of hydrogeological criteria for delineation of source protection zones as the basis for policy harmonisation and protection of karst springs used for public water supply in Bosnia & Herzegovina, Croatia, Montenegro and Albania	Una, Neretva, Cetina, Bilećko Lake, Cemi/Cijevna
4. Creation of future projections of water demands depending on socio-economic analysis	Una, Neretva, Trebišnjica, Cetina, Bilećko Lake
5. Definition of common criteria for: a) delineation of the sanitary protection zones and b) for setting cost-efficient measures for groundwater protection in karst areas	All TBAs
6. Definition of legal framework for establishment and law enforcement in sanitary protection zones	Una, Neretva, Cetina
7. Inventory of non-point and point sources of pollution (landfills, septic tanks, quarries, wastewater discharges, and others)	All TBAs
8. Establishment of regulations between the countries to set up regulatory frameworks concerning the discharge of wastewaters into the land stressing the importance of a unified policy	All TBAs
9. Fostering better control of the current agricultural and similar practices	Neretva
10. Promotion of eco-tourism	Neretva
11. Precise mapping of land and water usage	Cetina

With this vision in mind, the following objectives were planned:

- *Groundwater Quantity*
  Objective A: Ensure sufficient groundwater availability in dry periods, especially for water supply and to support environmental flow.
- *Groundwater Quality*
  Objective B: Maintain and improve (where required) quality of karst groundwater in the Dinaric region.
- *Protection of Groundwater Dependent Ecosystems (GDE)*
  Objective C: Ensure protection of GDE, specific features and their ecosystem services for the future.
- *Equitable Use*
  Objective D: Support equitable use of groundwater resources.
- *Capacity Building*
  Objective E: Raise awareness and capacity building related to karst water and dependant ecosystem.

Regional management actions and desired outcomes for reaching the expected Water Resources and Ecosystem Quality Objectives (WREQO) are concentrated on four classes of actions to provide structure to the future implementation:

- Policy, legislative and institutional actions;
- Monitoring and data management actions;
- Research training and awareness-raising actions, and;
- Investment actions.

Regular exchange of data represents one of the key steps toward sustainable transboundary aquifer management. This fact is recognised also in international law and acknowledged in the UN Resolution 63/124 the Law of Transboundary Aquifers (Article 8 Regular exchange of data and information). The DIKTAS countries agree and commit to exchange information among themselves at the regional or bilateral levels and prepare an instrument for the legal basis (e.g. Memorandum of Understanding, or other agreements). DIKTAS countries will exchange collected data through the national institutions responsible for managing of National Water Information System and under supervision of the Consultation and Information Exchange Body.[1]

The Consultation and Information Exchange Body is intended to create conditions for the development of sustainable and productive consultation mechanisms for information exchange and collaboration among the four DIKTAS countries aiming at, among other, ensure equitable use of groundwater resources in the whole DIKTAS region. The Consultation and Information Exchange Body will deal with, but not be limited to: elaboration of the aquifer acts, consideration of improved water supply in critical periods by proper management and technical solutions; prioritisation of the water end-users related to water availability and maintaining common information system.

It was concluded that all countries have legislation in place for groundwater protection which is, to a greater or lesser extent, implemented in each country. However, due to different criteria for delineation of protection zones and different prescribed protection measures within these zones, it is not possible to establish and implement efficient groundwater protection in the areas of transboundary aquifers based on the existing legislation. The Strategic Action Plan proposes to develop common criteria based on the existing ones for delineation of protection zones and common protection measures within these zones.

The establishment of a regional expert working group from experts in the fields of hydrogeology, water protection, economy and legislation is also envisaged. The regional expert working group's main tasks should be to determine the criteria for delineation of protection zones, define programmes for additional surveys, follow up monitoring network establishment and data exchange, and similar.

Finally, one of the most important objectives is to *raise awareness and capacity building related to karst water and dependant ecosystem*. The environmental

---

1 The signing of multilateral agreement between DIKTAS countries should provide a base for establishing the Consultation and Information Exchange Body. Signed Multilateral Agreement on establishment of the Consultation and Information Exchange Body will, among others, represent the acceptance by the DIKTAS countries to exchange data, which will be included in the Common Information System.

awareness campaign will be based on using the best practices and further elaboration of new public awareness concepts in accordance with the project objectives. The responsibility of the implementation of the global awareness campaign will be under the Consultation and Information Exchange Body Secretariat. Promotional materials in printed, and in electronic form and publications will be an important channel of communication. Formats will be various: leaflets, newsletters, brochures and should be distributed at all major events – press conferences, school lectures and meetings.

1.5   CONCLUSIONS

Following Cvijić's research, a large number of specialists have further improved the knowledge of the Dinarides in terms of hydrology, geomorphology, geology and hydrogeology. Today, more than a hundred years after the initial research, the Dinaric karst is relatively well investigated on a regional scale, but due to the complexity and intensity of the karst aquifer regime, detailed survey and systematic monitoring have to be further improved. The DIKTAS project contributed to better understanding of complex transboundary inter-linkages and the importance of sustainable and equitable use of water resources.

Due to its historical importance for the development of karst science an initiative has been taken to include the Dinaric region and its selected areas in the UNESCO list of World Heritage Sites. This reflets its exemplary karst development with numerous geo-heritage sites, several national parks and protected areas such as those under Ramsar's Convention, endemic species which inhabit underground world of caves and abundant groundwater resources. If this proposal comes to reality this will further strengthen activities on nature and water protection and raise awareness of the local population of the importance of sustainable use of natural resources.

REFERENCES

Bonacci O. (1987) *Karst Hydrology; with special reference to the Dinaric Karst*. Springer-Verlag, Berlin.

Božičević S. (1966) Caves, potholes and ponors with water in Dinaric karst area (in Croatian). *Krš Jugoslavije*, Jug. Akad. Zn. i Um. Zagreb, 6, 105–136.

Cvijić J. (1893) *Das Karstphänomen. Versuch einer morphologischen Monographie*. Geographischen Abhandlung, Wien, V(3), 218–329.

DIKTAS database; http://dinaric.iwlearn.org. Last visited on 10/05/2015.

DIKTAS, 2014: Strategic action plan, Trebinje.

Eftimi R. (2010) Hydrogeological characteristics of Albania. *AQUAmundi*. 1, 79–92.

Ford D. (2005) Jovan Cvijić and the founding of karst geomorphology. In: Stevanović Z & Mijatović B. (eds.): *Cvijić and karst*, Board on karst and spel. Serb. Acad. of Sci. and Arts, Belgrade, 305–321.

Herak M. (1972) Karst of Yugoslavia. In: Herak, M. and Stringfield, V.T. (eds.), *Karst: Important Karst Regions of the Northern Hemisphere*. Amsterdam, Elsevier. 25–83.

Hrvačević S. (2004) *Resources of surface water in Montenegro*. Elektroprivreda Crne Gore. Podgorica. 331 p.

Komatina M. (1983) Hydrogeologic features of Dinaric karst. In: *Hydrogeology of the Dinaric Karst*. Mijatovic B (ed.). Spec. ed. Geozavod, Belgrade. 45–58.

Krešić N. (1988) *Karst and caves of Yugoslavia*, Naučna knjiga, Belgrade, 149 p.

Kukuric N. (2011) Assessment of internationally shared Karst aquifers: example of Dinaric karst aquifer system, In: Polk, J. and North, L, *Proceedings of the 2011 International Conference on Karst Hydrogeology and Ecosystems*. Environmental Sustainability Publications. Book 2.

Meçe S., Aliaj Sh. (2000) *Geology of Albania*. Gebrüder Borntaeger. Berlin-Stuttgart.

Mijatović B. (1983) Karst poljes in Dinarides. In: Hydrogeology of Dinaric karst. Field trip to the Dinaric karst, Yugoslavia, May 15–28, 1983. (ed. Mijatović B). 'Geozavod' and SITRGMJ, Belgrade, p. 69–84.

Mijatović B. (1984) *Hydrogeology of the Dinaric Karst*. International Association of Hydrogeologists, Heise, Hannover. Vol. 4.

Milanović P. (1979) *Hydrogeology of karst and methods of investigation* (in Serbian), HET, Trebinje, 302 p.

Milanović P. (2000) *Geological engineering in karst*. Zebra Publishing Ltd., Belgrade. 347 p.

Milanović P. (2005) Water potential in south-eastern Dinarides. In: Stevanović Z. & Milanović P. (eds.): *Water Resources and Environmental Problems in Karst* CVIJIĆ 2005, Spec. ed. FMG. Belgrade, 249–257.

Milanović P. (2006) *Karst of eastern Herzegovina and Dubrovnik littoral*. ASOS, Belgrade, 362 p.

Milanović S. (2007) Hydrogeological characteristics of some deep siphonal springs in Serbia and Montenegro karst. *Environ. Geol.* 51(5), 755–759.

Pekaš Z., Jolović B., Radojević D., Pambuku A., Stevanović Z., Kukurić N., Zubac Z. (2012) Unstable regime of Dinaric karst aquifers as a major concern for their sustainable utilization. *Proceedings of 39 IAH Congress*, Niagara Falls, (CD publ.).

Radulović M. (2000) *Karst hydrogeology of Montenegro*. Sep. issue of Geological Bulletin, vol. XVIII, Spec. ed. Geol. Survey of Montenegro, Podgorica, 271 p.

Roglić J. (1972) Historical review of morphological concepts. In: Herak, M. and Stringfield, V.T. (eds.), *Karst: Important Karst Regions of the Northern Hemisphere*. Amsterdam, Elsevier Publishing Company. 1–17.

Stevanović Z., Mijatović B. (2005) Cvijic and karst / Cvijic et karst, Monograph: Spec. ed of Board of Karst and Speleology SANU, Belgrade, 405 p.

Stevanović Z., Eftimi R. (2010) Karstic sources of water supply for large consumers in south-eastern Europe – sustainability, disputes and advantages, *Geologica Croatica*, 63(2), 179–186.

Stevanović Z., Kukurić N., Treidel H., Pekaš Z., Jolović B., Radojević D. Pambuku A. (2012) Characterization of transboundary aquifers in Dinaric karst – A base for sustainable water management at regional and local scale. *Proceedings of 39 IAH Congress*, Niagara Falls, (CD publ).

Šarin A. (1983) Hydrogeologic regional classification of the karst of Yugoslavia. In: *Hydrogeology of the Dinaric Karst*. Mijatovic B (ed.). Spec. ed. Geozavod, Belgrade, 35–44.

UN Resolution 63(124), The law of transboundary aquifers.

Chapter 2

How confident are we about the definition of boundaries in karst? Difficulties in managing and planning in a typical transboundary environment

Mario Parise
CNR-IRPI, Bari, Italy

ABSTRACT

Karst aquifers may also be transboundary, i.e. crossing political or administrative borders. Typical karst hydrogeology includes a lack of correspondence between the surface and subsurface limits and difficulties in identifying hydraulic boundaries and volumes. A large amount of high-quality drinkable water comes from karst aquifers, with this volume increasing in the future. There is an urgent need to understand karst, and safeguard it and its resources, without being limited by administrative borders. Karst management requires cooperation between the people living on the land, to transfer and disseminate research outcomes, to create awareness of the fragile environment, to build resilience and readiness to cope with natural hazards and the changes they cause, and to minimise disturbance.

2.1 INTRODUCTION: THE KARST ENVIRONMENT

Karst is of huge importance for groundwater storage and transport in aquifers which have been used for many decades for supplying drinking water (Figure 2.1). About one quarter of the global population is completely or partially dependent on drinking water from karst aquifers, with karst regions representing about 10% of the Earth's continental area (Ford & Williams, 2007; Hartmann *et al.*, 2014). The percentage of drinking water from karst is expected to increase in the future. However, the valuable, high-quality, freshwater resources contained in karst aquifers are extremely vulnerable to contamination.

Karst aquifers are often transboundary, i.e. crossing political or administrative borders (Figure 2.2). To define the limits of a water catchment, and to understand whether the likely surface limits correspond to those that are present underground, is a very difficult task in karst lands, at the same time representing one of its main peculiarities (Gunn, 2007). A typical element of karst hydrogeology is the lack of correspondence between the surface and the subsurface limits, which also has consequences in terms of politics, planning and management of territories and of natural resources.

In the great majority of the natural environments, catchment boundaries are defined by surface morphology, by outlining ridges and watersheds, dividing a certain area into several basins. Knowing the amount of rainfall that falls in a specific

Figure 2.1 The Ombla spring, source of drinking water for the town of Dubrovnik, in Croatia.

Figure 2.2 The Reka River originates in the Classical Karst. After a surface course on flysch deposits, it sinks underground at the Škocjanske Jame, characterised by a 200 m high, collapsed doline (A), and follows a course of some 40 km, before coming to the surface in Italy, near Trieste, at the Timavo resurgences (B), to eventually flow into the Adriatic Sea.

watershed, as well as its size and geological characters, it is possible to estimate the amount of water collected in the corresponding discharge area.

Water circulation in karst differs from other types of terrain. This is due to limited surface runoff, balanced by rapid infiltration of water underground through the complex network of discontinuities within the rock mass, and regulated by the role of the epikarst. Water sinks at a certain site, and, due to complex underground systems made of conduits and caves, can be transported to another, nearby, watershed. The only way to be sure of the course of water in karst is to follow it underground, where possible, or using dye tracers (Goldscheider & Drew, 2007) (Figure 2.3). Locating the karst catchment boundaries represents a highly complex goal, not least because of the great variability that may be recorded in time. It has been proved that in conditions of high

Figure 2.3 Dye tracing test within a cave in southern Italy (photo: N. Damiano, Alburni Exploration Team).

groundwater levels, fossil and otherwise inactive karst conduits become activated, thus originating overflow from one catchment to those nearby (De Waele & Parise, 2013).

For the sustainable development of transboundary karst terrains, the main actions should address facilitating a fair and sustainable use and management of the transboundary water resources, and safeguarding the unique groundwater ecosystems that are present in karst from natural and man-made hazards (Parise & Gunn, 2007; Gutierrez, 2010; Parise, 2010; De Waele *et al.*, 2011). Such hazards now also include climate change. These goals should be pursued by taking advantage of the lessons learned from the ancient communities that used scarce available amounts of water (Parise & Sammarco, 2015) or from the practices for water use in arid and semi-arid lands (Stevanovic *et al.*, 2012).

Dagnino Pastore and co-workers (2013) point out several negative tendencies regarding the management of water resources:

- Non-sustainable use of aquifers, and lack of adequate management of groundwater.
- Presence of human settlements in areas potentially affected by flooding.

- Excessive use of agro-chemicals, with likely contamination of aquifers.
- Limited progress regarding environmental and economic education for improving awareness about the value of water and the social and environmental cost of poor water management, as well as the need to preserve water resources and the environment.

Karst aquifers, in particular, require application of specific hydrogeological methods. Key aspects are investigation and proper understanding of groundwater flow, and the identification and implementation of the most suitable management actions for protection of the resource.

There is a need for water management, to sustain water quality and to ensure rational use of the resource; in other words, to safeguard its natural cycle, as well as the related surface and subsurface ecosystems (Figure 2.4). At the same time, managing water resources is a complicated issue, with difficulties of different types and origin occurring both in industrialised and under-developed countries.

About 40% of the global population lives in river basins shared by more than one country, and many communities depend on drinking water from the same aquifers that have been over-exploited. Thus water scarcity resulting from its poor management frequently brings about the risk of conflict (Dagnino Pastore *et al.*, 2013). Furthermore, 90% of the world's population lives in 145 countries with shared hydrographic basins (UNDP, 2007), where water management could become a potential focus of conflicts.

2.2   THE MEANING OF TRANSBOUNDARY

Decision makers and land use planners, attempting to achieve sustainable development, must take into account the multiple aspects of water resources management (van Dijk & Zhang, 2005). The delicate balance existing in karst, in particular with regard to the transboundary karst water resources, is not easily maintained. There needs to be specific actions to safeguard the high-value biodiversity of the area (Culver & Pipan, 2009).

The extreme heterogeneity of karst, with a high number of variables in different fields, each one of them changing in both time and space, makes analysis complex. A multi-disciplinary approach is required, and understanding the interaction between groundwater and surface water, as well as their influence on biota, both at the ground surface and underground, are crucial points to be addressed.

Several examples can be recalled to outline the delicate and fragile importance of transboundary karst. In Europe, the lakes Ohrid and Prespa are located in the Balkan Peninsula (Figure 2.5). Lake Ohrid is shared by Macedonia and Albania, and Lake Prespa by Macedonia, Albania and Greece (Popovska & Bonacci, 2007). Many caves are present in the surrounding areas, and the lakes are a well known site for biodiversity and cultural heritage (Lake Ohrid was declared a World Heritage Site by UNESCO in 1980). Lake Ohrid is fed by the waters coming from Lake Prespa, that has no surface outflow (Amataj *et al.*, 2007). Hydrologically, the two lakes should be managed as a transboundary water resource. However, so far, no coordinated water resources management has been established.

Figure 2.4 Karst springs: A) the Auso, one of the main springs at the foothills of the Alburni Massif (Southern Italy); B) SyriiKalter, a beautiful spring in southern Albania; C) the main spring of the Unica River, coming out from the Planinska Jama at the southern edge of Planinsko Polje, in Slovenia.

The high importance of transboundary water resources has been recognised by UNESCO within specific programmes on sustainable water resources management, by coordinating the International Network of Water/Environment Centres for the Balkans (INWEB). 80% of the water resources in the Mediterranean region are shared between two or more countries, and in South Eastern Europe transboundary

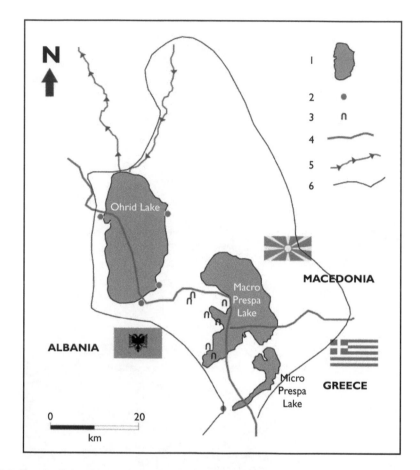

Figure 2.5 Sketch of the Ohrid and Prespa Lakes, at the boundaries among three countries (modified after Popovska & Bonacci, 2007). Key: 1) lake; 2) karst spring; 3) cave; 4) state boundary; 5) water course; 6) limit of the karst catchment (as identified by Popovska & Bonacci, 2007).

groundwaters are the most important source of freshwater. The collection and analysis of data and preparation of hydrogeological maps on transboundary aquifers are key objectives (INWEB, 2008).

Mexico is one country that is interested in transboundary water management, in karst as well in non-karst settings. At its north western borders, the state of Baja California is located within a large area characterised by arid and semi-arid climate. The area has become one of the most important economic regions in Mexico despite water scarcity. About 4 million people live there (the main municipalities are Mexicali, Tecate, Tijuana, and San Luis Río Colorado), using the Colorado River as the main water source by agreement with the American Government in accordance to the 1994 International Waters Treaty.

The situation of the transboundary watershed with the United States is of even greater concern in the face of the possible effects produced by climate change, especially because droughts are expected to occur in the future. The problem has even become more acute due to a significant increase in urban and industrial demand (Román Calleros *et al.*, 2010).

On the southern Mexican border (shared with Guatemala and Belize), there is another transboundary aquifer. Water availability is not a problem in terms of quantity and quality, but there are differences between the individual countries as regards the basic water and sanitation services. Although the area contains six transboundary river basins, shared cooperation of water among the countries is virtually absent.

In the Dominican Republic, the karst region of Los Haitises, covering an area of about 1.6 km², is an important geomorphologic landscape of limestone domes, with large volumes of good-quality water. This karst ecosystem is of extreme importance for water supply in the area, and potentially represents one of the main alternatives for the future water supply for part of the city of Santo Domingo and the eastern part of the country. However, the subterranean waters are at high risk of pollution from a number of planned anthropogenic actions.

About 67% of the Cuban territory is mostly karstified carbonate rocks. Being an island there is always risk of seawater intrusion (Hernandez *et al.*, 2013). The delicate balance among the different types of waters can easily be affected by anthropogenic actions (Molerio León & Parise, 2009).

2.3   APPLYING THE CONCEPTS OF RESILIENCE AND DISTURBANCE IN KARST

The term resilience was initially proposed in the field of ecology as a core concept within ecosystems (Holling, 1973) and has been widely used in recent years, as the ability of a system to absorb perturbation or disturbances. Resilience is the capacity to resist and recover from losses (Kleina *et al.*, 2003). One of the more widespread definitions of resilience nowadays includes: i) response to disturbance; ii) capacity to self-organise; and iii) capacity to learn and adapt.

Since the social and environmental consequences of degradation (including pollution events) are increasing, and cause high costs to society, there is the need to involve stakeholders and the population in strategies to manage the impacts. This is especially required in heavily populated areas, or in those sectors in the proximity of big cities and metropolitan areas, where the natural and anthropogenic factors are mixed (Serre & Barroca, 2013).

In karst, the concept of disturbance produced by man to the natural environment has been analysed by means of the Karst Disturbance Index (van Beynen & Townsend, 2005; North *et al.*, 2009). The Karst Distribution Index takes into account a number of indicators, subdivided into five different categories (geomorphology, atmosphere, hydrology, biota, cultural), to evaluate the disturbance to the karst environment deriving by anthropogenic actions. The index has been applied to different settings, from European karst areas, to selected states or regions in the Americas. As the main outcome from these applications, it appears that knowledge of the main features of the karst environments, encompassing many different fields and disciplines of interest, is fundamental for a proper understanding of the changes occurring, and for linking such changes to specific actions by man or to variability of other factors (i.e. climate change).

Among the other indices that have also been designed for use in karst environments are the Karst Sustainability Index (van Beynen *et al.*, 2012), and the Priority Management Index (Angulo *et al.*, 2013). The first is a standardised metric of sustainable development practices in karst settings. It uses predetermined targets to ascertain the overall sustainability of a karst region, by taking into account 25 indicators related to the fields of environment, economy, and society. The Karst Sustainability Index eventually provides a measure of sustainability in karst areas. The second index (by Angulo *et al.*, 2013) is a holistic approach to analyse the level of disturbance in a protected karst area. Use of two standardised indices (zonal Karst Significance Index, and zonal Karst Disturbance Index) allows the most disturbed sectors to be identified. The two indices can be combined into a Priority Management Index, in order to highlight the main management needs of the protected area.

Another is the evaluation of disturbance in caves (Harley *et al.*, 2011), where the Cave Disturbance Index addresses deficient cave management strategies. The index was based on cave resource data collected during an in situ inventory of 36 terrestrial caves in Florida. The approach was later followed by Van Aken *et al.* (2014) using a GIS-based model to predict disturbance for the same sample of caves. This approach aims at providing land managers or stakeholders with tools to conduct an in situ inventory of a cave, in order to better focus both management strategies and conservation efforts.

The main positive effect from all these approaches is that the peculiarity of karst has been partly acknowledged, together with the need of specific actions for study, management, and decision, based upon the main features of this highly vulnerable environment. Recognition of the uniqueness of karst is a crucial point to be reached in the process of creating an environmental awareness among the communities living in karst lands.

Repeated episodes of mismanagement in karst, with events of pollution and contamination of karst aquifers, both at the surface and underground, or as post-war scenarios (Calò & Parise, 2009), highlight how fragile karst is.

The concept of resilience could be applied to floods (Figure 2.6) in karst (Parise, 2003; Gutierrez *et al.*, 2014; Jourde *et al.*, 2014). In general, floods are the most

Figure 2.6 Effects of floods in karst environments: A) inundated area in the aftermath of a hurricane in Cuba in 2009 (photo courtesy of H. Farfàn Gonzalez); B) flash flood in the city of Bari (Apulia, Southern Italy) (after Mossa, 2009).

common type of natural hazards in the world (Liao, 2012). It has been estimated that they constitute 40% of the natural disasters, and have caused 8.6% of the total casualties, and 80% of injured people from all types of natural hazards in the last 25 years. In a typical tropical karst landscape, floods are often associated with changes in land use, through implementation of inadequate agricultural practices, expansion of urban areas, bad constructive practices, and other human actions inappropriate for the environment (Farfan Gonzalez *et al.*, 2009).

Floods are also an interesting issue as regards the possible different uses of water. Especially in regions where there is a high contrast between the wet and dry seasons (lack of surface water resources, with intervening short periods of abundant water, if not floods), a proper regulation and distribution of the amount of water available are needed. Some of the engineering works that promote constant use of water throughout the year are damming the rivers, and impounding reservoirs. In the case of transboundary rivers, and of rivers shared by more than one country, since dams modify the flow of rivers, they might create conflicts.

The regulatory capacity of a reservoir could be used to buffer the impact of exceptional rises in water level, thus contributing to flood amelioration. Such actions, however, should be carefully planned and managed, in order not to disturb the ecological need for flow of rivers.

The same applies to other engineering works, as outlined by the case of the Ombla spring, in Croatia. The Ombla spring is the source of drinking water for the town of Dubrovnik (Figure 2.1), with an average discharge of some 24 m³/s. The spring is located at the tectonic contact between karstified Mesozoic carbonates and impervious flysch sediments, with these latter acting as a hydrogeological barrier (Milanovic *et al.*, 2014). Construction of a hydroelectrical power plant has been planned, which raised several doubts in the scientific community and the local populations. The plant would consist of an underground dam (spanning from 280 m below sea level to 135 m above sea level) and a power plant.

An interesting example of the different uses of water, even linked to exploitation activities in karst, is represented by the Pertosa cave, in Campania, Southern Italy. The cave, about 3 km long, is located at the north east foothills of the Alburni Massifs, the most important karst area in Southern Italy, hosting several hundred caves (Del Vecchio *et al.*, 2013). It became a show cave in 1932, only two years after the opening of the first show cave in the area, the nearby Castelcivita Cave, situated on the other side of the same carbonate massif. The initial interest in the cave was archaeological, since at its entrance excavations carried out at the start of the twentieth century revealed Paleolithic and Neolithic age material (Larocca, 2015). The cave is an active resurgence, hosting a river with discharge in the range 350–600 l/s, and a dam for hydroelectric power was built at the site in the 1920s (Figure 2.7).

2.4 CONCLUSIONS

The peculiarities of karst environments require a specific approach that accounts for the interrelationships between the surface and the underground environments, as well as the time and space variability of many of the factors influencing karst ecosystems.

Figure 2.7 Double view of the entrance of the Pertosa Cave and its underground stream: on the left the cave with the artificial water basin at his highest level held by the dam; on the right with the open dam and natural water naturally flow (after Larocca, 2015).

Karst aquifers are a valuable natural resource in many transboundary areas, and since biodiversity in karst represents a remarkable and valuable character, it is necessary to safeguard this environment, both from the scientific and technical and the administrative and management standpoints.

Many karst aquifers are transboundary. Sustainable development in karst follows actions that allow growth of the local economy but, at the same time, are respectful of the peculiarities of this environment. Where mismanagement of karst and degradation of the natural resources have occurred, efforts should be made towards a resilience approach, aimed at fully understanding the disturbance caused by man, and at both recovering the karst landscape and exploiting the capacity to learn and adapt from past errors or situations.

REFERENCES

Amataj S., Anovski T., Benischke R., Eftimi R., Gourcy L.L., Kola L., Leontiadis I., Micevski E., Stamos A., Zoto J. (2007) Tracer methods used to verify the hypothesis of Cvijic about the underground connection between Prespa and Ohrid lake. *Environmental Geology* 51(5), 749–753.

Angulo B., Morales T., Uriarte J.A., Antigüedad I. (2013) Implementing a comprehensive approach for evaluating significance and disturbance in protected karst areas to guide management strategies. *Journal of Environmental Management* 130, 386–396.

Calò F., Parise M. (2009) Waste management and problems of groundwater pollution in karst environments in the context of a post-conflict scenario: the case of Mostar (Bosnia Herzegovina). *Habitat International* 33(1), 63–72.

Culver D.C., Pipan T. (2009) *The biology of caves and other subterranean habitats.* Oxford University Press.

Dagnino Pastore J.M., Sturzenegger A., Charreau E.H., Vardè O., Bauer C., Bereciartùa P. (2013) Considerations on the situation of water resources in Argentina. In: Jimenez-Cisneros B., Galizia-Tundisi J. (editors), *Diagnosis of waters in the Americas*, IANAS Water Programs, 17–89.

Del Vecchio U., Lo Mastro F., Maurano F., Parise M., Santo A. (2013) The Alburni Massif, the most important karst area of southern Italy: history of cave explorations and recent developments. In: Filippi M., Bosak P. (editors), *Proceedings 16th International Congress of Speleology*, Brno, 21–28 July 2013, 1, 41–46.

De Waele J., Parise M. (2013) Discussion on the article "Coastal and inland karst morphologies driven by sea level stands: a GIS based method for their evaluation" by Canora F., Fidelibus D., Spilotro G. *Earth Surface Processes and Landforms* 38(8), 902–907.

De Waele J., Gutierrez F., Parise M., Plan L. (2011) Geomorphology and natural hazards in karst areas: a review. *Geomorphology* 134(1–2), 1–8.

Farfan Gonzalez H., Corvea Porras J.L., Martinez Maquiera Y., Diaz Guanche C., Aldana Vilas C., de Bustamante I., Parise M. (2009) Impact of the hurricanes Gustav and Ike in the karst areas of the Viñales National Park, Cuba. *Geophysical Research Abstracts* 11, 1186.

Ford D., Williams P.W. (2007) *Karst hydrogeology and geomorphology.* Wiley and Sons.

Goldscheider N., Drew D. (editors) (2007) *Methods in karst hydrogeology.* International Contributions to Hydrogeology 26, International Association of Hydrogeologists, 264 p.

Gunn J. (2007) Contributory area definition for groundwater source protection and hazard mitigation in carbonate aquifers. In: Parise M., Gunn J. (editors), *Natural and anthropogenic hazards in karst areas: recognition, analysis and mitigation.* Geological Society of London, Special Publication 279, 97–109.

Gutiérrez F. (2010) Hazards associated with karst. In: Alcántara I., Goudie A. (editors), *Geomorphological Hazards and Disaster Prevention.* Cambridge University Press, Cambridge, 161–175.

Gutierrez F., Parise M., De Waele J., Jourde H. (2014) A review on natural and human-induced geohazards and impacts in karst. *Earth Science Reviews* 138, 61–88.

Harley G.L., Polk J.S., North L.A., Reeder P.R. (2011) Application of a cave inventory system to stimulate development of management strategies: The case of west-central Florida, USA. *Journal of Environmental Management* 92, 2547–2557.

Hartmann A., Goldscheider N., Wagener T., Lange J., Weiler M. (2014) Karst water resources in a changing world: review of hydrological modeling approaches. *Reviews of Geophysics* 52, doi:10.1002/2013RG000443.

Hernandez R., Ramirez R., Lopez-Portilla M., Gonzalez P., Antiguedad I., Diaz S. (2013) Seawater intrusion in the coastal aquifer of Guanahacabibes, Pinar del Rio, Cuba. In: Farfan Gonzalez H., Corvea Porras J.L., de Bustamante Gutierrez I., LaMoreaux J.W. (editors) *Management of water resources in protected areas.* Springer, Berlin, 301–308.

Holling C.S. (1973) Resilience and stability of ecological systems. *Ann. Rev. Ecol. Syst.* 4, 1–23.

INWEB (2008) Inventories of transboundary groundwater aquifers in the Balkans, UNESCO Chair and Network INWEB, Thessaloniki, Greece http://www.inweb.gr/.

Jourde, H., Lafare, A., Mazzilli, N., Belaud, G., Neppel, L., Doerfliger, N., Cernesson, F. (2014) Flash flood mitigation as a positive consequence of anthropogenic forcings on the groundwater resource in a karst catchment. *Environmental Earth Sciences* 71, 573–583.

Kleina R.J.T., Nicholls R.J., Thomalla F. (2003) Resilience to natural hazards: How useful is this concept? *Environmental Hazards* 5, 35–45.

Larocca F. (2015) La Grotta di Pertosa (Salerno) e il suo giacimento archeologico. In: De Nitto L., Maurano F., Parise M. (editors), *Proceedings XXII National Congress of Speleology – Euro Speleo Forum 2015 "Condividere i dati"*, Memorie dell'Istituto Italiano di Speleologia, 29, 651–656.

Liao K.H. (2012) A theory on urban resilience to floods – A basis for alternative planning practices. *Ecology and Society* 17(4), 48.

Milanović P., Stevanović Z., Čokorilo Ilić M. (2014) *Field trip guide: Trebinje, June 2014*. DIKTAS Project, Faculty of Mining and Geology, Department for Hydrogeology, Belgrade.

Molerio León L., Parise M. (2009) Managing environmental problems in Cuban karstic aquifers. *Environmental Geology* 58(2), 275–283.

Mossa M. (2007) The floods in Bari: what history should have taught. *Journal of Hydraulic Research* 45(5), 579–594.

North L.A., van Beynen P.E., Parise M. (2009) Interregional comparison of karst disturbance: West-central Florida and southeast Italy. *Journal of Environmental Management* 90(5), 1770–1781.

Palmer A.N. (2007) *Cave Geology*. Cave Books, Dayton.

Parise, M. (2003) Flood history in the karst environment of Castellana-Grotte (Apulia, southern Italy). *Natural Hazards and Earth System Sciences* 3(6), 593–604.

Parise M. (2010) Hazards in karst. In: Bonacci O. (editor), *Proc. Int. Sc. Conf. "Sustainability of the Karst Environment, Dinaric Karst and Other Karst Regions"*, Plitvice Lakes, Croatia, 23–26 September 2009, IHP-UNESCO, Series on Groundwater, 2, 155–162.

Parise M., Gunn J. (2007) *Natural and anthropogenic hazards in karst areas: an introduction*. In: Parise M. & Gunn J. (editors) *Natural and anthropogenic hazards in karst areas: Recognition, Analysis, and Mitigation*. Geological Society, London, Special Publication 279, 1–3.

Parise M. & Sammarco M. (2015) The historical use of water resources in karst. *Environmental Earth Sciences* 74, 143–152.

Popovska C., Bonacci O. (2007) Basic data on the hydrology of Lakes Ohrid and Prespa. *Hydrological Processes* 21, 658–664.

Román Calleros J., Cortez Lara A., Soto Ortiz R., Escoboza García F., Viramontes Olivas O. (2010) El aguaen el noroeste. In: Jiménez B., Torregrosa M.L., Aboites A. (editors), *El Agua en México: Cauces y encauces*. Academia Mexicana de Ciencias, 479–504.

Serre D., Barroca D. (2013) Natural hazard resilient cities. *Natural Hazards and Earth System Sciences* 13, 2675–2678.

Stevanović Z., Balint Z., Gadain H., Trivić B., Marobhe I., Milanović S. *et al.* (2012) *Hydrogeological survey and assessment of selected areas in Somaliland and Puntland*. Technical report no. W-20, FAO-SWALIM (GCP/SOM/049/EC) Project, (http://www.faoswalim.org/water_reports) Nairobi.

UNDP (2007) *Informe de desarrollo humano, 2006. Másallá de la escasez: poder, pobreza y la crisismundial del agua*. United Nations Development Programme.

Van Aken M., Harley G.L., Dickens J.F., Polk J.S., North L.A. (2014) A GIS-based modeling approach to predicting cave disturbance in karst landscapes: a case study from west-central Florida. *Physical Geography* 35(2), 123–133.

van Beynen P.E., Townsend K. (2005) A disturbance index for karst environments. *Environmental Management* 36, 101–116.

van Beynen P.E., Brinkmann R., van Beynen K. (2012) A sustainability index for karst environments. *Journal of Cave and Karst Studies* 74(2), 221–234.

van Dijk M.P., Zhang M. (2005) Sustainability indices as a tool for urban managers, evidence from four medium-sized Chinese cities. *Environmental Impact Assessment Review* 25, 667–688.

Chapter 3

WOKAM – The world karst aquifer mapping project, examples from South East Europe, Near and Middle East and Eastern Africa

Zoran Stevanović[1], Nico Goldscheider[2], Zhao Chen[2] & the WOKAM Team
[1]*University of Belgrade – Faculty of Mining and Geology, Centre for Karst Hydrogeology of the Department of Hydrogeology, Belgrade, Serbia*
[2]*Karlsruhe Institute of Technology (KIT), Institute of Applied Geosciences, Division of Hydrogeology, Karlsruhe, Germany*

ABSTRACT

The WOKAM project supported by the International Association of Hydrogeologists and UNESCO was established in 2012 in order to obtain a better global overview of karst aquifers, to create a basis for sustainable international and global water resources management, and to increase the awareness and knowledge concerning these special resources, both among the public and the decision-makers. The team of international experts is responsible for data collection and evaluation, definition of methodology and creation of the new karst aquifers map and the associated database. The map, to be completed in 2016, will not only show carbonate rock and evaporite outcrops, but also display deep and confined karst aquifers, large karst springs, including thermal and mineral springs, drinking water abstraction sites and selected caves. The World Karst Aquifer Map is based on the highly detailed Global Lithological Map (GLiM) and is intended to supplement the existing map of Groundwater Resources of the World (WHYMAP). The paper presents some examples and large differences between karstic aquifers of the Alpine system of South East Europe, the Near and Middle East, which have some of the most karstified and richest water reserves in the world, and the less productive and developed East African karst.

3.1 INTRODUCTION

According to an often-cited estimation by Ford & Williams (2007) surface and subsurface outcrops of potentially soluble karstic rocks occupy around 20% of the planet's ice-free land, but probably not more than 10–15% is extensively karstified. The same authors found that probably more than 90% of the evaporitic rocks anhydrite and gypsum do not crop out, while this percentage in the case of remarkably soluble salty rocks is almost 99%. About 25% of the global population is at least partly supplied by freshwater from these karst aquifers. Owing to their specific hydrogeological properties, karst aquifers are particularly vulnerable to contamination and difficult to manage. Furthermore, many karst aquifers are hydraulically connected over wide areas and thus require transboundary exploration, protection and management concepts.

For almost half a century, since the Karst Commission was established under the umbrella of the International Association of Hydrogeologists, through the publication of books, and organisation of conferences, field trips, seminars, and workshops, it has made many important contributions to the knowledge of karstic aquifers and groundwater. One of its most recent collaborative efforts is the implementation of the project WOKAM, the ultimate goal of which is to create a world map and database of karst aquifers, as a further development and refinement of earlier maps.

Numerous international karst experts contribute to this project by collecting data and working on an international Scientific Advisory Board that meets on a regular basis[1]. The first results of this project were presented by Scientific Advisory Board members at several global conferences, including 'Karst without boundaries' held in Trebinje in June, 2014. There were opportunities to discuss the draft legend, the design and preliminary content of the associated database, as well as the first sketch maps for Europe, Asia and North America.

3.2 METHODOLOGICAL APPROACH

Although karst phenomena were listed and presented in numerous books and encyclopedias (e.g. Gunn, 2004; Culver & White, 2005), and the most attractive and largest were included on many websites, several attempts were also made to present various karst features on the map and in the common database. These activities took place largely in the USA (Epstein *et al.*, 2001; Hollingsworth E. *et al.*, 2008; Veni, 2002). The sketch maps showing regional distribution of carbonate rocks are also presented in Ford and Williams' book *Karst Hydrogeology and Geomorphology* (2007). The same map with a somewhat better resolution is available at the web address of the University of Auckland, New Zealand http://www.sges.auckland.ac.nz/sges_research/karst.shtm.

After completion of WHYMAP, the map showing the groundwater resources of the World (Richts *et al.*, 2011), the Past-President of IAH, Willi Struckmeier, proposed during the karst conference held in Malaga in 2010 that a karst map should be prepared to complement the existing global groundwater resources map. The concept was accepted and an initiative team under the coordination of Nico Goldscheider from Karlsruhe Institute of Technology was formed and started to work. From the beginning it was agreed that one main output would be a GIS-based map in scale 1:10 Million, while final hard (paper) copy will be at a scale of 1:25 Million, with several layers and a connected database.

---

1 WOKAM Scientific Advisory Board consists of the following experts (in alphabetical order): Augusto Auler (Brazil), Michel Bakalowicz (France), David Drew (Ireland), Nico Goldscheider (Germany, Project leader), Guanghui Jiang (PR China), Jens Hartmann & Nils Moosdorf (Germany, GLiM Team), Andrea Richts (Germany, WHYMAP Team), Zoran Stevanovic (Serbia) and George Veni (USA). Zhao Chen, PhD student at KIT, is compiling the map, with support of a student assistant, Franziska Griger. Alexander Klimchouk, Art Palmer, Paul Williams and Daoxian Yuan contributed by exchange of ideas during the project preparation.

To make such a thematic map on a very small scale, appropriate answers to very specific issues are required. Some of them are:

- Classification of karst aquifers (including criteria such as lithology, porosity, structures, hydrodynamics);
- Distribution of fresh, thermomineral or saline karstic groundwater;
- The depth up to which the confined (buried) karst aquifers will be considered and shown on the map (shallow with fresh waters, deep with thermal waters, oil/gas reserves);
- The karst features that might be included on the map and criteria for their selection (main sources, cave systems, regional groundwater flow directions, etc.).

In addition, management of such a project is not an easy task because due to their specialisations (e.g. geomorphology, hydrogeology, chemistry) there are often diverse opinions among the experts. This also affects what the priorities are and how to present them in the final materials. However, during several Scientific Advisory Board meetings held in Karlsruhe between 2012 and 2014, common approaches were achieved on essential issues:

1. The mapping of rock distribution is based on the highly detailed Global Lithological Map (GLiM), which was assembled from 92 regional lithological maps of the highest available resolution (Hartmann & Moosdorf, 2012).
2. Deep and confined karst aquifers are identified on the basis of regional literature and geological sections and their approximate boundaries are shown on the map.
3. WOKAM GIS database includes large karst springs, thermal springs, submarine springs, major drinking water abstraction sites and important caves, which are all shown on the map as point symbols.

3.2.1   Rock complexes – Aquifer systems

Following the generalisation and reclassification of GLiM, four lithological units and subsequent aquifer systems are presented on the World Karst Aquifer Map: 1) igneous and metamorphic rocks, 2) non-carbonate sedimentary formations, which comprise unconsolidated sediments, siliciclastic dominate sedimentary rocks, and pyroclastic formations, 3) carbonate sedimentary rocks, which are further subdivided into continuous and discontinuous (when outcrops of karstified rocks are separated by non-karstic rocks), and 4) evaporites. In fact, only the last two belong to classical rock complexes resulting in two main karst aquifer systems. The draft version of the WOKAM lithology units for all of Europe is shown on Figure 3.1.

3.2.2   Deep and confined karst aquifers

Karst aquifers not exposed at the surface cannot be delineated precisely without detailed 3D geological information that is usually not available. Therefore, this work step is done manually in a pragmatic way, with geologic expertise, and concealed karstified rocks are presented as potential deep or confined karst water resources. For this purpose a 'thrust fault symbol' is used: the teeth point in the dipping direction, toward the deep aquifer.

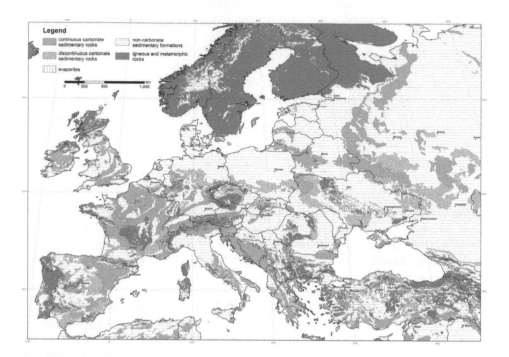

Figure 3.1 Lithology units in Europe and adjacent areas as a base for the draft WOKAM map.

3.2.3 WOKAM GIS database

The two main groups of objects to be presented on the map are water points (large karst springs, thermal springs, submarine springs, major drinking water abstraction sites), and caves. Although the former group is of primary importance considering that water resources are the focus of the WOKAM project (Goldscheider *et al.*, 2014), the caves are also important not only as the main morphological feature in karst, but as an indicator of karstification intensity.

Data on springs and caves are both separated in the two groups, those to be included in the database (larger group) and those to be shown on the map (selected as the largest). After discussion the Scientific Advisory Board made a preliminary decision on the following criteria:

3.2.3.1 Springs

The numbers in brackets indicate candidates to be included on the map in the database, while the presentation of selected springs on the final map will be much less detailed:

- Permanent karst springs with a minimum discharge > 200 l/s (for the map: 500 l/s²)

2 Later, the Scientific Advisory Board proposed a norm of 2,000 l/s, but still in a flexible manner, so that if a spring is not reaching the criterion, but it is unique and represents a regionally important karst aquifer it could be shown on the map. In this case expert's opinion is relevant.

- Temporary or highly variable springs with a maximum discharge > 10 m³/s (50 m³/s)
- Very important submarine springs (experts' opinion)
- Thermal water resources/systems with total discharge > 100 l/s that are >4°C warmer than the average air temperature (200 l/s and 10°C warmer than air temperature)
- Karst springs > 100 l/s with peculiar gas composition, such as CO_2 or H_2S (200 l/s)
- Important wells, well fields, drainage galleries or other water abstraction structures in karst with a discharge/pumping rate > 200 l/s (500 l/s).

The 'estimated annual minimum discharge' (low flow, baseflow) is thus a main criterion. It can be determined in three different ways: 1) One or several discharge measurements or estimations during low-flow conditions; 2) Estimated on the basis of minimum and mean discharge; 3) Calculation on the basis of long-term discharge data. But it was also agreed that because of the heterogeneity of the database and the heterogeneous global distribution of large springs, the legend and selection criteria could be slightly modified. Accordingly, not all the large springs have to be shown on the map in terms of their very high density (e.g. Dinaric karst), while some of the smaller springs that are regionally very important, such as the Jordan spring, although fulfilling none of criteria, can be shown on the map. Although it is not logical to exclude some larger springs because there are a lot of them the printing scale is such that overlapping of objects would be unavoidable. In contrast, some small springs can be included as an extra item in the legend because of their importance.

3.2.3.2 Caves

The Scientific Advisory Board has established basic ranking criteria by appointing the following values (total maximum of 10 points): up to 5 points for dimensions (3 for length and 2 for depth) and up to 5 points for significance (2 for hydrological significance, 1 for human use and ecosystems, 2 for regional significance). In principle, caves with less than 4 or 5 points would not be mapped, and will remain just a part of the database. The final goal is to have a short list of the most relevant caves for the map. These rankings also confirmed that caves that are significant in terms of archeology, cave paintings, fossil remnants, and tourism are less likely to be displayed on the map than, for instance, a 'short' cave with high hydrologic and regional significance.

3.2.4 Case example – Springs and caves of South East Europe, Near and Middle East, and Eastern Africa

Data were collected by the first author of this paper from a wide region shared by more than 40 countries. The northernmost part is Poland (up to 55° N), while the southernmost parts are India and the east (Asia) and Tanzania on the west (Africa, below the equator, Figure 3.2). For this exercise extensive work and the involvement of many local experts were required. Not all of these countries possess karst and karstic features (e.g. Rwanda, Seychelles, Bhutan) while in others with considerable karst (Yemen, Afghanistan) data collection was not possible mainly because of the absence of local expertise or literature. Some requests of local researchers for additional field survey were disregarded due to limited funds and the security situation, and because

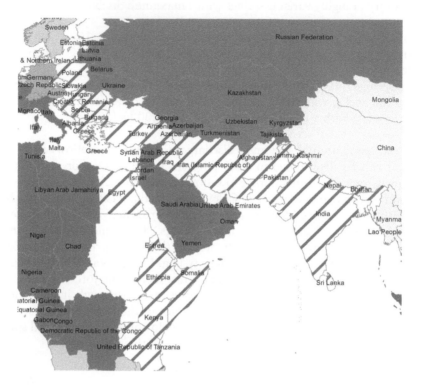

Figure 3.2 Study region – from Central and South East Europe to Indian sub-continent and eastern Africa. The lines show countries with information on water points and caves in the project stage I.

of the scale of the WOKAM map. However, until the end of 2014, involved local experts provided a significant amount of information and prepared inquiry lists for springs and caves completed by the Scientific Advisory Board. It might be concluded that the work has been done successfully, while in a 'live organism' such as the GIS database, additionally collected data will improve the overall quality of the study and enable further creation of regional maps at larger scales.

The karstic systems of the region might be divided into two major groups: one represents aquifers formed in large *geosyncline structures* (Tethys), recently transformed into a highly deformed orogenic belt consisting of high mountains intersected by large depressions, and the second, the *platform type of karst*, slightly deformed and karstified (Stevanović, 2015).

The first, the Alpine geosyncline karstic system, has several branches: Carpathians, Balkanides, Dinarides, Hellenides, Taurides, Zagros, and Himalayan mountain chains with adjacent areas. It is an area of highly developed karst with some 'classical' karst regions such as Dinaric karst. Creation of karst is associated with the transgression of the Tethys Ocean and later intensive orogenesis (intensive uplifting and folding) repeated in several stages: the two most intensive orogenic phases took place at the end of the Upper Cretaceous and during the Paleocene (Laramian tectonic phase), and in Eocene/Oligocene (Helvetian phase). The karst in the Northern Hemisphere is almost entirely carbonate (limestones and dolomites), while varieties of evaporitic karst are more frequent in the southern parts (Iraq, Iran, Somalia). The carbonate sedimentary complex can be very thick: for instance the thickness of Dinaric karstic rocks is often greater than 1000 m. The Alpine system is characterised by abundant water resources but they are unequally distributed throughout the year, due mostly to the specific climate and a high karstification rate (Bonacci, 1987). Some areas in Dinaric karst, such as southern Montenegro, are characterised by a very intensive water balance where the average specific yield is over 40 l/s/km^2 (Radulović, 2000). Apart from richness in water reserves the Alpine system is also known for various karstic features, and some of the deepest potholes and longest caves.

The second, platform karst system is developed in Africa and the Arabian Peninsula. It is less developed karst, first because of the younger age of sedimentary complex (shorter karstification), and secondly, and more important due to less tectonised and deformed carbonate and evaporitic rocks. In Eastern Africa (Somalia, Ethiopia) there are some outcrops of Jurassic carbonates, but widely exposed karstic aquifers are of Eocene and Miocene ages. Horizontal or slightly dipping layers did not allow deeper karstification and intense groundwater flow. As an example, the average specific groundwater yield is >1 l/s/km^2 for most of the surveyed karstic aquifer systems in Somaliland and Puntland provinces of northern Somalia.

The database created for the entire area of 47 countries including both Alpine and platform karstic systems consists of 124 major springs and 52 caves. The springs database contains information of spring coordinates, lithology, minimum/average/maximum discharges, annual low flow (as flow estimated based on one or several discharge measurements or estimations during low-flow conditions), average water and air temperatures, gas content, water chemistry (total dissolved soils and/or electrical conductivity), utilisation and main references. Based on expert judgment, of the 124 springs at least 53 are proposed to be shown on the printed map, while the rest are marked as 'perhaps yes'.

3.2.4.1   Springs

The largest number of springs occur in the Dinaric karst (Mijatović, 1984; Janež et al., 1997; Kranjc, 1997; Bakalowicz et al., 2003; Stevanović, 2010; DIKTAS database http://dinaric.iwlearn.org). Although a more restrictive approach has been applied in the selection of springs, 53, or 40% of all major springs belong to that aquifer system across seven countries which nowadays share the water resources. For instance, the four countries of former Yugoslavia have the highest density of springs: distribution of large karstic springs which achieve the discharge criterion against territory covered by karst in Croatia, Slovenia, Montenegro and Bosnia & Herzegovina is equal to or larger than 1 spring/2000 km^2. Probably additional restriction or adaptation of the legend (grouping of objects) will be required once a full version of the map with water points/caves is displayed (Figure 3.3).

A considerable number of springs are also characteristic of other countries which contribute in part to the Dinaric system: In the Albanian part of the Dinarides (Albanian Alps) there are 2 large springs (Eftimi, 2010), in the Dinarides of western Serbia, 4, and in FRY of Macedonia, 4. In the southern extensions of Alpides: Pindes and Hellenides, 4 Albanian and 2 Greek springs are included in the WOKAM database.

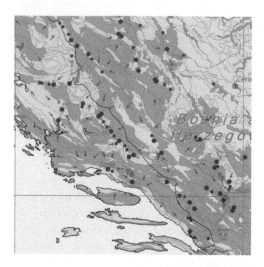

Figure 3.3 The segment of the digital Hydrogeological map of Dinaric karst prepared under the DIKTAS
project (http://dinaric.iwlearn.org). The boundary area between Bosnia & Herzegovina and
Croatia (between Zrmanja and Neretva rivers) is very rich in karstic springs discharging a
minimum of over 100 l/s (smaller dots) and over 1000 l/s (larger dots), not all of which can
be shown on the WOKAM map.

The Carpathian karst is also rich in groundwater reserves and large springs,
but to a much lesser extent than the southern Alpine branches. The number of large
springs or well-fields included in the WOKAM database is: 1 in Czech Republic, 1 in
Slovakia, 4 in Poland, 3 in Romania (Oraşeanu & Iurkiewicz, 2010) and 3 in Serbia
(Stevanović, 1994). In the Balkanides of Bulgaria there are four large springs proposed for the WOKAM database (Benderev *et al.*, 2005). In the Hungarian part of the
Pannonian basin and its margin there are several fresh water and thermal springs, four
of them are included in the WOKAM list.

The Taurus Mountains in central and southern Turkey are the most karstified
region in the Near East (Günay, 2010). Most of the 9 selected springs, which
issue from Mesozoic and rarely from Tertiary sediments, belong to that unit. The
six large springs selected are in Israel; among them is the famous Jordan River
source.

In Iraq, five springs met the criteria for inclusion in the WOKAM database. They
belong to the two major geo-structural units, the Taurides and Zagros Mountains,
both extending over the northernmost part of the country along the borders of Turkey
and Iran. They drain the two main karstic aquifer systems, 'Bekhme' of Cretaceous,
and 'Pila Spi' of Tertiary age (Stevanović & Iurkiewicz, 2004). Similarly, nine major
springs of Iranian karst were selected for inclusion on the WOKAM list. They also
drain limestones of Cretaceous (Sarvak Formation) and Oligocene-Miocene age
(Asmari Formation; Raeisi & Stevanović, 2010). The evaporitic rocks are connected
to younger formations (Mio-Pliocene, Quaternary).

In India, out of the twelve selected, nine springs are located in Kashmir. The other
three are on the Meghalaya – Shillong Plateau.

North eastern and eastern African and Arabian plateaus are less karstified and almost none of these karstic occurrences satisfied the established criterion of minimum discharge (Figure 3.4) for the database. However, 3 major spring groups (oases) have been selected in Egypt, along with one well-field in Somalia and one well-field in Ethiopia, as representatives of local karst aquifers. Although Eocene carbonate and evaporitic rocks have major extensions in Somalia and eastern Ethiopia, the two selected well fields are linked to Jurassic limestones (Stevanović et al., 2012).

Concerning the spring discharges, of the 124 springs in this wide region, 28 or 1/5 have a minimal discharge larger than 2000 l/s which represents key evidence of water availability. The largest is the famous Dumanli spring in Turkey (38 000 l/s),

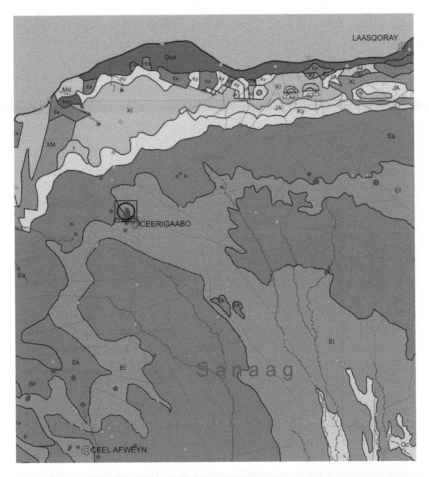

Figure 3.4 The segment of digital Hydrogeological map of Sanaag karstic area between Somaliland and Puntland provinces (northern Somalia) prepared under project SWALIM, IV Phase (http://www.faoswalim.org/water_reports). Karstic springs or well fields situated in Eocene karstic rocks (all formations on map labeled by E) are relatively dense, but springs that discharge more than 10 l/s (big dots on the map) are very rare.

nowadays impounded by the Oymapinar reservoir. The second ranked is the Bistrica group of springs in Albania (including the well-known Blue eye spring) with a minimum discharge of 12 000 l/s. The largest number of springs discharging regularly over 2 000 l/s are in Bosnia & Herzegovina (8) and Turkey (8) followed by Montenegro (5). Milanović (2006) noted that springs along the Neretva Valley and in eastern Herzegovina (Buna, Bunica) are the largest in the Adriatic basin.

3.2.4.2 Caves

The distribution of caves is not always the same as that for springs. Some 'karstic' countries possess larger speleological objects than water objects. For instance, the Carpathian karst is specific in its development of large underground cavities formed along preferred tectonic pathways. As such, 18 caves, or 1/3 of the total proposed for the WOKAM list, are from that geo-structural unit: Romania is a leader in this group with 9 caves, followed by Poland (Tatra Mountain) with 3, Czechia (Moravsky kras) and Hungary with 2, and Slovakia and Serbia with one each.

Slovenia has the most caves in the southern Alpine group with 9 caves proposed for the WOKAM list. Among them is the famous touristic cave Postojna. Some other well known caves, or world's largest potholes, are also selected to be displayed on the map: the Humpleu-Poienita cave system (Romania, 40 km of explored channels in length), Demänovské jaskyne system (Slovakia, 36 km), Amaterska cave (Czechia, 35 km), Wielka Śnieżna Cave (Poland, 23 km), Čehi 2 pothole (Slovenia, 1500 m in depth), and Slovačka jama (Croatia, 1300 m deep).

The largest African cave, Sof Omar Holluca in Ethiopia, with 15 km of explored channels, is included in the list. The largest Indian cave among five proposed for the WOKAM list (all in Meghalaya) is Krem Liat Prah with 30 km of explored channels.

Of 52 WOKAM caves from the studied region, 11 have more than 20 km of explored channels, while 6 potholes are deeper than 1000 m.

3.3   DRAFT MAP OF THE STUDIED REGION

The WOKAM project will be completed in 2016 and the map is scheduled to be printed in late 2016. Because this chapter was written before the final evaluation of the submitted data and the Scientific Advisory Board's review, all information is of a preliminary nature. Figure 3.5 shows a part of the draft map with selected springs of the south eastern Europe, eastern Mediterranean basin and adjacent areas.

3.4   CONCLUSIONS

The WOKAM project will be completed in 2016 and will result in the World Karst Aquifer Map, one of the layers of, and supplement to, the existing map of Groundwater Resources of the World (WHYMAP). It will provide a better global overview of karst aquifers, and is expected to become a useful tool for sustainable global water resources management, and to increase public awareness of the importance of these resources, currently utilised for drinking purposes by some 25% of the world's population.

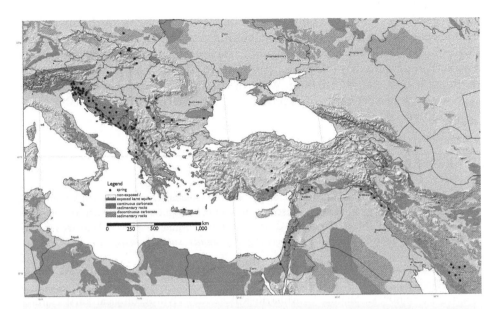

Figure 3.5 A part of preliminary WOKAM map showing selected springs of the South East Europe and adjacent areas.

The team of international experts, members of the Scientific Advisory Board working on the project since 2012, has defined the methodology for data collection and evaluation, and for the creation of the map and its legend as well as the associated database. As the most feasible solution to present the lithology and corresponding aquifer systems, the existing Global Lithological Map was chosen. Selection criteria for water points (springs, thermal springs, well fields, extraction intakes) and caves as single geomorphologic karstic forms to be included in the map were defined in accordance with the preferred scale of 1: 10 Million. However, considering the large differences between highly karstified areas and rich karstic aquifers, and poorly developed karst with small water quantity and availability, only two approaches to the selection of objects for the map and database were found to be feasible: the more restrictive approach in the former case and the more flexible in the latter. This concept is demonstrated by the examples from South East Europe, and the Near and Middle East where karstic aquifers belong to the Alpine system and are the richest in water reserves in the world, and in contrast with examples from less productive and developed Eastern African karst. In eastern Africa karstic formations have a large area, but the yield of springs or drilled wells is relatively low and rarely exceeds 10 l/s. Thus, although they are the main source for local water supply, they do not meet the WOKAM criteria for display on the map.

ACKNOWLEDGEMENTS

Authors and members of WOKAM Scientific Advisory Board gratefully acknowledge the efforts by the consultants and colleagues from South East Europe, the Near and Middle East: Romeo Eftimi (Albania), Boban Jolović (Bosnia & Herzegovina), Aleksey Benderev (Bulgaria), Judit Mádl-Szőny (Hungary), Želimir Pekaš (Croatia), Jerome Perrin (BRGM-France, data for India), Ezzat Raeisi (Iran), Milan Radulović (Montenegro), Jacek Rózkowski (Poland), Iancu Orăşeanu (Romania), Peter Malik (Slovakia), Nataša Ravbar (Slovenia) and Gültekin Günay (Turkey) to collect and provide valuable data and information. Without the help of all those involved, this work would not have been complete and productive. We also thank IAH and UNESCO for provided financial support.

REFERENCES

Bakalowicz M., Fleury P., Dörfliger N., Seidel J.L. (2003) Coastal karst aquifers in Mediterranean regions. A valuable ground water resource in complex aquifers. In: Instituto Geologico y Minero de Espana Publ., Technologia de la Intrusion de Agua de mar en Acuiferos Costeros: paises mediterraneos (TIAC). *Hydrogeologia y aguas subterraneas,* no. 8, Alicante, 2, p. 125–140.

Bonacci O. (1987) *Karst Hydrology; with special reference to the Dinaric Karst.* Springer-Verlag, Berlin.

Benderev A., Spasov V., Shanov S., Mihaylova B. (2005) Hydrogeological karst features of the Western Balkan (Bulgaria) and the anthropological impact. In: *Water Resources and Environmental Problems in Karst* – CVIJIĆ 2005, eds. Z. Stevanović and P. Milanović. Spec. ed. FMG. Belgrade, pp. 37–42.

Culver D.C., White W.B. (2005) *Encyclopedia of Caves.* Burlington, MA, Elsevier Academic Press, 654 p.

Eftimi R. (2010) Hydrogeological characteristics of Albania. *AQUAmundi.* 1, 79–92.

Ford D.C., Williams P.W. (2007) *Karst Hydrology and Geomorphology.* London, Wiley Chichester, 2nd ed., 576 p.

Epstein J.B., Orndorff, R.C.,Weary, D.J. (2001) U.S. Geological Survey National Karst map [abs.]: National Speleological Society Convention Program Guide, 87 p.

Goldscheider N., Chen Zh., WOKAM Team (2014) The world karst aquifer mapping project – WOKAM. [abs.] In: Kukurić N, Stevanović Z, Krešic N (eds.) *Proceedings of the DIKTAS Conference: 'Karst without boundaries',* Trebinje, June 11–15 2014, p. 391.

Günay G. (2010) Geological and hydrogeological properties of Turkish karst and major karstic springs. In: Kresic N., Stevanovic Z. (eds.) *Groundwater hydrology of springs: Engineering, theory, management and sustainability,* Elsevier Inc., BH, Burlington-Oxford, 479–497.

Gunn J. (2004) *Encyclopedia of Caves and Karst Science.* New York, Fitzroy Dearborn, 902 p.

Hartmann J., Moosdorf N. (2012) The new global lithological map database GLiM: A representation of rock properties at the Earth surface. *Geochemistry, Geophysics, Geosystems* 13 (12): 1–37. doi:10.1029/2012GC004370.

Hollingsworth E., Van Brahana, Inlander E., Slay M. (2008) Karst Regions of the World (KROW) (2008): Global karst datasets and maps to advance the protection of karst species and habitats worldwide. http://pubs.usgs.gov/sir/2008/5023/pdf/06hollings.pdf

Janež J., Čar J., Habič P., Podobnik R. (1997) *Vodno bogastvo visokega krasa.* Geologija d.o.o., Idrija, 167 p.

Kranjc A. (1997) Karst hydrogeological investigations in south-western Slovenia. *Acta Carsologica* 26(1), 388 p.

Mijatović B. (1984) *Hydrogeology of the Dinaric Karst.* International Association of Hydrogeologists, Heise, Hannover. Vol. 4.

Milanović P. (2006) *Karst of eastern Herzegovina and Dubrovnik littoral*. ASOS, Belgrade, 362 p.

Oraşeanu I., Iurkiewicz A. (2010) (eds) *Karst hydrogeology of Romania*. Belvedere Publ. Oradea.

Radulović, M. (2000) *Karst hydrogeology of Montenegro*. Sep. issue of Geological Bulletin, vol. XVIII, Spec. ed. Geol. Survey of Montenegro, Podgorica, 271 p.

Raeisi E., Stevanović Z. (2010) Springs of Zagros mountain range (Iran and Iraq). In: Kresic N., Stevanovic Z. (eds.) *Groundwater hydrology of springs: Engineering, theory, management and sustainability*, Elsevier Inc., BH, Burlington-Oxford, 498–515.

Richts A., Struckmeier W.F., Zaepke M. (2011) WHYMAP and the Groundwater Resources of the World 1:25,000,000. In: Jones, J.A.A. (Ed.) *Sustaining Groundwater Resources*, pp. 159–173.

Stevanović Z. (1994) Karst ground waters of Carpatho – Balkanides in Eastern Serbia. In: Stevanović Z, Filipović B. (eds) *Ground waters in carbonate rocks of the Carpathian – Balkan mountain range*. Spec. ed. of CBGA, Allston, Jersey, pp. 203–237.

Stevanović Z., Iurkiewicz A. (2004) *Hydrogeology of Northern Iraq, Vol. 2. General hydrogeology and aquifer systems*, Spec. Edition TCES, FAO/UN, Rome, 175 p.

Stevanović Z. (2010) Major springs of southeastern Europe and their utilization, In: Kresic N., Stevanovic Z. (eds.) *Groundwater hydrology of springs: Engineering, theory, management and sustainability*, Elsevier Inc., BH, Burlington-Oxford, 389–410.

Stevanović Z, Balint Z, Gadain H, Trivić B, Marobhe I, Milanović S *et al.* (2012). Hydrogeological survey and assessment of selected areas in Somaliland and Puntland. Technical report no. W-20, FAO-SWALIM (GCP/SOM/049/EC) Project. (http://www.faoswalim.org/water_reports) Nairobi.

Stevanović Z. (2015) Characterization of Karst Aquifer. In: Stevanović Z. (ed) *Karst Aquifers – Characterization and Engineering*, Series: Professional Practice in Earth Science. Springer International Publishing Switzerland, 47–126.

Veni, G., April 2002, Revising the karst map of the United States. *Journal of Cave and Karst Studies*, 64(1), 45–50.

Some useful web sites on the topic:
http://www.sges.auckland.ac.nz/sges_research/karst.shtm
www.karstportal.org
http://dinaric.iwlearn.org
www.speleogenesis.info
www.caverbob.com
www.caves.org

Chapter 4

Groundwater flow in the Orontes River basin and the Syria–Lebanon water sharing agreement

François Zwahlen[1], Michel Bakalowicz[2],
Raoul Gonzalez[3], Ahmed Haj Asaad[3],
Myriam Saadé-Sbeih[3] & Ronald Jaubert[3]
[1]*Chyn, University Neuchâtel, Switzerland*
[2]*HydroSciences, Montpellier University, France*
[3]*Graduate Institute of International and Development Studies,*
Geneva, Switzerland

ABSTRACT

This chapter analyses groundwater flow in the Orontes River basin and changes which occurred in the past forty years as a result of the massive expansion of irrigated lands using the groundwater resources. The region contains significant karstic water resources supplying springs in the upper and middle reaches of the basin. Although variations in annual flows are difficult to assess precisely, there has been a significant decrease since the 1960s. The Syrian–Lebanese agreement on sharing of the Orontes water, signed in 1992, focused almost exclusively on surface water resources. The drilling of wells in Lebanon was restricted only near the Orontes River. Amendments to the agreement, in 1997 and 2002, restricted the drilling of wells in the Lebanese section of the basin as a whole. However no restrictions were imposed on groundwater withdrawals in Syria. Because of the continuity of the main Jurassic-Cretaceous aquifer between the two countries the latter withdrawals are likely affect groundwater resources in Lebanon.

4.1 INTRODUCTION

The extensive development of irrigation in the Orontes River basin, since the 1970s, has substantially disorganised subsurface flows feeding the water resources of this region, particularly within the area of the Lebanese-Syrian border.

A literature review located the approximate recharge areas of the main springs and showed the impact of the recent intensive exploitation of the Jurassic-Cretaceous aquifer on subsurface water flows.

From the upstream Lebanese section of the watershed, surface water carried by the Orontes river flows toward Syria, as well as significant amounts of groundwater from the Jurassic-Cretaceous aquifer complex extending at depth into to the Syrian territory.

This chapter analyses the effects of irrigation development on the dynamics of the groundwater resources and how this issue was taken into account in the 1994 water agreement and the 1997 and 2002 amendments. The study was conducted as part of a research programme supported by the Global Program Water Initiatives of the Swiss Agency for Development and Cooperation.

4.2 ORONTES RIVER BASIN: GENERAL HYDROGEOLOGY AND SUPPLY OF MAIN SPRINGS AND PRODUCTION WELLS

The Orontes basin contains significant karstic water resources. These are largely fed the Orontes River before the extensive development of the irrigation in the past four decades. In the 1960s, the downstream discharge rate reached almost 100 m³/s at the Syrian – Turkish border, but in the 2000s, it fell to less than 15 m³/s.

The large aquifers supplying the main springs located in the upstream reach of the basin are thick limestone formations of Jurassic and Cretaceous age (Figure 4.1). Although these formations are not equally fractured and karstified, they contain groundwater flowing through them and from one to the other via faulting or fracturing, even if they are separated by lower Cretaceous impermeable strata. The scale of the basin allows the Jurassic and Cretaceous formations to be considered as a unique complex aquifer, a very large reservoir in hydraulic continuity (Figure 4.2).

In the southern and central parts of the basin, this large karstic aquifer supplies many springs (Figure 4.3). The annual flow of the main springs is of several m³/s. Their regime is more or less stable throughout the year because of the very large water reserves, the high hydraulic conductivity, the well-developed internal drainage and the extended confinement of this complex aquifer. Recharge to the Jurassic and Cretaceous aquifer is particularly important in the highest areas of the basin, especifically on Mounts Lebanon and Anti-Lebanon. Where limestone formations outcrop, the recharge can reach up to 60% of the total precipitation, varying between 750 and 2000 mm per year on the mountain range (Droubi, 2012). Recharge takes place mainly during the winter season, and lasts until spring due to snow-melt water.

4.3 INVENTORY OF GROUNDWATER SOURCES, EVOLUTION OF THEIR FLOW AND OF THE ORGANISATION OF THE SUBSURFACE FLOWS SUPPLYING THEM

The inventory of groundwater sources and resurgences of major interest of the Orontes basin (Figure 4.1), amounts to around 30 springs, including several major ones located close to the Lebanese-Syrian border.

The evolution of their annual average flow between the 1960s and the 2000s is difficult to assess. There has been a significant decrease because of the intensive use of water, more specifically groundwater pumped from deep wells, due to recent and rapid extension of irrigated lands, mainly in Syria.

The main springs in Lebanon are Ayn ez Zarqa (Orontes spring) and Ayn el Laboue close to the border and in Syria Ayn at Tannur, Uyun as Samak and Ayn al Damamel (Kloostermann, 2008). They are supplied by groundwater in the Jurassic an Cretaceous strata.

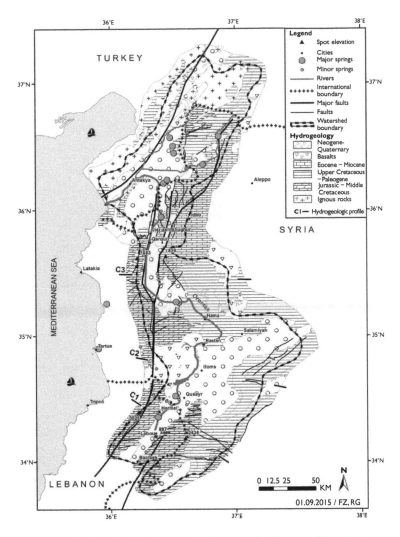

Figure 4.1 Simplified hydrogeological map of the Orontes River basin.

Remote groundwater discharge from the Jurassic and Cretaceous aquifer has been much more affected downstream, in particularly in the East of the Al Ghab plain. Several springs have dried up, as shown in Table 4.1, which provides references of the sources (Figure 4.1).

4.3.1 Schematic diagram of the organization of the subsurface flow in the 1960s

Observations presented above allow a schematic diagram of the organisation of the main subsurface flow lines in the Orontes basin to be created for the 1960s. Figure 4.4 shows the groundwater flow oriented northward, mainly affected by the

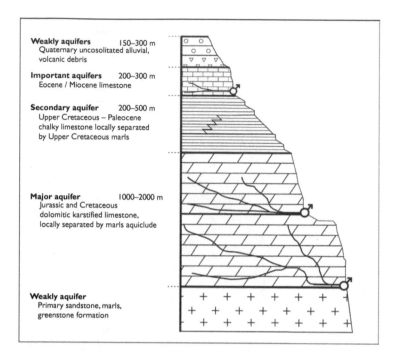

Figure 4.2 Hydro-litho-stratigraphical figure of the Orontes River basin formations.

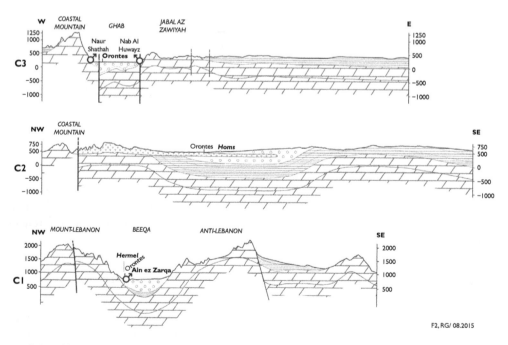

F2, RG/ 08.2015

Figure 4.3 Hydrogeological cross-sections of the Orontes River basin (legend see Figure 4.1).

Table 4.1 Estimated flows of the main springs or group of springs, in the 1960s and in the late 1990s, south and central parts of the basin.

Main springs or groups of springs	Flow l/s years 60	Flow l/s years 1990–2000	Sources	ID	Elevation
Ayn ez Zarqa	**13 000**	**13 000**	Ayn ez Zarka	1.01	676
Ayn el Laboue	**1 400**	**700**	Ayn el Laboue	1.02	903
			Ayn at Tannur	2.01	510
Ayn at Tannur region	**2 300**	**1 500**	Uyun as Samak	2.02	512
			Ayn al Damamel	2.03	518
			Tall al Uyun	2.11	193
			Ayn Qalat al Madiq	2.12	172
Al Ghab east region	**13 000**	**virtually zero flow**	Ayn at Taqah, Ash Shariah	2.13	170
			Nab an Nasiriyah	2.14	172
			Nab al Huwayz	2.15	171

longitudinal hydrogeological structure of the basin. South of Mount Anti-Lebanon, however, the flow is oriented in the opposite direction southward toward the Syrian spring of Figueh.

4.3.2 Schematic diagram of the organisation of the subsurface flow in the years 1990–2000

In response to the intensive development of irrigation in the Syrian part of the basin, more particularly in areas with no surface water, a proliferation of legal and illegal wells has taken place during the late 1990s.

Deep well pumping of the Jurassic and Cretaceous groundwater induces signifiant drawdown of groundwater heads especially in the artesian areas. The TNO numerical simulation, (Kloostermann & Vermooten, 2008), which roughly simulates the extension of groundwater drawdown, clearly delineates three heavily impacted areas: Qusayr, north of the Lebanese-Syrian border, Homs in the middle section of the basin and Asharneh ahead of the Al Ghab plain. At Qusayr, the drawdown dropped several tens of metres (Figure 4.5).

This large-scale drawdown caused a sharp reduction in discharge from numerous springs in the Orontes basin, more particularly in the central part of the basin where some of them dried up. In the vicinity of the Lebanese-Syrian border, the springs have also been seriously affected but to a lesser degree.

The direction of the nearby groundwater flow lines have in some areas significantly changed, because they have been attracted to main drawdown areas, especially in the Asharneh plain area (Figure 4.5).

In the border area, the intense water pumping around Qusayr, creates significant local drawdown and may increase the groundwater flow passing through the border. This could ultimately lead to reduced groundwater resources in the Lebanese part of the Orontes basin.

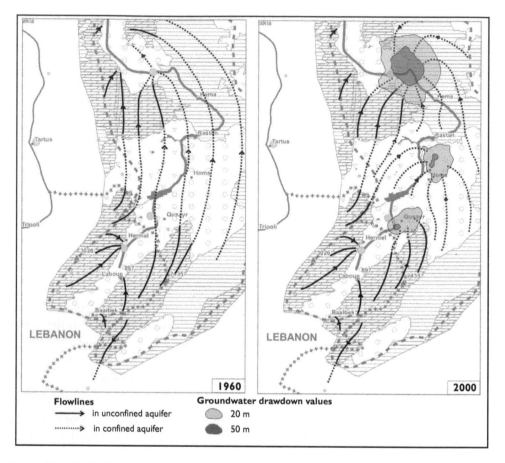

Figures 4.4 and 4.5 Schematic diagram of the organisation of the subsurface flow, in the 1960s (left) and 2000s (right).

4.4 THE SYRIA-LEBANON AGREEMENT AND GROUNDWATER FLOW INVOLVEMENT

4.4.1 History of the agreements

Discussions between the two countries, Syria and Lebanon, for establishing a water resources sharing agreement in the upstream part of the basin, started in the 1940s. A first agreement was signed in 1994, modified by an addendum in 1997 and followed in 2002 by a new agreement which is still in force.

4.4.1.1 The 1994 Agreement

The 1994 'Agreement of the Distribution of the Orontes River Water Originating in Lebanese Territory' defined 'a fixed amount of 80 million cubic meters (MCM/year) *(scheduled10 MCM for each following period, Sept–Oct, Nov–Feb, Mar–Apr and*

50 MCM for May–Aug period one) allocated to Lebanon if the river flow exceeds 400 MCM/year at the Hermel Bridge gauging station and 20% of the annual flow if the discharge volume is less than 400 MCM/year' (Sarraf, 2013).

Comair (2013) specified that 'the 1994 Agreement addressed the issue of groundwater withdrawals with a provision of authorizing pumping from wells drilled before the signature of 1994 agreement, but prohibiting new wells. The wells allowed were the ones located within a radius of 1500 m from the center of the source and 500 m of either banks of the river'.

In the absence of dams or diversion works (not mentioned in the agreement), a large part of the water allocated to Lebanon, in particularly during winter season, cannot be used. Moreover, because of groundwater withdrawal restrictions, the Lebanese border village of Qaa and Hermel could not undertake any irrigation development.

4.4.1.2   The 1997 Addendum

Because of the weaknesses of the 1994 Agreement, considered as not equitable by Lebanon, an addendum with two new clauses. The utilisation of the waters of four small closed basins (Yammoune, Marjhine, Joubab el Homor and Ouyoun Orgosh) shall be equal to the quantity of renewable water of these basins. In addition the Lebanese party may benefit, for the Baalbek-Hermel region, from all the waters deriving from the Laboue springs during the irrigation season (end of April till October 15) as well as from the drinking water in use in the neighboring villages.

This means that the quantity of waters mentioned in the two clauses are no longer included in the discharge volume of 400 MCM/year referred in the 1994 Agreement.

4.4.1.3   The 2002 Agreement

This new agreement comprises the 1994 Agreement, the 1997 Addendum and approved minutes of the Joint Lebanese–Syrian meetings held in the 2000s. These minutes include the construction of a derivation dam with a storage capacity of 27 MCM (located directly after the Ayn ez Zarqa sources that will serve two sides of the river supplying an area of 3000 ha) and a multipurpose dam with a storage capacity of 37 MCM upstream of the Hermel bridge (for supplying drinking water, irrigation water for 3800 ha and for power production). Finally, the proposed irrigation schemes should cover a total of 6800 ha in the Hermel and Al Qaa area.

This last Syrian-Lebanese Agreement is seen as equitable for both parties and complying with international law, in particularly the 1997 'Convention on the Law on the Non-Navigational Uses of International Watercourses (UNWC)'. However, the non-involvement of Turkey in the negotiations remains an obvious negative aspect.

4.5   HOW GROUNDWATER IS TAKEN INTO ACCOUNT IN THE AGREEMENTS

4.5.1   General aspects

At present, the use of the water from the Orontes River is limited in Lebanon to small scale farming, fish farms and tourism. Total use of water is estimated, according to

Sarraf (2013), at only 21 MCM/year of which 23% is for domestic purposes and the rest for irrigation (irrigated areas in the Lebanese part of the Orontes basin are officially 1.703 ha in area).

As the fixed amount of water allocated to Lebanon is of 80 MCM/year, Lebanese withdrawals are much less than the volume attributed and the situation so far does not require any specific discussion between the two countries regarding the shared water part as established in the Agreement. However, sharing the water according the Agreement is quite complex, involving a calculate for each month, in dry years, of the water deficit regarding the monthly flow of an average year which will be taken into account in the following months for reduced withdrawal.

4.5.2   Hydrogeological aspects

Water sharing between Syria and Lebanon, based to the flow of the Orontes river at the Hermel Bridge gauging station, only apparently encompasses the surface flow. Most landscape of the Lebanese part of Orontes basin is karstified outcrops without any surface runoff. The episodic runoff flowing after heavy rains in the middle of the Bekaa valley, on the Neogene formation, is almost negligible in comparison to the Orontes discharge flowing at Hermel.

Thus the Syrian-Lebanon agreement exclusively covers the groundwater issues.

In the case of the deep wells pumping from the Jurassic and Cretaceous rocks, the Agreement is very clear, prohibiting any new wells and allowing only the wells drilled before 1994 close to the sources or the Orontes River to be used.

This means that the groundwater flow which does not supply directly or indirectly the Orontes river is attributed to Syria, all over the Lebanese part of the basin, except the four "closed" basin and partly the Laboue sources mentioned in the 1997 Addendum.

No restriction is mentioned regarding the Syrian exploitation of groundwater in the vicinity of the border, even though the hydrogeological situation clearly shows the continuity of the main Jurassic and Cretaceous aquifer between the two countries. In reality, the important quantity of groundwater pumped close to the border, as it does exist now south of Qusayr, will increasingly affect the Lebanese groundwater head in the long term, and in that way will indirectly impacting the 400 MCM annual river flow taken as reference in the Agreement.

A rough evaluation of the subsurface annual flow passing the border in the Jurassic – Cretaceous aquifer is of note. A rough calculation using Darcy's law can be applied and gives an order of magnitude of annual groundwater volume entering Syria of 65 MCM, (about 2 m³/s with K 5×10^{-4} m/s, hydraulic gradient 10^{-3}, width of the water flow 28 km, average thickness of hydraulic flow 150m). This volume can be compared to the annual volume of 30 MCM corresponding to the over exploitation of groundwater in the irrigated area of Qusayr, according to the 'Annual Effective Storage use' calculated in the TNO numerical simulation (Kloosterman, 2008).

4.6   CONCLUSIONS

This chapter provides an update on recent issues from the hydrogeology of the Orontes basin and seeks to show how the existing large groundwater resources have been taken into account in the Lebanese-Syrian Agreement.

Unlike river flow rate, sharing of groundwater resources cannot be directly based on fixed observation points. It requires taking into account the groundwater flow for large areas and long periods of time, involving a great deal of uncertainty. In such a situation it is especially difficult to reach a common approach between the negotiating parties.

The history of the Lebanese-Syrian Agreement is a good example of how groundwater flow has eventually been taken in account. It also makes clear how far groundwater diversity is taken into account and remains partly absent because of the difficulty of determining simple and effective rules based on observations on the subsurface environment.

The creation of new numerical procedures for water flow simulation provides a useful aid for the negotiators. The outcomes nevertheless depend on robust data and the establishment of a coherent conceptual model, with which underpin the basis for the negotiations.

REFERENCES

Comair G.F., McKinney D.C., Scoullos M.J., Flinker R.H., Espinoza G.E. (2013) *Transboundary cooperation in international basins: Clarification and experiences from the Orontes River Basin agreement: Part 1*. Environmental Science & Policy 31, 133–140.

Comair G.F., McKinney D.C., Scoullos M.J., Flinker R.H., Espinoza G.E. (2013) *Transboundary cooperation in international basins: Clarification and experiences from the Orontes River Basin agreement: Part 2*. Environmental Science & Policy 31, 141–148.

Droubi A. (2012) Communication interne. Meeting du projet IHEID Orontes River basin Liban-Syrie. Novembre Bekaa, Lebanon.

ESCWA13 (2013) Inventory of shared water resources in Western Asia. *Chapter 7, Orontes River Basin, UN, ESCWA/SDPD/2013/WG13/REPORT.*

FAO (2011) Aquastat database. *Syria Arab Republic.* http:www.fao.org/nr/water/aquastat/countries/syria/index.stm; last visit 10.01.2012.

Hamade S., Tabet, Ch. (2013) *The impacts of climate change and human activities on water resources availability in the Orontes watershed: Case of the Ghab Region in Syria.* Journal of Water Sustainability 3(1), 45–49.

Kloosterman F.H., Vermooten, J.S.A. (2008) Final Report, development of a numerical groundwater flow model for the Larger Orontes Basin. *Dutch-Syrian Water Cooperation, TNO report.*

Kloosterman F.H. (2009) Notes on the Origin of the Groundwater issuing from The Ain Altnour, Ain Asamak and Ain Alzzarka Springs. *Mission Report EVD Syria Bridging Phase. TNO/Deltares.*

Maalouf F. (1999) Approche du fonctionnement de l'aquifère de l'Oronte. DEA Univ. St Joseph, Beyrouth.

Sarraf S. (2013) *Connecting transboundary water resources management with national visions and plans in Lebanon.* Near East & North Africa Land and Water Days, Amman, 1–9.

Chapter 5

Hungarian–Slovakian transboundary karstic groundwater management under the scope of ENWAT and TRANSENERGY EU projects

Peter Malík[1], *Radovan Černák*[1] *& György Tóth*[2]

[1]*Štátny geologický ústav Dionýza Štúra – Geological Survey of Slovak Republic, Bratislava, Slovakia*
[2]*Geological and Geophysical Institute of Hungary, Budapest, Hungary*

ABSTRACT

Within the framework of European Union projects ENWAT and TRANSENERGY, Hungarian and Slovakian hydrogeologists cooperated on common descriptions, modelling and formulating proposals for the groundwater management of transboundary karstic aquifers, one of them a deep-seated geothermal aquifer and the other a classical plateau karst structure. Some notes from lessons learned during the work on both projects may be useful to other hydrogeologists trying to establish cross-border cooperation especially in Europe, such as maintaining two independent web portals for public use and for specialists. Keeping units to be constantly linked with the data in common databases of input parameters, and adequate (possibly metric) coordinate systems seems to be useful. Choice of software tools should be left for individual selection by working teams, while basic standards of data inputs and outputs seem to be the most effective common understanding for hydrogeologists managing transboundary aquifers.

5.1 INTRODUCTION

Groundwater bodies, delineated according to the EU Water Framework Directive (2000/60/EC) by both countries along the Hungarian–Slovakian border, form interconnected systems, which supply both countries with drinking water. Surface waters, rivers and wetland ecosystems are groundwater dependent. The EU Water Framework Directive deals with the quantitative and qualitative status of groundwater, and protection of the ecosystems, which depend on groundwater. Due to the EU regulations, the groundwater body is considered as transboundary only after its approval by both respective countries. After bilateral negotiations, several groundwater bodies were recognised along the Hungarian/Slovakian border. Within the framework of the European Union INTERREG IIIA project 'Environmental state and sustainable management of Hungarian–Slovak transboundary groundwater bodies (ENWAT)', three transboundary aquifers were investigated in the Hungarian–Slovakian border region: Ipoly/Ipel' Valley, Bodrog region (both of them with porous aquifers) and Aggtelek-Slovak Karst region, with prevailing karstic type of permeability (Brezsnyánszky *et al.*,

Figure 5.1 Location of the two discussed transboundary aquifers between Hungary and Slovakia (1: Komárom-Štúrovo geothermal karstic body, 2: Aggtelek/Slovenský kras transboundary aquifer).

2008; Malík *et al.*, 2012; No. 2 on Figure 5.1). Another project, dealing with transboundary groundwater is TRANSENERGY (Transboundary Geothermal Energy Resources of Slovenia, Austria, Hungary and Slovakia). The project was targeted at geothermal waters (geothermal resources) in several regions: the pilot area of Komárom-Štúrovo (Komárno-Párkány) is a deeply seated geothermal karstic aquifer (Nádor *et al.*, 2013; No. 1 on Figure 5.1).

5.2   ENWAT TRANSBOUNDARY GROUNDWATER PROJECT

One of the three pilot areas of the ENWAT project was the Aggtelek Mountains and the Slovenský kras Mountains (Slovak Karst). The area represents a large common karstic aquifer system in the Eastern part of both countries (Figure 5.2). Both countries considered it as a highly important transboundary water body. The area is protected as a National Park on both sides of the border, the Aggtelek National Park (Aggteleki Nemzeti Park) in Hungary and National Park of Slovak Karst (Národný park Slovenský kras) in Slovakia. In 1995, all the caves of the Aggtelek and Slovak Karst were designated World Heritage Sites by UNESCO. Significant drinking water resources in Slovakia and regionally important resources in Hungary, although still preserving sufficient quality, are located in vulnerable karstic area and require preventive protection.

The groundwater body is in a Mesozoic complex with morphologically visible karstic plateau and canyon-like river valleys, separating different hydrogeological units. Outcropping rocks are very different according to the character of permeability, character of groundwater circulation, type of groundwater regime, and also in the resulting yield of springs. From the hydrogeological point of view, the most

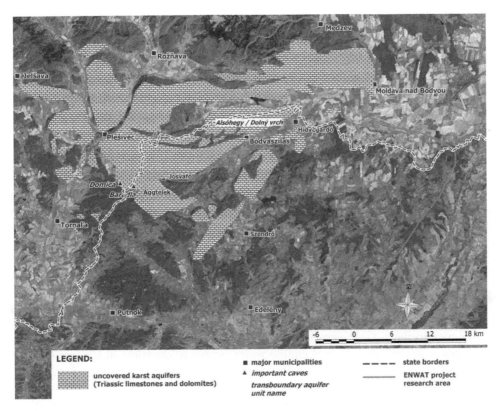

Figure 5.2 Aggtelek/Slovenský kras transboundary aquifer.

important tectonic unit in the area is the Silicikum unit, mainly its Middle Triassic and Upper Triassic part. The most important aquifer with karst-fissure type of permeability is formed by the Middle and Upper Triassic limestones and dolomites. Similarly there are important hydrogeological units on the Hungarian side such as Alsóhegy, Nagyoldal, Haragistya and Galyaság, which contain the Aggtelek-Domica cave system. Tertiary filling of sedimentary basins (mostly clays and clayey sands) act as a regional impermeable barrier for the groundwater accumulated in Triassic carbonates.

Natural groundwater circulation in karstic aquifers is typified by intensive drainage towards the major springs. Their infiltration areas are not easy to delineate, but results of the water balance suggest that at many places groundwater infiltrating in one country is draining towards a spring located in the neighboring country. For example, the underground hydrological system of the Domica-Baradla cave system is fed by water recharged in Slovakia, while the major drainage element here is the Jósva spring at Jósvafő in Hungary. Another example is the Alsóhegy/Dolný vrch karstic aquifer unit, elongated along an east to west axis, with a national boundary crossing over it. The eastern part of this aquifer drains towards springs in Hungary, while natural groundwater outlets are not present on the Slovak territory. By contrast, the western part of the Alsóhegy/Dolný vrch karstic aquifer drains towards north, to springs located on the Slovak territory.

Groundwater chemical composition or quality originates mainly from water-rock interaction. Groundwater circulates predominantly in limestones and dolomites of the Mesozoic rock formations of Middle and Upper Triassic age. The $Ca-HCO_3$ and $Ca-Mg-HCO_3$ types of chemical content dominate in groundwater that is not affected by anthropogenic activities. The natural character of groundwater circulation is at the moment still able to preserve good properties of high quality groundwater and most of the area provides high quality groundwater for regional waterworks. This groundwater fully meets the criteria for drinking water. In most cases, concentrations of trace elements are low and mostly below the limits of detection. Anthropogenic contamination was found only rarely, usually close to settlements. It was indicated by raised concentrations of nitrate, chloride, sulphate, chemical oxygen demand and potassium. High iron and manganese concentrations may be found at the adjoining Tertiary artesian aquifers which have a reducing environment.

In the Aggtelek/Slovak Karst area the chemical status of groundwater is good, but possible future climate change may affect this with increased occurrence of extreme hydrological events. In the worst scenario, from the water management point of view, higher and more rapid flood peaks will apply pressures to water channels by erosion while drought periods damage the ecology of the fragile karst area. Small-scale water management arrangements buffering hydrological extremes, such as building of minor water storage systems enable water supply to continue in dry periods, and polders, dry reservoirs to collect high water stages are the best strategy. A significant threat in this economically less developed part of Slovakia and Hungary is uncontrolled land use and building on flood-prone areas. [More details are found on the project output website http://www.all-in.sk/enwat, maintained in Hungarian, Slovak and English languages.]

5.3   TRANSENERGY TRANSBOUNDARY GROUNDWATER PROJECT

The second multinational project, TRANSENERGY – Transboundary Geothermal Energy Resources of Slovenia, Austria, Hungary and Slovakia (Nádor et al., 2013), was aimed at providing implementation tools based on a firm geoscientific basis for enhanced and sustainable use of geothermal resources linked to CEU Programme, Area of Intervention 3.1. (developing a high quality environment by managing and protecting natural resources). The project was focused on the key problem of using natural resources – here the geothermal energy – shared by different countries in a sustainable way. Natural resources, such as geothermal energy, whose main transporting medium is groundwater along regional flow paths, are strongly linked to geological structures that reach across the state borders. Therefore, only a transboundary approach can handle the assessment of geothermal energy and water use in sustainable way. Results of the project were addressed at the needs of decision makers and stakeholders by providing a user-friendly web-based decision supporting tool (an interactive web portal) as a main core output of the project. Expert knowledge about geothermal resources and sustainable reservoir management, gained during the project, was transferred to end-users to provide an overview and for them to make simple estimations on geothermal reserves within the project area. This publicly accessible implementation tool shows all the relevant information on the potential, vulnerability and sustainability of the geothermal system in the transboundary regions.

Results integrate all of the activities carried out within the TRANSENERGY project, such as screening utilisation needs with special respect to national, EU and international legislation, collecting and harmonising geoscientific data, performing additional measurements, organising all data in harmonised, multi-lingual joint databases, producing various cross-border geoscientific models and performing scenario modelling for different extraction of geothermal heat/water.

In addition to the web-based decision planning tool, the project also delivered a methodology for joint groundwater management and utilisation maps summarising the legal steps and actions towards a harmonised management strategy of transboundary geothermal resources, and a best practice on geothermal use. The project partners were the Geological Surveys of the four countries – Hungary, Slovakia, Austria, Slovenia, which are involved in several international projects concerning natural resources. They also have a long tradition of bilateral cooperation with each other in several projects.

Five pilot areas were treated within the TRANSENERGY project, while one of them – the Komárom-Štúrovo (Komárno-Párkány) pilot area is a deeply seated karstic groundwater body (Figure 5.3). The Komárom-Štúrovo pilot area belongs to the Komárno block, in Slovakia Komárno high block and Komárno marginal block

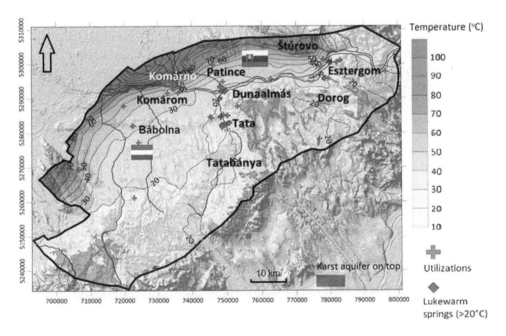

Figure 5.3 Position of the transboundary Komárom-Štúrovo karstic geothermal aquifer (after Gáspár and Tóth 2013).

are distinguished (Remšík *et al.*, 1992) comprising a subsided northern blocks of the Gerecse and Pilis Mountains (Hungary).

The main and most important aquifers in the pilot area are the Upper Triassic platform limestones and dolomites (Dachstein Limestone and Main or Haupt Dolomite). The Middle Eocene denudation caused strong karstification in the more than 1500 m thick carbonate sequence. These well karstified conduits and fractures along the tectonic elements determine the karstic groundwater flow; the karstification in the upper part of the system has higher permeability and this is where the main part of the groundwater flow takes place. From the area of outcropping Upper Triassic rocks (North-Bakony, Vértes, Gerecse, Pilis mountains in Hungary) the recharged precipitation descends and flows towards the deeper regions to the north west and west. From the north west edge of the aquifer the water turns towards north-north east and in the Slovakian parts towards the east.

Along the margin of the mountains (Tata, Dunaalmás, Patince and Esztergom) natural discharge areas of the lukewarm karst springs (~20–27°C) discharge in: Esztergom springs at 26–29°C, Tata springs at 20–22°C, Dunaalmás-Patince sprigs higher at 23–24°C and 25–27°C. The higher (~40°C) water temperatures in Štúrovo (against ~28°C in Esztergom) can be explained by longer flow paths from the north west and west (Gáspár & Tóth, 2013). The marginal (west, north west and north) and deeper part (more than 600 m below the sea level) of the Upper Triassic carbonate aquifer is characterised by higher temperatures and belongs to the Komárno marginal block, is without hydraulic contact with the Komárno elevated block and has no recharge area (Remšík *et al.*, 1992). This thermal karst water (40–60°C) is produced by deep wells in the north west and north of the area (near Bábolna, Komárom and Komárno).

The main utilisation is for health purposes in both countries in the north east part of the area (Esztergom and Štúrovo). In Patince and Dunaalmás (historical) balneological and drinking water uses exist. Near Komárom and Komárno balneological and agricultural use prevails. Most of the users utilize the lukewarm or thermal water of the Triassic karstic aquifer, but some Miocene and Cretaceous local aquifers near Komárom and Komárno are also exploited.

Because of the long-term brown coal mining coupled with intensive water abstraction (Tatabánya – Dorog area), the whole cold- and thermal karst system was affected by a regional depression which caused the drying out of most of the lukewarm springs. After the mining ceased the water level has been rising since the beginning of the 1990s (Alföldi & Kapolyi, 2007). In this dynamically changing system it is hard to estimate the actual dimension of the impacted area in karst (difference between natural groundwater level in karstic aquifer existing prior to mining activities and its current state), but in the south and south west part of the area in 2013 it was about 30 m, and decreases toward the norh. Along the Danube river from Komárom-Komárno to Esztergom-Štúrovo the actual drop in the water level remaining as a consequence of the past mining activities in the karst was about 10 m. Existing abstraction for drinking water supply in the area is much smaller than the mining abstraction in the past and has a potentially low influence on the main lukewarm springs. In the area of Tata the rising karst water level results in seepages on the surface and affects surface infrastructure such as buildings and garages. Therefore, solutions have to be found to utilize and/or drain the (surplus) seepage waters. In this area there is also a competition between the water demand of balneological users, the drinking water abstractions and protection of the groundwater dependent ecosystems, which represents a very important environmental aspect with high priority in the EU Water Framework Directive. Therefore, ranking of different needs and an integrated assessment of their impacts is of vital importance. The coordinated utilisation and planning of further development is needed in both countries to maintain the current situation and the historical heritage. In the area of the natural discharges (Tata, Esztergom in Hungary, Patince in Slovakia) the thermal pollution of the surface waters is minimal and nature has adapted to lukewarm karst waters.

The assessment of sustainability was based on an overview of 34 geothermal wells, 8 on the Slovakian side and 26 on Hungarian side (in the Komárno high block). No over-exploitation was reported. Due to abandoning mining areas and halting of water pumping, water levels are rising. However, some negative effects on infrastructure were reported and a continuous karstic water level monitoring programme was proposed.

The hydrogeological structure of the Komárno marginal block that is closed to water circulation and has no hydraulic contact with the surrounding structures does not have reinjection of geothermal water (Gáspár & Tóth, 2013, Švasta et al., 2013).

Although water has no borders and flows freely between the countries, the categorisation and delineation of the water bodies is not harmonised across the border. Both countries have a different way and philosophy of dealing with the delineation of geothermal water bodies. For a mutual assessment and management of the groundwater bodies, harmonised rules are needed in the future.

5.4 CONCLUSIONS AND USEFUL NOTES

Results of both projects, based on evaluations and models, local needs, costs and best practices, are a step forward in the creation of a joint Hungarian–Slovakian water management plan by supplying basic data and new information on transboundary groundwater bodies. The work also indicated the need to continue the process of potential pollutant characterisation, risk assessment and common regulations to control water abstraction in both countries. Some notes from lessons learned during the work on both the projects may be useful to other hydrogeologists trying to establish cross-border cooperation especially in Europe. (a) Maintaining two independent web portals has proved to be useful to keep the project agenda for both specialists (working portal as a place of exchange information during project between working groups) and public (public portal to publish preliminary and final results of the projects via lay reports, as well as by more technical and scientific outputs). (b) Common database of input parameters that should allow simple manipulation and visualisation of data and contain forms to browse and edit the data: the role of showing units for every component as frequently as possible is inevitable! Majority of misunderstandings and time consuming discussions and explanations were created by different units used by respective working groups. (c) Coordinate system: practically all countries keep the geospatial data in their national (and usually mutually incompatible) systems; the Hungarian EOV system and Slovakian/former Czechoslovakian S-JTSK Křovák system. Popular and frequently used Longitude/Latitude WGS 84 shows distances in vector format. These are impracticable in frequently used hydrogeological software models. Universal Transverse Mercator (UTM) for the respective zone was a solution, although German metric system Gauss-Krüger (Pulkovo, 1942) was also historically applied in both countries. (d) Space of freedom: after many time-consuming discussions about the software that should be used by both sides (usually intended to force the partners to use 'our' software tools that we are used to) the simplicity and basic standards of keeping data inputs and outputs seems to be the most effective way of common understanding and managing of transboundary aquifers by hydrogeologist acting on both sides of the borders.

REFERENCES

Alföldi L., Kapolyi L. (2007) Bányászati karsztvízszint-süllyesztés a Dunántúli-középhegységben, MTA Földrajztudományi Kutatóintézet, Budapest.

Brezsnyánszky K., Malík P., Gaál G., Szőcs T., Tóth G., Bartha A., Havas G., Kordík J., Michalko J., Bodiš D., Švasta J., Slaninka I., Leveinen J., Kaija J., Gondár-Sőregi K., Gondár K., Kun É., Pethő S., Ács V. (2008) ENWAT: Hungarian–Slovakian transboundary ground-water bodies. *European Geologist* 26, 37–41.

Gáspár E., Tóth G. (2013) Report on Komárom – Štúrovo Pilot Area scenario modelling. Project TRANSENERGY internal report, 1–40, http://transenergy-eu.geologie.ac.at/results; last visited 14/2/2014.

Malík P., Brezsnyánszky K., Gaál G., Szőcs T., Tóth G., Bartha A., Bottlik F., Havas G., Kordík J., Michalko J., Bodiš D., Švasta J., Slaninka I., Leveinen J., Kaija J., Gondár-Sőregi K., Gondár K., Kun É., Pethő S., Ács V. (2012) Evaluation of environmental state of Hungarian–Slovakian transboundary groundwater bodies within the "ENWAT" EU project. In: T. Nałęcz (ed.) Transboundary Aquifers in the Eastern Borders of the European Union. Regional Cooperation for Effective Management of Water Resources. Springer Science+Business Media Dordrecht 2012, NATO Science for Peace and Security Series – C: Environmental Security, 149–163.

Nádor A., Szőcs T., Rotár Szalkai A., Goetzl G., Prestor J., Tóth G., Černák R., Švasta J., Kovács A., Gáspár E., Rman N., Lapanje A., Fuks T. (2013) Strategy paper on sustainable cross-border geothermal utilization. Project TRANSENERGY internal report, http://transenergy-eu.geologie.ac.at/results; last visited 14/2/2014.

Remšík A., Franko O., Bodiš D. (1992) Geothermal resources of Komárno block. *Západné Karpaty, séria hydrogeológia a inžinierska geológia* 10, Geologický ústav Dionýza Štúra Bratislava, 159–199.

Švasta J., Remšík A., Černák R., Gregor M. (2013) Report on steady state hydraulic model of the Danube basin pilot area. Project TRANSENERGY internal report, http://transenergy-eu.geologie.ac.at; last visited 14/2/2014.

Chapter 6

Development and protection of transboundary karst and karst aquifers in West Stara Planina Mountains (Bulgaria–Serbia)

Aleksey Benderev[1], Zoran Stevanović[2], Boyka Mihaylova[1], Vladimir Živanović[2], Konstantin Kostov[1], Saša Milanović[2], Stefan Shanov[1] & Igor Jemcov[2]

[1]*Geological Institute, Bulgarian Academy of Sciences, Sofia, Bulgaria*
[2]*Faculty of Mining and Geology, Department of Hydrogeology, University of Belgrade, Belgrade, Serbia*

ABSTRACT

The West Stara Planina Mountains are situated across two countries – Serbia and Bulgaria. The karstification is developed mainly in two carbonate complexes: Triassic and Upper-Jurassic, while in the western extension, i.e. in Vidlič Mountain, karstified carbonate rocks of Cretaceous age prevail. This is a typical mountain karst with wide distribution of classical karstic landforms: dolines, small poljes, and blind valleys. There are also hundreds of caves in both countries, many of them well explored. Due to high infiltration of rainfall, abundant groundwater reserves are formed. The drainage takes place over numerous large karst springs. There are overflow sources while other sources are from upwelling water, draining the deeper saturated zones. Their discharge regime varies from relatively constant to highly changeable. Some are tapped and used for potable water supply, as in the case of Pirot, Dimitrovgrad (Serbia) or Svoge (Bulgaria). Most of the karst springs have relatively clearly defined inner catchment areas and are not subject to transboundary discharge. The water budget analysis at the bordering territories of Serbia and Bulgaria requires more hydrological, hydrogeological and climatic data, field survey and water tracings.

6.1 INTRODUCTION

The Balkan Mountains (*Stara Planina Mountains*) are a mountain range in the eastern part of the Balkan Peninsula and represent a part of the Alpine-Himalayan orogenic belt. The western part, located on the territory of Bulgaria and Serbia, has, for various reasons, not been well explored. In the absence of a common survey and exact data there is considerable speculation concerning water flows from one side of the border to the other. During recent years, the joint research of Serbian and Bulgarian scientists, both sponsored by their Academies of Sciences in the fields of geology, geomorphology and hydrogeology, has been active. One of their most interesting and topical problems having an important ecological and practical significance was to standardise the research for cross-border areas to solve the problems of karst aquifer distribution and karst water flow directions.

6.2   PRECONDITIONS FOR DEVELOPMENT OF KARST AND KARST WATERS

The target of the study is the part of the Western Balkan Mountains where karstified rocks are continuously distributed throughout the territories of the two countries. Such areas are located south of the main ridge of the mountain, between the Temska River (in Serbia) and the Iskar River (in Bulgaria) (Figure 6.1). The southern boundary of the area is marked by rivers draining surface and underground waters of this part of West Stara Planina – the Nišava River in Serbia and the Elovitsa and Blato Rivers in Bulgaria. The total surface of the transboundary region is about 2000 km².

The landscape is typically mountainous. The highest elevations – between 1300 and 2015 m a.s.l. (peak of Kom) – are on the main ridge of Stara Planina. South of and parallel to the main ridge are series of ridges and valleys. The lowest parts are located along the southern border – from about 300 to 700 m a.s.l. The relief provides various climatic conditions. The annual average air temperature fluctuates from 3.4°C to more than 10°C according to the altitude. The annual rainfall varies widely from about 300 to 1000 mm (Koleva & Peneva, 1990; Stevanović, 1991; Ristić, 2007). According to the graph based on data from the rainfall stations in Bulgaria and Serbia, there is a good correlation between rainfall and altitude (Figure 6.2).

The formation of karst and karst aquifers depends mainly on the geological conditions. They have been studied by several researchers and the results are summarised in geological maps at scale 1:100 000, map sheet Pirot (Geological Survey Serbia, 1970) on Serbian territory and map sheets Pirot (Haidutov, 1992), Berkovitsa (Haidutov & Dimitrova, 1992) Vlasotince & Breznik (Zagorchev & Kostadinov, 1991) and Sofia (Yanev, 1992) on Bulgarian territory. Between the sheets on both sides of the state border there are some discrepancies in the geological boundaries and rock outcrops

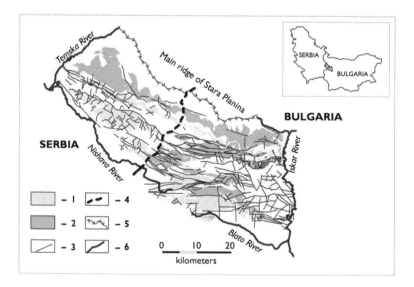

Figure 6.1 Geological sketch map of West Stara Planina Mountains. 1 – Upper Jurassic – Lower Cretaceous limestones; 2 – Triassic limestones and dolomites; 3 – Main faults; 4 – State boundary; 5 – Main ridge of Stara Planina; 6 – Rivers.

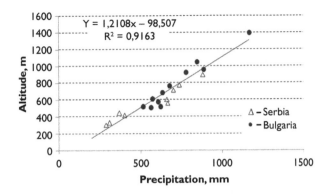

Figure 6.2 Correlation between rainfall and altitude in Western Stara Planina.

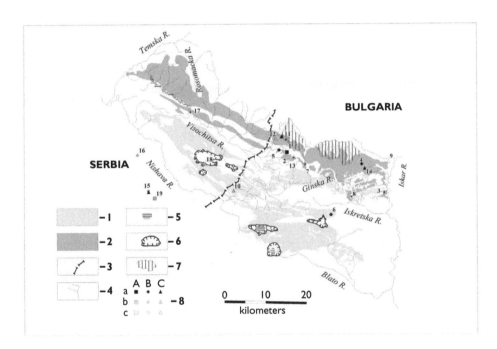

Figure 6.3 Karst in West Stara Planina Mountains. I – Karstified rocks of Southern zone; 2 – Karstified rocks of Northern zone; 3 – State boundary; 4 – Rivers; 5 – Lakes; 6 – Poljes; 7 – Blind valleys; 8 – Caves (Length: A – >3000 m, B – 1000–3000 m, C – <1000 m; Height difference: a – >100 m, b – 50–100 m, c – <50 m),

of different ages and lithology. To solve these problems a joint Serbian-Bulgarian research project (Tchoumatchenco *et al.*, 2011a, b) has been launched in recent years.

The karst and karst aquifers are formed in Triassic limestone and dolomite and Upper Jurassic and Lower Cretaceous limestone (Figure 6.1). Their outcrops and spatial distribution are associated with the complex tectonic structure in the area. The two zones are separated (Figure 6.3). In the first, the Northern zone, covering the higher parts of the mountain, the two carbonate complexes build monoclinal structures dipping to the south and separated by non-karstic Lower–Middle Jurassic rocks. To the south it is bordered by Lower Triassic sandstones and Paleozoic rocks in the higher parts of the mountains. The southern boundary is a thrust belt of east-west direction passing through the study area.

The second, the Southern zone, is characterised by more complex block structures due to a series of horst and graben structures also oriented east-west. The complex tectonics are reflected on the landscape of this zone with a series of depressions linearly oriented in the same direction separated by ridges.

The river network also has an important impact on the karst and karst waters. Rivers are usually formed in the highest parts of the mountain and flow to the south. Some of them (mainly on the territory of Bulgaria) completely lost their flow entering the karst terrains. Some of the rivers cross the limestone and dolomite in the first zone and flow periodically after rainfall: for example the Gintsi River, the upper reaches of the Visochka (Visočica) River, and the upper reaches of its right tributaries in Serbia (Rosomačka, Jelovička). The Visočica and Nišava Rivers represent the major drainage arteries of the second (southern) zone and the Iskar River drains only the eastern part of the northern zone. Several rivers such as the Iskrets and Blato in Bulgaria begin from large karst springs. The western boundary of the area – the Temska River – has almost no connection with karst.

The Nišava River crosses the border between Bulgaria and Serbia near Dimitrovgrad with a total length of 202 km to its confluence with Južna Morava River and a total catchment area of 4068 km^2 (Dukić, 1975). About 70% of the region falls within the catchment area of its upper reaches, while the remaining eastern part of the region is in the catchment area of the Iskar River and its tributaries. The Nišava originated from the higher parts of the Stara Planina Mountains from numerous springs. In the upper part of the watercourse it is called River Ginska. Entering karst carbonate rocks, it lost a substantial part of its flow only to reappear in Godech where it had already been named Nišava.

The Visočica River is also a cross-border river, which after its confluence with the Temska River flows into the Nišava River. This river also originates from the high parts of the Stara Planina in Bulgaria and after entering Serbia is a major drainage artery of the northern, high part of the study area. Only a small tributary south of Visočica River passes from Serbia to Bulgaria and flows into Nišava River. The characteristics of the quantitative parameters of the Nišava and Visočica Rivers, before their entry into Serbia, are described in detail from Hristova (2010), as the average annual water quantities of the Nišava vary from 0.91 to 3.47 m^3/s. According to Dukić (1975) the surface runoff for the territory of Bulgaria at Dimitrovgrad station amounted to 5.47 l/s/km^2, while in Pirot, where the Nišava River flow out of the investigated region is 6.63 l/s/km^2.

6.3 KARST AND KARST LANDFORMS

About 60% of the total area of the Western Balkans is occupied by karstified rocks.

The first mention of the distribution of karst landforms in the West Stara Planina Mountans was by the Austrian geologist Franz Toula (1882). Toula reported dolines and karren fields in the area between Vidlič Mtn and Nišava River (Toula, 1882; Radev, 1915).

Some notes about karst landforms in the studied area are published by Jovan Cvijić in his first karstological monograph 'Caves and Underground Hydrography in Eastern Serbia' (1895) (Ćalić, 2007). The karst relief is described in the book 'Karst Landforms in West Stara Planina' (1915) by the Bulgarian geomorphologist Zheko Radev. Radev surveyed 11 caves and performed a detailed study of the surface karst landforms: poljes, uvalas, dolines and karren fields, supported with photos (Radev, 1915).

During the last several decades the karst geomorphology and speleology in West Stara Planina have been described by several authors: Petrović (1974, 1976), Benderev (1989), Angelova et al. (1995, 1999), Beron et al. (2006), Zlatkova (2006), Kostov (2008), Szeidovitz et al. (2008), Mihaylova et al. (2008, 2009), Shanov & Kostov (2015).

From a morphological point of view the karst is an exposed, mountain type with widespread surface karst landforms. The intensity of karst processes is different in the Northern and Southern zones of the studied area. In the Northern zone the karrenfelds, dolines and uvalas are widespread. The significant blind valleys play an important role especially on the territory of Bulgaria. Compared to the Southern zone, a large number of caves and potholes, for instance 184 in Bulgaria, are established here. The caves are widely distributed and the morphology has determined the monoclinal dipping to the south of the layers of carbonate rocks with old or active ponors of surface water. There are also caves representing active and temporary springs (Vodnata Cave, Krivata Cave, Dushnika Cave, etc.). In this zone most of the deepest and longest caves are located in the Western Balkans (Table 6.1, Figure 6.3).

Table 6.1 The longest and the deepest caves in the studied area.

No	Cave	Country	Zone	Length (m)	Depth (m)
1	Balabanova Dupka	Bulgaria	Northern	4800	80
2	Tizoin	Bulgaria	Northern	3599	320
3	Vodnata Cave	Bulgaria	Northern	3264	85
4	Katsite Cave	Bulgaria	Northern	2560	205
5	El Saguaro	Bulgaria	Northern	2217	135
6	Golyamata Temnota	Bulgaria	Southern	2100	106
7	Krivata Cave	Bulgaria	Northern	1500	75
8	Dushnika Cave	Bulgaria	Northern	827	27
9	Kozarskata Cave	Bulgaria	Northern	709	12
10	Temnata Dupka	Bulgaria	Southern	493	95
11	Malata Balabanova	Bulgaria	Northern	400	125
12	Granicharskata Cave	Bulgaria	Northern	344	72
13	Malkoto Saguaro Cave	Bulgaria	Northern	338	44
14	Kolkina Dupka Cave	Bulgaria	Northern	68	236
15	Peshterica pothole	Serbia	Southern	0	155
16	Velika Pecina	Serbia	Northern	1440	>50
17	Vladikina plocha	Serbia	Northern	300	20
18	Golema dupka ponor	Serbia	Southern	480	>50
19	Vetrena dupka	Serbia	Southern	4150	280

The karst distribution in the Southern zone varies in the different areas, depending on the specific geomorphological and hydrogeological conditions. There are both exposed karst areas and sections covered with thick soil layers. The distribution of uvalas and dolines is different: there are both areas with significant dolines density (for example 164 dolines on the flattened ridge part of Tri Ushi Ridge in Bulgaria over an area of 35 km^2 – Figure 6.4) and areas with relatively low levels of karstification. Typical of the zone are the large karst poljes: Rayanovsko, Dragomansko and Aldomirovsko poljes in Bulgaria, and Odorovačko in Serbia (Figure 6.5). According to the polje classification scheme of Gams (1994), the poljes in the Bulgarian part are of the overflow type. In Dragomansko and Aldomirovsko poljes permanent marshes are formed.

The number of caves in the Southern zone is much lower, only about 60 in the Bulgarian part. Most of the caves are small being up to 100 m long and 25 m deep. The exceptions are the Vetrena dupka Cave in Serbia (4150 m), and the Golyamata Temnota Cave (2100 m) and Temnata Dupka Cave (493 m) in Bulgaria. The latter two are active ponor caves.

Figure 6.4 Karst dolines on the flattened ridge part of Tri Ushi Ridge in Bulgaria.

Figure 6.5 Odorovačko karst polje in Serbia.

6.4   KARST WATERS

The data on the karst aquifers and waters in the area are summarised separately for Bulgaria (Antonov, 1963; Antonov & Danchev, 1980; Benderev *et al.*, 2005) and Serbia (Stevanović, 1991, 1994; Živanović, 2011). Along with the karst waters in the transboundary region, there are both fracture and intergranular aquifers (Figure 6.6). Fracture aquifers are formed in different aged rocks, with different lithological and petrographic features. Tectonic conditions determine complex relationships between karstic rocks and the other pre-Quaternary massive rocks. In the North, these rocks are relatively poorly permeable and underlie the karst aquifers. In the South, because of the block tectonics, the karst and fracture aquifers are characterised by different permeability, and their recharge depends on contacts (lithological and tectonic) with impervious rocks. Intergranular aquifers are formed in unconsolidated Quaternary and Pliocene deposits. Pliocene and Quaternary sediments in the southern part of the area have more significant hydrogeological importance and form independent hydrological units. In their underlying formations blocks of karst rocks, which outcrop in the uplands, and have hydraulic connection with them, are included.

The karst aquifer is primarily recharged by rainfall. The rainfall quantity increases from south to north (Figure 6.2), due to high precipitation and the converse reduction of the evapotranspiration in the higher parts of the mountains. In the northern zone an important component of the water balance is temporary and permanent infiltration

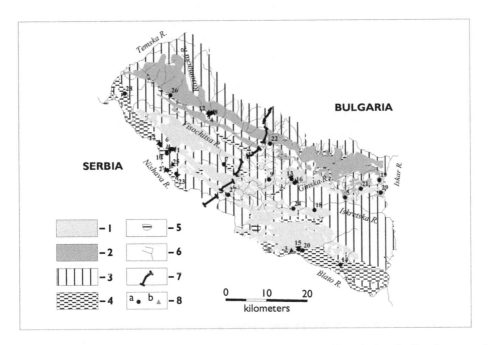

Figure 6.6 Hydrogeological map of West Stara Planina Mountains. I – Karstified rocks (Southern zone); 2 – Karstified rocks (Northern zone); 3 – Rock with fissure water; 4 – Quaternary and Pliocene sediments with porous groundwater; 5 – Lakes; 6 – Rivers; 7 – State boundary; 8 – Main spring (Temperature: a –T < 15°C, b –T > 15°C).

of surface rivers descends from the main ridge of Stara Planina. In the southern part the infiltration rate of surface water is much lower. It is related to those watercourses passing through karst poljes.

According to the water budget calculations by Stevanović (1994) for the Serbian part of Stara Planina Mountain, 18% of the precipitation recharges the karst groundwater. Water balance calculations for Bulgaria were made only at two isolated areas. The first occupies the eastern part of the northern area of the distribution of carbonate rocks (east of Ginska River) called Ponor Mountain (Benderev, 1989). The altitude for this area varies from 500 to 1400 m while the rate of evapotranspiration varies from 70 to 36% of rainfall (Turc method). Typical of this region is that almost the entire surface runoff recharges the karst groundwater and thus forms around 20% of the water reserves (regime data of the measurement of river quantity). The second area coincides with the catchment areas of the most southerly situated springs Opitsvet, Bistrets and Bezden, located at elevations between 550 and 900 m. Conducted studies used water balance methods showed that 22% of the precipitation recharges the groundwater system (Mihaylova & Benderev, 2010). This value is relatively close to the average for the Serbian territory where low altitudes prevail.

The general direction of groundwater movement is from north to south, flowing to the lowest parts of the terrains. One exception is in the easternmost part of the region, where the karst water is directed to the Iskar River.

The drainage is mostly by springs (Table 6.2, Figure 6.6), by groundwater abstraction and by underground flows into lateral permeable intergranular aquifers. There are both small gravity springs of local catchment areas with low flow rates and typical ascending karst springs with high flows. In the northern zone are springs with relatively high and variable discharges – including Peshta (Iskrets spring), Jelovičko vrelo and Skaklya. These springs are characterised by conduit flow and very fast response to precipitation events (Eftimi & Benderev, 2007). Springs of this type exist also in the southern zone at Bistrets and Berende izvor. Fault tectonics associated with the rise of some and the sinking of other blocks in many places create conditions for the formation of saturated zones which are drained by springs with more stable flow rates – Staroplaninsko oko, Kavak, Ropotski. Interesting are the larger drainage areas where there are several springs draining different hydrodynamic zones on the border of Stara Planina with Sofia and Pirot plains; for example Krupac 1 and 2, Topli izvor in Serbia and Bistrets, Bezden, Opitsvet in Bulgaria. In these areas both springs drain the upper part of the saturated zone with highly variable flow and the springs at lower elevations have a smaller range of flow rates. Characteristic is the presence of ascending springs with higher temperatures (19–22°C) draining areas with deep circulation – the Topli izvor and Toplo vrelo Springs in Serbia and the Opitsvet in Bulgaria.

6.5   VULNERABILITY AND ANTHROPOGENIC IMPACT

The specific characteristics of the hydrogeological conditions and the existing information for the transboundary region determine the approach used for the evaluation of groundwater vulnerability to pollution. The need to use the karst waters leads to

Table 6.2 Main springs.

	Spring	Country	Zone	Discharge, l/s Minimum	Maximum
1	Krupac 1	Serbia	Southern	220	1500
2	Gradište	Serbia	Southern	180	
3	Opitsvet*	Bulgaria	Southern	147	761
4	Staroplaninsko oko	Serbia	Northern	119	150
5	Peshta (Iskrets spring)	Bulgaria	Northern	90	54600
6	Kavak	Serbia	Southern	65	130
7	Vrelo Protopopinci	Serbia	Southern	60	
8	Jelovičko vrelo	Serbia	Northern	50	6000
9	Ropotski	Bulgaria	Southern	50	150
10	Krupac 2	Serbia	Southern	50	
11	Topli izvor*	Serbia	Southern	50	
12	Group of springs (Vrelo)	Serbia	Northern	50	
13	Zli dol	Bulgaria	Southern	40	1500
14	Dragovishtitsa	Bulgaria	Southern	40	120
15	Bistrets	Bulgaria	Southern	31	1860
16	Godech	Bulgaria	Southern	30	500
17	Klok	Serbia	Southern	30	
18	Buchin prohod	Bulgaria	Southern	24	180
19	Skaklya	Bulgaria	Northern	20	2164
20	Bezden	Bulgaria	Southern	20	370
21	Buchiloto	Bulgaria	Northern	20	300
22	Komshtitsa	Bulgaria	Northern	20	120
23	Spring (village Ciniglavci)	Serbia	Southern	20	
24	Vreloto (Lopushna)	Serbia	Southern	15	100
25	Toplo vrelo*	Serbia	Southern	10	
26	Spring (village Bela)	Serbia	Northern	7–10	
27	Berende izvor	Bulgaria	South	5	850
28	Spring (village Sopot)	Serbia	South	5–7	
29	Vodnata peshtera	Bulgaria	North	1	630

*Spring with temperature T > 15°C.

ecological problems because of the water quality. One of the most important problems in the area is the correct recognition of the sites of potential pollutants to the karst waters, as well as the need for zones of increased protection.

The EPIK method is a GIS-based, multi parametric method which takes four parameters into consideration: Epikarst, Protective cover, Infiltration condition and Karst network development (Doerfliger *et al.*, 1999; Zwahlen, 2003). It has been chosen as the most suitable approach for describing the vulnerability of the karst areas. It was created especially for areas of mountain karst and has been successfully applied for different karst basins (Gogu & Dassargues, 2000; Iurkiewicz *et al.*, 2005; Goldscheider, 2005; Awawdeh & Nawafleh, 2008). Maps have been drawn separately for the adjacent territories of Serbia and Bulgaria (Živanović & Dragišić, 2013; Mihaylova *et al.*, 2009), and they are unified and adjusted in the present study.

The rates of significance of the different EPIK parameters are determined by their weight coefficients. Each of these parameters, taking into account its peculiarities, is divided into a number of classes. These classes consider the quantitative and the qualitative indicators impacting the protection of the groundwater.

The weight coefficients of the parameters and the numerical values of the classes have been argued and determined in Doerfliger *et al.* (1999); Zwahlen (2003). Data for 'EPIK Vulnerability Index' are categorised into four groups – Very high, High, Intermediate and Low Vulnerability. They can be visualised on GIS prepared maps. Figure 6.7 presents the terrains of Very high and High Vulnerability for the Stara Planina Mountain region in Bulgaria and Serbia. The zone of Very High Vulnerability covers 157 km², or about 7.5% of the whole area, while the zone of High Vulnerability occupies 605 km² or 29% of the whole.

The distribution of the zones of Very High and High Vulnerability has significant importance for protection and management of the karst groundwaters. The zones are situated in areas of relatively less important anthropogenic impact. The most significant urban areas are the towns of Pirot, Dimitrovgrad, Dragoman and Slivnitsa, situated on low relief parts at the periphery of the region, and outside the karstic areas including the zone of intensive groundwater recharge. Industrial units exist only in the town areas. The towns and the villages are connected by roads but the overall traffic is low.

The Neogene sediments filling one of the east-west oriented grabens across the national border contain lignite beds. Excavation of these coals is carried out near the village of Staniantsi in Bulgaria and in the past also at the village of Mazgosh in Serbia. These two areas are situated outside the established zones of Very High and High Vulnerability, and they do not impact directly the karst waters.

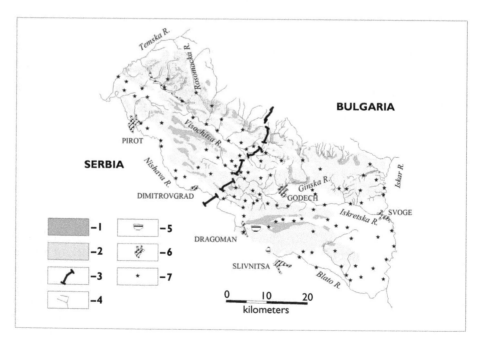

Figure 6.7 Zones of Very High and High Vulnerability. I – Very High Vulnerability; 2 – High Vulnerability; 3 – State boundary; 4 – Rivers; 5 – Lakes; 6 – Towns; 7 – Villages.

The town of Godech is situated in the centre of the studied area in Bulgaria, but it is also built on non-karstic rocks, on the bank of the Nišava River. Most of the other less significant inhabited sites are also situated outside the outcrops of karstified rocks, predominantly on terrains used for agriculture. A limited number of small villages, situated mainly in the lower parts of the river valleys, are within the zones of Very High and High Vulnerability.

There are a few villages situated in the zones of outcropping karst, where karst aquifers are recharged. An example is the village of Ponor in Bulgaria, built inside an large surface karst form. On the territory of Serbia few villages are situated on large karst poljes.

The inhabited sites are surrounded normally by cultivated terrains, and animal breeding is also important for the local population. Agricultural cultivation in the zones where karstified rocks exist is possible only where uninterrupted soil cover on these rocks is present. Such conditions exist on the flattened bottoms of the karst forms, as karst poljes or ponors. These cultivated terrains are one factor for the increase in karst water pollution.

Many quarries exist both in Bulgaria and Serbia where limestone and dolomites from the vadose zone are exploited. It is commonly accepted that they do not affect karst waters, even though some of them are close to the discharge zones and karst springs. One example is Iskrets Karst Springs which have a variable discharge from 280 l/s to more than 50 000 l/s (Benderev, 1989). They drained more than 80% of the territory of Ponor Mountain, a part of Stara Planina Mountain inside the area studied in Bulgaria. Precipitation over the capture area of the springs (about 140 km^2) provides 62% of the average annual discharge of the springs.

The springs have dried up several times during the twentieth century. After the Vranchea Earthquake in 1977 (M = 7.2) and at a distance of about 400 km from the site, the discharge rate dropped from 5.5 to 500 l/s in 7.5 hours (Paskalev et al., 1992). Later, the discharge rate rose abruptly to 13 500 l/s, and then began to decrease gradually. Similar events happened during the local Svoge Earthquake of 1979, as well as on 11 April 1982, when no significant earthquake was recorded at the localities or in the whole Balkan Region.

The question of the seismic impact is of importance for the local authorities because of the significance of the springs as a major source of fresh water for the town of Svoge. The quantities normally used are at the rate of 150 l/s. It was proved that the blasts in the quarries at the nearest vicinity of the springs do not have the potential to disturb the normal discharge of the springs (Shanov & Benderev, 2005). The only factor for such disturbances can be the local tectonics and the dynamics of the processes inside the karst system (Shanov & Kostov, 2015).

The most important hydro-technical facility built on the Bulgarian side is a canal designed for redirecting a part of surface waters, which recharges naturally in high parts of the mountains, towards the neighboring watershed for their use for hydroelectric power generation.

The most important hydro-technical facility in Serbia is Zavojsko Lake which dams the Visočica River. The biggest part of the lake is outside the area of karstified rocks, except for a limited zone near the dam. This lake was formed in 1963 after a spectacular landslide on the right river bank which dammed the Visočica River. A 50 m high natural dam was formed and caused total flooding of several upstream villages. The landslide activated as an effect of abundant water infiltration from snow melt water. Recently this dam was replaced by an artificial one providing the power for a hydro-electric power plant.

Karst springs on both sides of the border are an essential source of water supply for the settlements. Karst waters are tapped mostly at the springs with gravity transport towards consumers. Most of the existing springs are used, as the large water quantities are taken to supply the larger settlements, such as Krupac springs for Pirot, Vrelo Protopopinci for Dimitrovgrad, Opitsvet for Slivnitsa, Zli dol for Godech, Peshta (Iskrets spring) for the town of Svoge. However, karst groundwater is not directly affected by the exploitation and there are sufficient amounts of water for dependent eco-systems. In contrast, significant impact on the balance of karst waters may be caused by exploitation wells drilled in karst aquifers covered by Quaternary-Pliocene deposits in the Sofia and Pirot basins. At this stage that impact is negligible because of the small number of wells using water from karst aquifer and small rates of abstraction (Directorate for the Danube River Basin, 2015).

6.6 POTENTIAL TRANSBOUNDARY PROBLEMS

The specific geomorphological, geological and hydrogeological conditions greatly reduce cross-border flow in the areas near the state border where transboundary impacts can be found. In the Northern zone there is a possibility that water is transferred from the Komshtitsa River to the Kamenička River (Figure 6.8), but this assumption must be proven by detailed hydrometric measurements and tracing experiments.

In Bulgaria the area is characterised by very high karstification and vulnerability. The Komshtitsa River loses its flow partially or completely into the outcrops of the Triassic limestones and dolomites, and its bed is located above the saturated zone of karst groundwater. Proof of this is Granicharska Cave, whose entrance is located next

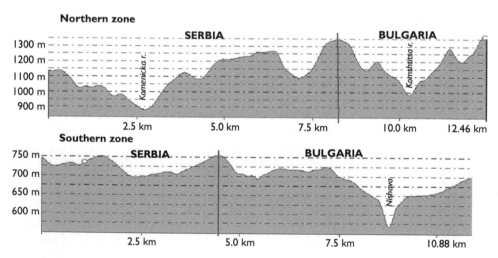

Figure 6.8 Schematic terrain elevation profile of transboundary areas in the Northern and Southern zones.

to the river about 3–4 m above it (Beron *et al.*, 2006) and where the river waters enter. River waters percolate down to the saturated zone but because of a fault against the elevated south block which passes through Serbian territory, the water probably flows to the west, where the lowest possible drainage areas are.

In the Southern zone the main drainage artery is the Nišava River but the probable influence of the transboundary impact is minimal due to the small catchment areas, absence of a significant river recharge into the area and lower level of karstification.

The zone represents a low mountain hill with leveled ridge areas with karst formations formed on both Bulgarian and Serbian territory. Precipitation can be drained away in different directions; to Berende spring in Bulgaria and to the springs in Dimitrovgrad and Krupac in Serbia, but there is no information on the spatial position of the underground watershed.

Neither of the transboundary regions are considerably affected by human activities.

6.7   CONCLUSIONS

The preliminary analysis of the existing information indicates that karst aquifers of the Northern and Southern zones along the border area of Serbia and Bulgaria are still not sufficiently studied, and cross-boundary groundwater flow is probably limited. A number of obstacles exist to understanding of the cross-border relationships because of limited access historical data due to the pre-existing border regime and the scarcity of available information. The main difficulties are related to:

1. Discordance of some geological boundaries and distribution of rock formations on both sides of the state border;
2. Absence of observations of the rivers and springs regime near the border;
3. Difficult access to comparable hydrological and meteorological data.

The continuation of the joint research by Serbian and Bulgarian hydrogeologists and karst scientists is essential. The common water balance of the border territories of Serbia and Bulgaria needs more hydrological, hydrogeological and climatic data, field survey and water tracings.

Future research will help to clarify the hydrogeological conditions and the quantitative and qualitative aspects related to groundwater for the sustainable development of cross-border areas and protection of karst water and landscapes.

REFERENCES

Angelova D., Benderev A., Baltakov G., Ilieva I., Nenov T. (1995) On the Karst Evolution of the Stara Planina Iskar Gorge. *Review of the Bulgarian Geological Society* 56(3), 111–124 (in Bulgarian with English abstract).
Angelova D., Benderev A., Kostov K. (1999) On the age of the caves in the Stara Planina Iskar gorge, NW Bulgaria. *Proc. "Karst 99" Europ. Conference*, 10–15. 09. 1999, Grand Causses – Verkors, France. 29–35.

Awawdeh M., Nawafleh A. (2008) A GIS-based EPIK Model for Assessing Aquifer Vulnerability in Irbid Governorate, North Jordan. *Jordan Journal of Civil Engineering* 2(3), 267–278.

Antonov H. (1963) *Karst waters in Western part of the Sofia Stara Planina.* Annual University of Mining and Geology, Sofia, 1961–1962, v.VII (in Bulgarian).

Antonov H., Danchev D. (1980) *Ground waters in Bulgaria.* Sofia, Tehnika, 360 p. (in Bulgarian).

Directorate for Water Management of Danube region (2015) URL: http://www.bd-dunav.org/content/Registar-na-izdadenite-razreshitelni-ot-podzemni-vodi-42/; last visited 28/4/2015.

Benderev A. (1989) Karst and karst waters of Ponor Mountain. PhD Thesis, NIPI-KG, 157 p. (in Bulgarian).

Beron P., Daaliev T, Jalov Al (2006) *Caves and Speleology in Bulgaria.* PenSoft, Bulgarian Federation of Speleology and National Museum of Natural History, Sofia. Bulgaria. 507 p.

CORINE Map. (2010) The European Topic Centre on Land Use and Spatial Information. European Environment Agency. Retrieved from: http://www.eea.europa.eu/data-and-maps/data/corine-land-cover-2006-raster-3.

Cvijić J. (1895) *Caves and underground hidrography in Eastern Serbia.* Glas. Serbian Royal Academy of Sciences, XLVI, Belgrade, 1–101 (in Serbian).

Ćalić J. (2007) Karst research in Serbia before the time of Jovan Cvijić. *Acta Carsologica* 36(2): 315–319.

Doerfliger N., Jeannin P. Y., Zwahlen F. (1999) Water vulnerability assessment in karst environments: a new method of defining protection areas using a multi-attribute approach and GIS tools (EPIK method), *Environmental Geology* 39 (2), 165–176.

Dukić D. (1975) Characteristics of hydrology Serbia. *Journal of the Geographical Institute "Jovan Cvijić"*, SANU, vol. 26, 23–94 (in Serbian).

Eftimi R., Benderev Al (2007) Utilization of hydrochemical data for characterization of the karst system & Example of Iskrets springs, Bulgaria. *Review of the Bulgarian Geological Society*, part 1–3, 167–174.

Gams I. (1994) Types of the poljes in Slovenia, their inundations and land use. *Acta Carsologica*, XXIII: 285–302.

Geological Survey of Serbia (1970) *Basic geological map of Yugoslavia. Sheet Pirot.* M 1:100 000, Belgrade.

Gogu R. C., Dassargues A. (2000) Sensitivity analysis for the EPIK method of vulnerability assessment in a small karstic aquifer, southern Belgium. *Hydrogeology Journal* (8), 337–345.

Goldscheider N. (2005) Karst groundwater vulnerability mapping: application of a new method in the Swabian Alb, Germany. *Hydrogeology Journal* 13(4), 555–564.

Haidutov I., Dimitrova R. (editors) (1993) *Geological map of Bulgaria.* M 1:100 000. Sheet Berkovitsa, VTS, Troyan.

Haidutov I., (ed.) (1993) *Geological map of Bulgaria.* M 1:100 000. Sheet Pirot, VTS, Troyan.

Hristova N. (2010) Hydrological picture of Nišava trans-boundary catchment. *Journal of the Geographical Institute Jovan Cvijić* SANU 60(2), 1–14.

Iurkiewicz A., Horoi V., Popa R. M., Dragusin V., Vlaicu M., Mocuta M. (2005) Groundwater vulnerability assessment in a karstic area (Banat Mountains, Romania) – Support for water management in protected areas, In: Stevanović Z. and Milanović P. (eds) *Proceedings of International Conference "Water Resources and Environmental Problems in Karst – CVIJIĆ"*, Beograd-Kotor, p. 127–133.

Koleva E., Peneva R. (1990) *Climatic guide. The rainfalls in Bulgaria.* BAS publishing house, Sofia, 169 p. (in Bulgarian with English summary).

Kostov K. (2008) Indices of paleoseismicity in karst terrains. PhD thesis, Sofia, 191 p. (in Bulgarian).

Mihaylova B., Kostov K., Danailova M., Benderev A. (2008) Characteristics and development of the karst in the catchment area of the Opitsvet-Bezden springs. *Proc. Conference*

"Investigation and protection of the Karst and Caves", Sofia, October 24, 2008, 96–103 (in Bulgarian with English abstract).

Mihaylova B., Benderev A., Kostov K. (2009) Application of GIS for localization of agricultural land within the vulnerable areas of karst terrains (the example of Western Stara Planina). *GEOSCIENCES, Bulgarian Geological Society*, Sofia. 125–126 (in Bulgarian).

Mihaylova B., Benderev A. (2010) Characteristics of formation of karst water regime in the southern slopes of the western part of Balkan Mountain (Bulgaria). In: *Groundwater resources. Current problems of learning and use. International Conference for the 100th anniversary of the birth of B.I.Kudelin*, Moscow, 13–14 May 2010. 331–337.

Paskalev M., Benderev A., Shanov S. (1992) Tectonic conditions of the region of the Iskrets Karst Springs (West Stara-Planina). *Review of the Bulgarian Geological Society* 53(2), 69–81 (in Bulgarian).

Petrović J. (1974) Karst of eastern Serbia. *Posebna izdanja Srpskog geografskog drustva*, Belgrade, 96 p. (in Serbian).

Petrović J. (1976) Potholes and caves in Serbia. *Vojnogeografski zavod, Belgrade*, 511 p. (in Serbian).

Radev J. (1915) Karst forms in Western Stara Planina. *Ann. of Sofia University, Historical and Philological Fac.*, v. 10–11, 149 p. (in Bulgarian).

Ristić V. (2007) Development of simulation model for daily spring discharge estimation of karst springs, PhD thesis, University of Belgrade, Faculty of Mining and Geology, Belgrade, 319 p. (in Serbian).

Shanov S., Benderev A. (2005) Seismic impact from earthquakes and from grouped blasts in quarries on the discharge of karst springs. In: Stevanović Z. and Milanović P. (eds) *Proceedings of International Conference "Water Resources and Environmental Problems in Karst – CVIJIĆ"*, Beograd-Kotor, p. 145–150.

Shanov S., Kostov K. (2015) Dynamic tectonics and karst. Springer-Verlag Berlin Heidelberg, 123 p.

Stevanović Z. (1991) *Hydrogeology of Carpathian-Balkan karst of Eastern Serbia and water supply opportunities*. University of Belgrade – Faculty of Mining and Geology, Belgrade, 245 p.

Stevanović Z. (1994) Karst ground waters of Carpatho-Balkanides in Eastern Serbia. In: Stevanović Z., Filipović B. (eds.) *Ground waters in carbonate rocks of the Carpathian-Balkan mountain range*. Spec ed. CBGA, Jersey-Belgrade, 203–237.

Szeidovitz Gy., Paskaleva I., Gribovszki K., Kostov K., Surányi G., Varga P., Nikolov G. (2008) Estimation of an upper limit on prehistoric peak ground acceleration using the parameters of intact speleothems in caves situated at the western part of Balkan Mountain Range, northwest Bulgaria. *Acta Geodaetica Geophisyca Hungarica* 43(2–3), 249–266.

Tchoumatchenco Pl., Rabrenović D., Radulović Vl., Malešević N., Radulović B. (2011) (a) Trans-border (north-east Serbia/north-west Bulgaria) correlations of the Jurassic lithostratigraphic units. *Geološki anali Balkanskoga poluostrva* 72, 1–20.

Tchoumatchenco Pl., Rabrenović D., Radulović Vl., Malešević N., Radulović B. (2011) (b) Trans-border (east Serbia/west Bulgaria) correlation of the morpho-tectonic structures. *Geološki anali Balkanskoga poluostrva* 72, 21–27.

Toula F. (1882) Grundlinien zur Geologie des Westl. Balkan. Denkschr der K. Ak. D. Wiss. Wien, 44.

Yanev S. (ed.) (1992) *Geological map of Bulgaria*. M 1:100 000. Sheet Sofia, VTS, Troyan.

Zagorchev I., Kostadinov V. (ed.) (1991) *Geological map of Bulgaria*. M 1:100 000. Sheet Vlasotice & Breznik, VTS, Troyan.

Živanović V. (2011) Pollution vulnerability assessment of groundwater – examples of karst in Serbia, Master thesis, University of Belgrade, Faculty of Mining and Geology, Belgrade (in Serbian). 215 p.

Živanović V., Dragišić V. (2013) Application of EPIK method for vulnerability assessment of groundwater of south part of Stara Planina Mt. *Proceedings of 7ᵗʰ Symposium on Karst Protection* (in Serbian), J. Ćalic (ed), Beograd, p 13–21.

Zlatkova M. (2006) Region 205 – Bezden. Geological-geographical notes. Golyamata Temnota Cave, village of Drenovo, Sofia district. *Caves and Cavers*, Sofia, 55–56.

Zwahlen F. (ed) (2003) Vulnerability and risk mapping for the protection of carbonate (karst) aquifers. Final Report COST Action 620, available at: http://capella.unine.ch/chyn/php/publica_intro.php.

Chapter 7

An assessment of territory participation in transboundary karst aquifer recharge: A case study from the Skadar Lake catchment area

Milan Radulović[1], *Goran Sekulić*[2],
Momčilo Blagojević[3], *Jelena Krstajić*[4] *& Entela Vako*[5]
[1]*University of Montenegro, Podgorica, Montenegro*
[2]*University of Montenegro, Podgorica, Montenegro*
[3]*Ministry of Agriculture and Rural Development, Podgorica, Montenegro*
[4]*University of Belgrade, Belgrade, Serbia*
[5]*Polytechnic University, Tirana, Albania*

ABSTRACT

Equitable sharing of groundwater resources between countries is a major challenge, especially in highly karstified terrains. In the karstic terrains of the External Dinarides water division takes place mainly below the surface, so maps of the spatial distribution of groundwater recharge, created by multi-parameter GIS methods, could be useful as one of the bases for water management. Such maps can be used for assessing the percentage in which the territories of bordering countries participate in recharging the shared aquifer. For the purpose of equitably sharing the karst aquifer on the south western edge of Skadar Lake (shared by Montenegro and Albania), the KARSTLOP method was applied. The method had previously been calibrated according to the nearby catchments of four terrestrial karst springs. The Recharge map obtained for the catchment area of sublacustrine springs allowed assessment according to which 291.7 million m³ of water infiltrates annually in Montenegrin territory, and 19.2 million m³ in the Albanian part.

7.1 INTRODUCTION

There are many countries that share karst aquifers. Political, cultural, socio-economic and other differences between bordering countries make groundwater management one of the most challenging issues today. Conflicts over the use of groundwater sometimes arise, even between municipalities within a single country. An absence of coordinated data sharing between municipalities and countries using the same aquifer can lead to deterioration in qualitative and quantitative groundwater status. In addition, disputes between countries over transboundary water resources often escalate, in some cases into armed conflicts. Adequate hydrogeological research and sustainable water management can help reduce the extent to which such instances arise. The growing problems of water scarcity and the unequal sharing of water resources between countries have prompted a number of authors to deal with this issue (Bittinger, 1972; Hayton & Utton, 1989; Yamada, 2004; Eckstein & Eckstein, 2005; Jarvis *et al.*, 2005; Puri & Aureli, 2005; Earle *et al.*, 2010; Puri & Struckmeier, 2010; Ganoulis & Fried, 2010; IGRAC, 2009; Stevanović *et al.*, 2012; Jarvis, 2014).

The hydrogeological research of transboundary aquifers in highly karstified terrain is an especially complex task. The aim of this chapter is to present an approach that will facilitate an assessment of groundwater resources in those areas. The External Dinarides, to which the pilot site belongs, are characterised by holokarst (highly karstified terrains), which demands a specific approach to hydrogeological research.

The approach presented is based on assessment of the spatial distribution of groundwater recharge using GIS techniques. The focus of the approach is on the application of the KARSTLOP method (Radulović *et al.*, 2012) as a tool for mapping the spatial distribution of recharge in karst terrains. The approach is applied to the transboundary catchment area of the sublacustrine springs of the south western coast of Skadar Lake, which is shared by Montenegro and Albania. Previously, the method was calibrated through analysis of the nearby catchment areas of terrestrial karst springs (Podgor, Karuč, Slatina and Crnojevića springs) (Figure 7.1).

7.2 METHODOLOGY

In highly karstified terrains, with numerous surface and subsurface karst landforms, and with the absence of surface runoff, it is not possible to determine the spatial distribution of aquifer recharge using standard methods (e.g. lysimeter, seepage meter, soil mass balance, zero-flux plane, tracer methods, method based on the Darcy's law, water budget methods). The standard methods can only provide assessment of recharge rates that relate to specific points or a whole catchment area of a karst spring.

Maps of the spatial distribution of groundwater recharge (recharge maps) created by multi-parameter GIS methods (Shaban *et al.*, 2005; Andreo *et al.*, 2008;

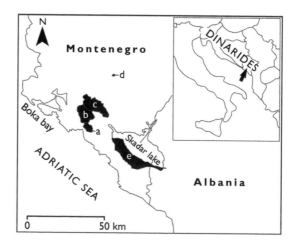

Figure 7.1 Position of catchment areas: a) Podgor springs; b) Crnojevića springs; c) Karuč springs; d) Slatina springs; e) the catchment area of the southwestern edge of Skadar Lake.

Radulović et al., 2012) could be useful for investigating transboundary karst aquifers. It is necessary to take into account those natural factors that have the greatest impact on groundwater recharge, such as climate, topography, hydrography, vegetation, soil and geology. Through overlapping maps of selected factors, and using an appropriate algorithm, a final groundwater recharge map can be obtained. Such a map could be used for rough assessment of the percentage in which territories participate in the recharge of a transboundary aquifer. Figure 7.2 shows the principle behind an assessment of the percentage in which the territories of bordering countries participate in the recharge of a shared karst aquifer.

The method had previously been calibrated according to the nearby catchment areas of four karst springs: Podgor springs, Crnojevica springs, Karuč springs and Slatina springs (Figure 7.1). The average recharge rates that relate to the whole catchment areas of these springs, are assessed by standard methods such as Cl⁻ mass balance, water budget (runoff/rainfall ratio), and assessment of evapotranspiration by empirical equations (Živaljević & Bošković, 1984; Živaljević, 1991; Zogović, 1992; Radulović, 1994, 2000, 2012; IJC, 2001). Those average values have been compared with average recharge rates computed by twenty potential algorithms of the KARSTLOP method. The algorithm that provided the best match between calculated and previously assessed average recharge rates was chosen for the further analyses (Radulović, 2012; Radulović et al., 2012).

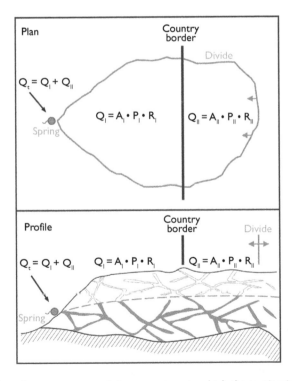

Figure 7.2 Conception of an assessment of the percentage in which the territories of bordering countries participate in the recharge of a shared karst aquifer; Q – discharge (L³/s), A – size of catchment area (L²), Precipitation (m/year), R – recharge (%).

In the process of obtaining the recharge map, the eight most important natural factors affecting the recharge rate are analysed. The initials for the selected factors make up the acronym in KARSTLOP method (where, K – karstfication; A – atmospheric conditions; R – runoff, S – slope; T – tectonics; L – lithology; O – overlying layers; P – plants).

K *map* (Karstification) is created by analysing surface (K_{sf}) and subsurface karstification (K_{ss}). Surface karstification (K_{sf}) is assessed on the basis of the area of karst landforms per surface unit (subfactors K_{sf1} and K_{sf2}). When estimating subsurface karstification (K_{ss}), each side of a speleological object (swallow holes, caves, pits, etc.) is marked in the range of 200 m, while the rest of the catchment area is assessed on the basis of indirect parameters such as: the discharge amplitude of karst springs (K_{ss1}), the mean registered velocity of artificial tracers at catchment areas (K_{ss2}) and the primary mineral saturation index of spring water (K_{ss3}).

A *map* (Atmospheric conditions) is obtained using the Digital Elevation Model (DEM), which is first used to generate an Altitude map A_1 and then a Shaded relief map A_2 showing factors affecting the intensity of solar radiation. An altitude (A_1) indirectly reflects the air temperature (and thus also evapotranspiration) and precipitation (an increase in altitude leads to an increase in precipitation and more intense groundwater recharge). The intensity of solar radiation (A_2) also influences the amount of evapotranspiration and recharge. Sides of mountains that are in shadow (surfaces with relative reflection lower than 0.5) for the mean annual sun position (at noon during the equinox) are characterised by lower intensity of solar radiation, i.e. by lower evapotranspiration and higher recharge rate.

R *map* (Runoff) is obtained based on the distribution of permanent streams at the catchment area, which may indicate the permeability of karstified terrains.

S *map* (Slope) is obtained using DEM. Terrains with smaller slopes are described as being more suitable for groundwater recharge.

Two subfactors are considered for the preparation of T *map* (Tectonics): the density of faults T_1 and dip of strata T_2.

L *map* (Lithology) is obtained based on the following lithological data: the type of carbonate rock (subfactor a), bedding (subfactor b) and mineral-petrographic impurities (subfactor c).

O *map* (Overlying layers) is obtained based on data relating to the type and thickness of geological cover and soil.

P *map* (Plants) is obtained based on data relating to terrain vegetation cover.

By overlapping these eight maps according to the established algorithm a recharge map is obtained (Figure 7.3, Table 7.1).

7.3   APPLICATION AND RESULTS

The method was applied to the catchment area of the southwestern edge of Skadar Lake shared by Montenegro and Albania.

7.3.1   Description of the study area

The study area is represented by the catchment area of the south western edge of Skadar Lake. It is located in the south eastern part of the Dinarides (Figure 7.1). The total size

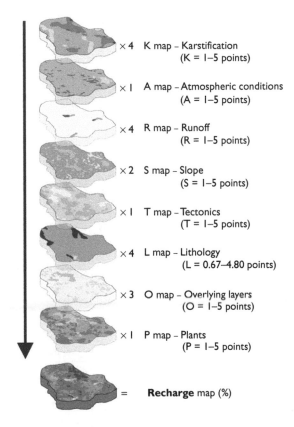

Figure 7.3 Flow Chart of the KARSTLOP overlapping procedure.

of the catchment area is approximately 185 km². Around 173 km² (93.5%) belongs to the territory of Montenegro, with the rest in the territory of Albania (12 km², i.e. 6.5%).

The area of Skadar Lake has a modified Mediterranean climate characterized by hot and dry summers and cold winters. Mean annual precipitation in this area is around 2500 mm, and the mean annual air temperature is around 10°C.

Vegetation over the catchment area is represented by forest, especially at higher altitudes. As elevation decreases, the area covered by forest gradually transforms into one characterised by the low vegetation typical for karst terrain (scrubby vegetation). A significant part of the catchment area is covered by pasture, but vegetation is generally absent from the coastal karst belt. The catchment area is characterised by the presence of residual soil, the thickness of which is below 15 cm, with the exception of areas with smaller slopes, where the thickness reaches 60 cm.

Geomorphologically, the area can be divided into two parts: the north eastern slopes of Rumija Mountain and the Basin of Skadar Lake (Figure 7.4). The north eastern slopes of Rumija are made up of carbonate rock masses in which numerous surface and subsurface karst landforms have developed. The most prevalent landforms are karrens that cover a significant portion of the surface. Sinkholes and uvalas are also widespread, and the bottoms are generally slightly inclined in a strata dip

Table 7.1 Classification matrix for the KARSTLOP method (Radulović *et al.*, 2012).

K – KARSTIFICATION

Area of degraded zone (karren fields, ruine-like relief etc.) per unit square (10^3 m^2/km^2)	Ksf_1	Area of karst depressions per unitsquare (10^3 m^2/km^2)	Ksf_2	$Ksf = (Ksf_1 + Ksf_2)/2$
<60	1	<25	1	1
60–120	2	25–50	2	>1–2
120–180	3	50–75	3	>2–3
180–240	4	75–100	4	>3–4
>240	5	>100	5	>4–5

$$K = \frac{Ksf + Kss}{2}$$

$Qmsx/Qmin$	Kss_1	v (cm/s)	Kss_2	d > 4 km SI	d > 1–4 km SI	d <1 km SI	Kss_3	$Kss = (Kss_1 + Kss_2 + KSS_3)/3$
<5	1	<0.1	1	>0.3	>0	>-0.3	1	1
5–50	2	0.1–1	2	0 – 03	-0.3 – 0	-0.6 – -0.3	2	>1–2
5–100	3	1–10	3	-0.3 – 0	-0.6 – -0.3	-0.9 – -0.6	3	>2–3
> 100	4	>10	4	<-03	<-0.6	- 0.9	4	>3–4
zone of 200 m from all sides of speleologically explored subsurface objects								5

A – ATMOSPHERIC CONDITIONS

Altitude (m)	A_1
<300	1
300–800	2
800–1300	3
1300–1800	4
>1800	5

Reflectance value	A_2
>0.5	1
<0.5	5

$$A = \frac{A_1 + A_2}{2}$$

R – RUNOFF

Surface within the catchment area of referred spring	R
Zone of 200 m from both sides of perennial streams formed by discharge at contact of karstificated limestone or dolomite with more compact carbonate sections	1
Zone of 200 m from both sides of perennial streams formed at the edge of karst depressions (polje, uvala etc.), as well as around perennial streams which runs from non-karstic terrain etc.	3
Remaining of catchment area	5

S – SLOPE

Slope (°)	S
0–5	5
5–15	4
15–25	3
25–35	2
>35	1

T – TECTONICS

Lengths of faults (km/km²)	Tf	Dip angle of stratum (°)	Td
0–1	1	<30	1
1–2	2	30–60	3
2–3	3	>60	5
3–1	4		
>4	5		

$$T = \frac{Tf + Td}{2}$$

L – LITHOLOGY

Carbonate rocks	a
Limestone	1
Dolomite	0.7

Bedding	b
Massive, Thickly bedded	2.4
Bedded	1.8
Laminated, Thinly laminated	1.2

Mineralogical-petrographic ingredients	c
Calcified	2
Dolomitic	1.6
Sandy, Silificated	1.2
Marly, Clayish, Bituminous	0.8

$$L = a \times b \times c$$

O – OVERLYING LAYERS

O_1 Soil	Thickness				
	<15 cm	15–30 cm	30–60 cm	60–100 cm	>100 cm
Calcomelanosols, Euthric cambisols, Rendzina	5	4	3	2	1
Calcocambisols, Terra Rossa	4	3	2	1	1

O_2 Geological cover	Thickness		
	<3 m	3–6 m	>6 m
Alluvial, Glacial-fluvial, Moraine sediments, Scree	5	3	1
Diluvium	3	2	1
Glacial-limnic sediments	2	1	1

$$O = \frac{O_1 + O_2}{2}$$

P – PLANTS

Plants	P
Bare rocks	5
Sparsely vegetated areas	4
Pastures	3
Transitional woodland-scrub	2
Forests	1

Recharge = 4xK + A + 4xR + 2xS + T + 4xL + 3xO + P

Figure 7.4 The south western coast of Skadar Lake photographed from Montenegrin territory.

direction, i.e. to the north east. Also, a large number of caves can be found in this area. Skadar Lake represents a crypto-depression whose floor in the areas of sublacustrine springs descends below sea level.

Many researchers both at home and abroad have studied the geological structure of this region (Tietze, 1884; Baldacci, 1886; Cvijić, 1899; Nopcsa, 1916; Bourcart, 1926; Waisse, 1948; Milovanović, 1965; Bešić, 1969; Grubić, 1975; Mirković *et al.*, 1985).

The main base for exploring the geology of this area is a Geological map of Montenegro 1:200 000 (Mirković *et al.*, 1985). The Skadar Lake area belongs to the tectonic unit of the Visoki Krš. Mesozoic rocks show dominant distribution and they are represented mainly by carbonates, i.e. limestones and dolomites. The total thickness of these carbonate rocks can be over 3000 m.

Numerous hydrogeological studies have been conducted in the Skadar Lake catchment area (Torbarov & Radulović, 1966; Radulović, 1989, 2000, 2012; Radulović *et al.*, 1979, 1989, 1998, 2013, 2015; Zogović, 1992; Radulović & Radulović, 2004; Stevanović *et al.*, 2008; Djordjević *et al.*, 2010; Sekulić & Bushati, 2013).

The surface drainage network in this area is poorly developed, with a small number of streams that flow only in the rainy period of the year. In the study area the karst aquifer is developed and it is mainly recharged by diffuse infiltration of rainwater (autogenic recharge). The hydraulic conductivity of the carbonate aquifer is relatively high. Groundwater flows mainly in privileged directions marked by faults and joints.

Discharge zones are spread along the south western edge of Skadar Lake. Sublacustrine springs (vruljas), that occurring along the coastal part of the lake (Figure 7.5), are actually underwater dolines through which a carbonate aquifer discharges. Around 40 sublacustrine springs are registered in the bottom of Skadar

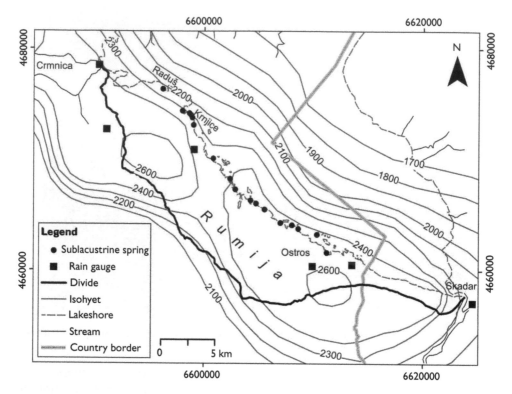

Figure 7.5 Map of known sublacustrine springs on the south western edge of Skadar Lake with the isohyets and the locations of the rain gauges.

Lake. The bottom of the underwater dolines is relatively deep, ranging from 10 to 70 m. One feature of karst aquifers in this area is a significant fluctuation in yield. Some sublacustrine springs, such as the Raduš and Krnjice springs, show a very high yield in the rainy period (above 30 m³/s). Groundwater quality is relatively good, with a high concentration of calcium and hydrocarbons.

7.3.2   Application of the KARSTLOP method

The final recharge map has been obtained through the overlapping the eight maps previously explained, according to an established algorithm. The recharge rate is expressed on the map as a percentage (%) of mean annual precipitation (Figure 7.6).

From the recharge map (Figure 7.6) it can be seen that the catchment area is heterogeneous according to its potential for aquifer recharge. Based on these results it can be concluded that the greatest potential for groundwater recharge is the areas of the strongest sublacustrine springs (Raduš and Krnjice springs). Slightly lower potential for recharge occurs in terrains in the central part of the catchment area, which are drained by temporal springs with a lower yield. The minimum values of the recharge rate relate to the south eastern and north western parts of the catchment area, from which strong sublacustrine springs are entirely absent.

Through statistical analysis of the digital recharge map (Figure 7.6) the average recharge rate for the entire catchment area was obtained (68.5%). Using the same map, in combination with other water balance components, the percentage of territory participation in recharging the transboundary karst aquifer was assessed (Table 7.2). The results indicate that 93.8% of water that outflows through sublacustrine springs infiltrates via Montenegrian territory (291.7 million m³/year), and 6.2% on that belonging to Albania (19.2 million m³/year).

7.4 DISCUSSION

The results obtained through applying the KARSTLOP method represent one of the bases for further planning and sustainable management of the transboundary karst aquifer shared between Montenegro and Albania.

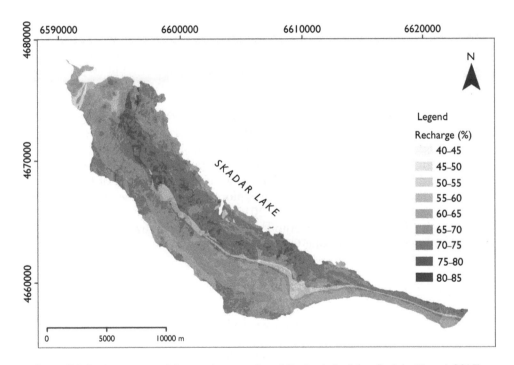

Figure 7.6 Recharge map of the south west edge of Skadar Lake (after Radulović *et al.*, 2015).

Table 7.2 Balance components for the catchment area of the south western edge of Skadar Lake on the territories of Montenegro and Albania.

Country	Catchment area (km²)	Precipitation (mm)	Average recharge (%)	Outflow (m³/s)	Specific yield (l/s/km²)	The percentage of the total outflow (%)
Montenegro	173	2461	68.5	9.25	53.5	93.8
Albania	12	2422	66.2	0.61	50.8	6.2

The calibration of the method from the catchment areas of nearby terrestrial springs proved applicable for assessing the spatial distribution of the mean recharge rate. The calibration process shows that the error in assessing the recharge rate could be up to around 5%. The discrepancy between calculated and previusly assessed values was in range from 0.1 to 5.1% (Radulović et al., 2012). However, the method has been tested only on the four catchment areas, so that the possible error could exceed the value.

However, as all methods of this type involve approximation, it is necessary to have supporting data to reduce the subjectivity factor to a minimum. As the main limitations of the KARSTLOP method, Radulović et al. (2012) identified the following:

– the method cannot assess allogenic recharge (inflows from non-karstic terrains must be assessed through the application of other hydrological methods),
– the method cannot assess temporal variations of the recharge rate (only the mean annual value).

To apply the approach described it is important first to determine the spatial position of the divides (watersheds), which is a difficult task in highly karstificated terrain. However, if data related to the discharge of springs and the amount of precipitation exists, the method can be used inversely, i.e. the recharge map can be used to find the spatial position of divides.

In applying the method described it is not possible to assess static groundwater reserves (storage), only dynamic reserves, i.e. the amount of water which inflows into the system through the process of infiltration (recharge) and outflows via springs (discharge).

Based on results obtained through previous application, use of the method is recommended for highly karstic areas, but the weighting coefficient of factors should be adapted to the characteristics of a given exploration area, i.e. the method should be additionally calibrated (Radulović et al., 2012).

The main advantages of this method are that the input data are readily available and application is facilitated using GIS techniques.

7.5 CONCLUSIONS

The application of the approach has proven to be useful for the purpose of equitably sharing water resources in highly karstified terrains. For applying the method in other karst areas, it is necessary to bear in mind the details of the terrain and customise the appropriate algorithm accordingly.

Use of the KARSTLOP method has generated one of the bases for the delineation and rational management of the transboundary karst aquifer shared between Montenegro and Albania.

The approach could find a wider application in assessing participation in karst aquifer recharge for other territories. Potential areas for the application of this approach could be found in the Balkans, where the distribution of carbonate rocks and the fact that a large number of countries exist within a relatively small area caused the existence of transboundary karst aquifers.

Examples can be found in the cross-border areas of Croatia and Slovenia, Croatia and Bosnia and Herzegovina, Bosnia and Herzegovina and Montenegro, Serbia and Montenegro, Serbia and Bosnia and Herzegovina, and Montenegro and Albania. It is necessary to analyse karst aquifer potential in order to avert conflicts over water use. The application of the approach could help to improve water management among these countries.

It always should be kept in mind that the method provides assessed recharge rates on a regional scale, so results should be used with caution, and preferably compared with values obtained by comprehensive hydrogeologic assessments.

Since water management for transboundary karst aquifers is a highly important and sensitive issue, this topic should be considered carefully in future research, with the participation of experts from various fields.

REFERENCES

Andreo B., Vías J., Durán J., Jiménez P., López-Geta J.A., Carrasco F. (2008) Methodology for groundwater recharge assessment in carbonate aquifers: application to pilot sites in southern Spain. *Hydrogeology Journal* 16, 911–925.

Baldacci L. (1886) Ricognizione geologico–mineraria del Montenegro. *Bollettino del Reale Comitato geologico d'Italia* 17, 9–10.

Bešić Z. (1969) *Geology of Montenegro–Karst of Montenegro* (in Serbian). Geological Survey of Montenegro, Podgorica.

Bittinger M.W. (1972) A survey of interstate and international aquifer problems. *Ground Water* 10(2), 44–54.

Bourcart J. (1926) Nouvelles observations sur la structure des Dinarides adriatiques. In: *Actas del XIV Congreso Geológico Internacional*, Madrid, 1899–1941.

Cvijić J. (1899) Glacial and morphological studies of the mountains of Bosnia, Herzegovina and Montenegro (in Serbian). *Voice of the Serbian Royal Academy* 57, 1–196.

Djordjević B., Sekulić G., Radulović M., Šaranović M., Jaćimović M. (editors) (2010) *Water potentials of Montenegro* (in Serbian). Montenegrin Academy of Sciences and Arts, Podgorica.

Eckstein Y., Eckstein G.E. (2005) Transboundary aquifers: Conceptual models for development of international law. *Ground Water* 43(5), 679–690.

Earle A., Jägerskog A., Öjendal J. (editors) (2010) *Transboundary water management: principles and practice.* Earthscan, London.

Ganoulis J., Fried J. (2010) Transboundary water resources managment – Needs for a coordinated multidisciplinary approach. In: Earle A., Jägerskog A., Öjendal J. (editors) *Transboundary water resources managment: principles and practice.* Earthscan, London, 9–25.

Grubić A. (1975) Tectonics of Yugoslavia. *Acta Geologica* 41, 365–384.

Hayton R., Utton A.E. (1989) Transboundary ground waters: The Bellagio draft treaty, *Natural Resources Journal* 29, 663–722.

IJC (Institut Jaroslav Cerni) (2001) *Water master plan of Montenegro.* Government of Montenegro, Podgorica.

IGRAC (2009) Transboundary aquifers of the world, 1:50 000 000. Special edition for the 5th World Water Forum, Istanbul.

Jarvis T., Giordano M., Puri S., Matsumoto K., Wolf A. (2005) International borders, ground water flow, and hydroschizophrenia. *Ground Water* 43(5), 764–770.

Jarvis T. (2014) *Contesting hidden waters-confilct resolution for groundwater and aquifers.* Routledge, New York.

Milovanović B. (1965) Epeirogenic and orogenic dynamics in the area of External Dinarides and the problems of paleokarstification and geological evolution of holokarst (in Serbian). *J Geol Surv Serbia* 4(5), 5–44.

Mirković M., Živaljević M., Đokić V., Perović Z., Kalezić M., Pajović M. (1985) Geological map of Montenegro 1:200 000. The Republic Self Managing Community of Interest for Geological Exploration of Montenegro, Titograd.

Nopcsa F. (1916) Begleitwortzur geol. Karten von Nordalbanien, Rasien und Ostmontenegro. *Földtani Közlöny* 46, 7–12.

Puri S., Aureli A. (2005) Transboundary aquifers: a global programm to asses, evaluate, and develop policy. *Ground Water* 43(5), 661–668.

Puri S., Struckmeier W. (2010) Aquifer resources in a transboundary context: a hidden resource?– Enabling the practitioner to 'see it and bank it' for good use. In: Earle A., Jägerskog A., Öjendal J. (editors) *Transboundary water management: principles and practice.* Earthscan, London, 73–90.

Radulović V. (1989) *Hydrogeology of Skadar Lake watershed* (in Serbian). Geological Survey of Montenegro.

Radulovic M (1994) Report on hydrogeological exploration of Sjenokos locality in Crmnica area – I phase (in Serbian). University of Montenegro, Podgorica.

Radulović M. (2000) *Karst hydrogeology of Montenegro* (in Serbian). Geological Survey of Montenegro, Podgorica.

Radulović M.M. (2012) Multi-parameter analysis of groundwater recharge in karstic aquifers– case examples from the Skadar Lake basin (in Serbian). PhD thesis, University of Belgrade, Belgrade

Radulović M., Radulović V. (2004) Hydrogeological map of Montenegro 1:200 000 (in Serbian). Geological Survey of Montenegro, Podgorica.

Radulović M., Radulović V., Popović Z. (1979) Final report on the task of preparation for the Basic hydrogeological map of SRY 1:100,000, "Titograd" sheet (in Serbian). Geological Survey of Montenegro, Titograd.

Radulović M., Popović Z., Vujisić M., Novaković D. (1989) Basic hydrogeological map of SRY 1:100,000, "Bar" sheet (in Serbian). Geological Survey of Montenegro, Titograd.

Radulović M., Popović Z., Vujisić M., Novaković D. (1998) *Guide book for the Basic hydrogeological map of SRY 1:100,000, "Bar" sheet* (in Serbian). Geological Survey of Montenegro, Titograd.

Radulović M.M., Stevanović Z., Radulović M. (2012) A new approach in assessing recharge of highly karstified terrains–Montenegro case studies. *Environmental Earth Sciences* 65(8), 2221–2230.

Radulović M.M., Novaković D., Sekulić G. (2013) Geological and hydrogeological characteristics of the Montenegrin part of the Skadar Lake catchment area (in Serbian). In: Sekulić G, Bushati S (editors) *Development of hydrogeological and hydraulic study of regulation of Skadar Lake and Bojana river water regime (IPA Project)–Volume 1.* Montenegrin Academy of Science and Arts, Podgorica, 9–115.

Radulović M.M., Radulović M., Stevanović Z., Sekulić G., Radulović V., Burić M., Novaković D., Vako E., Blagojević M., Dević N., Radojević D. (2015) Hydrogeology of the Skadar Lake basin (Southeast Dinarides) with an assessment of considerable subterranean inflow. *Environmental Earth Sciences* 74(1), 71–82.

Sekulić G., Bushati S. (editors) (2013) *Development of hydrogeological and hydraulic study of regulation of Skadar Lake and Bojana river water regime (IPA Project)–Volume 1.* Montenegrin Academy of Science and Arts, Podgorica.

Shaban A., Khawlie M., Abdallah C. (2005) Use of remote sensing and GIS to determine recharge potential zones: the case of Occidental Lebanon. *Hydrogeology Journal* 14, 433–443.

Stevanović Z., Radulović M., Shammy P., Radulović M.M. (2008) Karstic source 'Bolje Sestre': Optimal solution for regional water supply of the Montenegro coastal area (in Serbian). In: *Zbornik radova Odbora za kras i speleologiju (Recueil des raports du Comite pour le karst et la speleologie)*, Serbian Academy of Science and Arts, Belgrade, 9, 33–64.

Stevanović Z., Kukurić N., Treidel H., Pekaš Z., Jolović B, Radojević D., Pambuku A. (2012) Characterization of transboundary aquifers in Dinaric karst–a base study for sustainable water management at regional and local scale. 39th International Association of Hydrogeologists Congress, Niagara Falls, Canada, September 16–21, 2012.

Tietze E. (1884) *Geologische Uebersicht von Montenegro.* Hölder, Vienna.

Torbarov K., Radulović V. (1966) Regional hydrogeological research of Montenegro and Eastern Herzegovina (in Serbian). Geological Survey of Montenegro, Titograd.

Waisse J.G. (1948) Les bauxites de l'Europe centrale (Province dinarique et Hongrie). PhD thesis, University of Lausanne, Lausanne.

Yamada C. (2004) Second report on shared natural resources: transboundary groundwater. UN International Law Commission, UN Doc A/CN.4/539.

Zogović D. (1992) Final report of hydrogeological exploration of Karuc source (in Serbian). Energoprojekt, Beograd.

Živaljević R. (1991) Hydrological analysis of karst water flow–Crnojevica River case study (in Serbian). PhD thesis, University of Montenegro.

Živaljević R., Bošković M. (1984) The results from hydrological measurements in the catchment area of Orahovstica (in Serbian). Hydrological and Meteorological Service of Montenegro, Podgorica.

Part 2

Karst aquifer characterisation and monitoring

Optimal water management – Prerequisite for regional socio-economic development in the karst of the south-eastern Dinarides

Petar Milanović
President of IAH Serbian Chapter, Belgrade, Serbia

ABSTRACT

In the south-eastern Dinarides, with the highest precipitation in Europe, people still suffer from water shortages. The Dinaric Karst Aquifer System (DIKTAS) belongs to one of the largest karst areas extending along the coast of the Adriatic Sea from Slovenia to Albania. The large water potential of the DIKTAS area is mostly concentrated on the catchment areas of the Cetina, Neretva, Trebišnjica, Zeta and Bojana rivers. This potential underpins an advanced economic development programme for the region was recognized long ago. However, the karst phenomena always presented a significant barrier for groundwater use. The karst presents a variety of hazards and risks associated with human activities, particularly the construction of dams, reservoirs, tunnels and canals. Successful solutions require serious and complex approaches and close co-operation with a wide spectrum of scientists and engineers to define causes and consequences between human activities and impact. The area of south-eastern Dinarides, particularly Eastern Herzegovina, represents the optimal strategy for water resources management in karst areas which is a key requirement for regional socio-economic development.

8.1 INTRODUCTION

The area of south-eastern Dinarides is one of the most karstified regions in the world. It has an average annual precipitation of 1250–2450 mm, locally more than 5000 mm, and maximum of 8000 mm (Figure 8.1).

Sinking rivers, underground flows, temporary flooded karst poljes and lack of arable land are the main properties of the region. Arable land, karst poljes, uvalas and sinkholes are surrounded by bare rocks (Figure 8.2).

The people of this region have always had to cope with two kinds of misfortune: flood and drought (Figure 8.3).

Under natural conditions a large portion of the arable land is flooded 150 to 250 days each year. Sowing and harvesting are not determined by man, but by water. During dry summer periods villagers rely on rainwater collected during winter or on water from siphon lakes in deep natural karst shafts and caves. For centuries people have emigrated from this region in search of a better life. The population density per km^2 varies from 21 to 29. The only natural resource of importance for the regional socio-economic development is water. However, distribution of precipitation is

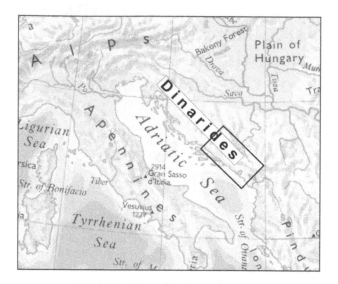

Figure 8.1 The south-eastern Dinarides.

Figure 8.2 Settled and cultivated uvala (Photo P. Milanović).

uneven during the year. More than 70% occurs during the wet season (October–March). The aquifer retardation capacity is extremely poor. Ninety percent of groundwater turnover occurs annually. The groundwater potential of all springs in the area of Eastern Herzegovina and Boka Kotorska Bay is estimated with a yearly

Figure 8.3 Fatnica Polje under water (upper). Dry Trebišnjica river bed (lower). Photos P. Milanović.

average of 300 m³/s. No doubt, this region is the largest and most important fresh water source in the Mediterranean area. However, only a small amount of this water is successfully tapped.

8.2 GENERAL GEO-STRUCTURAL FEATURES

More than 90% of the entire area consists of soluble Mesozoic carbonate formations. The depth of these rocks exceeds 3000 m. The Eocene flysch does not represent a significant lithostratigraphic unit, however, according to its hydrogeological role and location along reverse faults, it has a huge effect on karst aquifers at many karst poljes. The existing geo-structural features of the entire Dinaric Karst Aquifer System (DIKTAS) area are the result of subduction movement by the Adriatic microplate beneath the Dinarides mountain chain. The dominating tectonic stress is oriented from the south west to north east and subsequently the regional structures are orientated north west to south east. The main structural features are the thrust faults locally separating the flysch from the carbonate complex limestones and dolomites. Inclination of the reverse fault planes, dipping toward the north east, ranges between 55° and 75°. Strong tectonic movements cause folding and cascade lifting of the terrain parallel to the present-day coastline. Starting at an altitude of over 1000 m, these stepwise depressions, which erosion and sedimentation have turned into spacious karst valleys known as poljes, are the outstanding feature of the Dinaric, and of the entire DIKTAS karst region.

8.3   NEEDS FOR WATER MANAGEMENT – OLD SOLUTIONS

Organisation of the water regime of this part of south east Dinarides is the only way forward for regional economic development. To minimise floods and to keep water at the surface as long as possible is a basic requirement for economic progress. Protection of ponors (swallow holes) against natural plugging, construction of small surface water storage ponds and construction of water-driven mills at ponors have been ongoing for centuries. The aqueduct constructed in Roman times for the water supply of Epidaurus (Cavtat) and the tapping of three karst springs for Dubrovnik water supply are some of the ancient examples. The first dam in the DIKTAS area (Klinje Dam, in Herzegovina) was constructed in 1896 for the Gatačko polje irrigation system (Figure 8.4). By excavation of many dewatering tunnels, after World War II, a number of karst poljes in the DIKTAS region have been protected against floods.

Intensive construction of dams and reservoirs, mostly in Europe and the United States, started in the first half of the 20th century. One of the first successful projects was the American Tennessee Valley Project. In that time understanding of karst and karstification, from an engineering view point, was in its infancy. A number of large structure failures were registered in this period. Dried reservoirs or reservoirs with unacceptable heavy leakage were frequent (e.g. Hales Bar, US; Montejaque, Spain; Vrtac, Montenegro). Failure in karst occurred despite extensive investigation

Figure 8.4 Klinje Dam, Gatačko Polje (Photo P. Milanović).

programmes and sealing treatment. However, experience gathered during construction and operation of many large structures in karst, eventually enabled the construction risk to be reduced to an acceptable level.

8.4 OPTIMAL WATER MANAGEMENT IN KARST – TREBIŠNJICA MULTIPURPOSE HYDROSYSTEM

Large and successful reservoirs constructed in the Dinaric karst area after World War II (Peruća at Cetina River, Bileća at Trebišnjica River and Krupac + Slano at Zeta River) promoted the Yugoslav school of engineering karstology. The Grančarevo Dam and Bileća Reservoir, as part of the Trebišnjica Multipurpose Hydrosystem (TMH), are excellent examples of successful construction of large structures in hazardous a and unpredictable karst geological environment (Figure 8.5). This Hydrosystem is situated in the Eastern Herzegovina, in the middle of the DIKTAS project area.

Stepwise placement of karst poljes in Eastern Herzegovina allows optimal multipurpose use of the great water potential from an elevation 1000 m down to the sea level. The water storage at the surface from the rainy period of year allows continued use during the drier months. This goal can be achieved only by construction of dams and reservoirs interconnected by tunnels and canals, and the construction of an integrated regional hydrosystem. The Trebišnjica Multipurpose Hydrosystem was initiated in early 1950s. The idea is to construct a huge surface 'water way' in the shape of letter 'S' (Milanović, 2010) (Figure 8.6).

In this way part of the water potential will be accessible to the majority of the population in the region, for different purposes, through the year. The impacts from

Figure 8.5 Grančarevo Dam and Bileća Reserovir (photo P. Milanović).

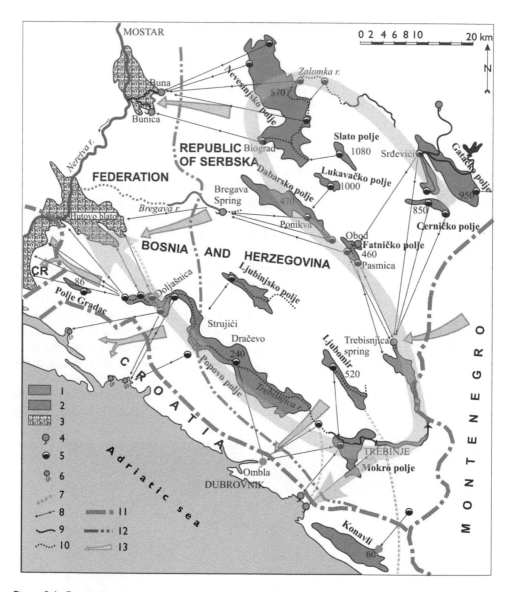

Figure 8.6 General concept of integral water management in Eastern Herzegovina. 1. Reservoirs, 2. karst polje, 3. alluvium and wetland, 4. permanent spring, 5. ponor (swallow hole), 6. submarine spring, 7. large fault zone, 8. established underground connection, 9. permanent river flow, 10. temporar river flow, 11. state border, 12. entity border, 13. general direction of underground flows. Light blue stripe – general direction of the man-made water route.

floods and droughts will be considerably reduced. The first governmental document for the construction of this integral reclamation and hydropower system was accepted in 1960. All structures of the Hydrosystem have been designed in such a manner that the water potential could be used for food production; water supply; flood reduction; irrigation; hydropower production; industry; fish farming; recreation; and prevention

of deforestation. It was also designed to have a positive influence on the critical minimal flows in urban and nature protected areas with a number of secondary benefits,
including the most important benefit to decrease the strong emigration trend away
from the region.

The TMH (Figures 8.7 and 8.8) consists of seven dams, six reservoirs, six tunnels
(with a total length of 74 km) and four canals (with a total length of 74 km). A large
part of TMH (Phase I) is already operational.

The Phase I includes waters from the Trebišnjica springs which are at sea level.
The Trebišnjica River is the most important in the area as it is the longest sinking river
in Europe with a total length of 90 km of which about 30 km has permanent flow.
The TMH aims to harness the potential energy of this river. The main structures of the
Phase I are: Grančarevo Dam (123 m high); 'Bileća' Reservoir (V = 1280 hm³); Gorica
Reservoir (V = 15 × 10⁶ m³); Gorica Dam (33.5 m high); tunnel Gorica – Dubrovnik PP

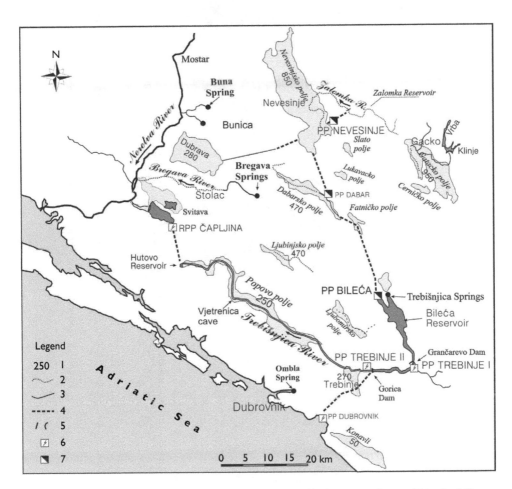

Figure 8.7 General scheme of the Trebišnjica Multipurpose Hydrosystem, layout. Altitude; 2. Temporary flow; 3. Permanent flow; 4. Tunnel route; 5. dam; 6. power plant (operational); 7. power
plant (designed).

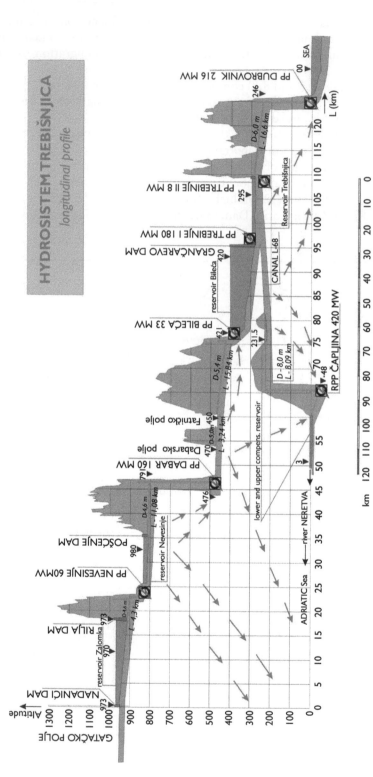

Figure 8.8 General scheme of the Trebišnjica Multipurpose Hydrosystem, longitudinal profile.

(16.57 km); Trebišnjica river bed, paved by shotcrete (65 km); Hutovo Reservoir (V = 5×10^6 m³); tunnel for reversible PP Čapljina (8.093 km), and power plants Trebinje I, Trebinje II, Dubrovnik and Čapljina.

The Phase II (so-called 'Upper Horizons') is now under construction. This part of the project requires the transfer of 30% of the water from the upstream part of the Neretva River (springs Buna + Bunica + Bregava, Q_{av} = 59.6 m³/s), to be used at the existing operational power plants (Phase I), and is then returned to the downstream part of Neretva River. As part of the water re-routing three temporary flooded karst poljes will be flood free and available for agriculture. According to the l hydrogeological and hydrological analysis, including four mathematical models, it was confirmed that the consequences of water transfer would be negligible on the Buna Spring discharge. In the case of the Bunica Spring the maximum flow (high precipitation period of year) would be reduced to about 70%, however in the period of low flow (dry period of year) a negative influence of water transfer is not possible. A similar result occurs in the Bregava Springs and Bregava sinking river. According to the design, the geotechnical measures are intended to improve the watertightness of the pervious Bregava river bed, with the consequence that the flow in the dry period would increase 3 to 5 times over the natural conditions (Milanović, 2006).

Once completed the TMH will enable irrigation of about 240 km² of arable land and an average annual hydropower output of 856 GWh. The 'Bileća' Reservoir, situated in karstic carbonates (volume 1.3×10^9 m³), without leakage, is one of largest and most successful reservoirs in karst. The question of reservoir watertightness was the key question of the entire project feasibility.

After construction the first phase of the TMH, the surface and underground water regime at one of world famous karst phenomena Popovo polje has been drastically changed (Figure 8.9). The possibility of massive floods (up to 40 m deep), which frequently happened under natural conditions, are now possible only at the very end of polje, but the flood water is shallow and of short duration.

Figure 8.9 Popovo polje before and after construction of first phase of the Trebišnjica multipurpose hydrosystem.

No doubt Trebišnjica Multipurpose Hydrosystem is one of the world most successful projects in karst. Important lessons learned during the investigation, design, construction and operation are contributed to the development and promotion of scientific karstology and engineering karstology.

8.5 NECESSITY FOR BALANCE BETWEEN REGIONAL DEVELOPMENT AND PRESERVATION OF NATURE

From the beginning the project was strongly supported by people from all communities from this part of the Dinarides and approved by responsible governmental institutions. At the same time the important question of possible environmental impact arises as a consequence of natural water regime changes. Some large springs, tapped for water supply (Trebišnjica Spring and Oko Spring) were submerged by reservoirs (Figure 8.10). The possible impact of Trebišnjica Spring submergence on reservoir integrity was also one of important questions to be answered in the stage of design.

Concern over the possible negative impact on the flow regime downstream from springs often creates conflicts between the owners of the reservoirs and the user of the springs. This too needs addressing.

Karst is rich with various fauna including a number of endemic species. Changes in the underground and surface water regimes can cause a negative effect on the fauna of subterranean karst. The important issue was how to keep the balance between the necessity for regional development and preservation of the sensitive and complex karst ecological system (Milanović, 2002).

Subterranean terrestrial and aquatic species exist in karst systems all over the world. As a consequence of the natural underground and surface water regime disturbance, a very distinct impact on a wide spectrum of biodiversities is expected. Often, construction of any structure in karst, particularly dewatering of temporarily flooded karst depressions and construction of dams and reservoirs has resulted in negative influence on fauna in the caverns.

In the East Herzegovina karst area a number of aquatic and terrestrial species have been found. Particularly well known is Vjetrenica cave (Popovo Polje) with several endemic species. A well-known cave-dwelling aquatic species is the blind salamander

Figure 8.10 Trebišnjica Spring, dry and submerged by a 75 m depth of water, in period of high discharge (Photos P. Milanović).

Proteus anguinus known as 'human fish' is common in the Dinaric karst. At Popovo Polje *Proteus* was found at more than 40 localities (Čučković, 1978). A problem arises in the case of the fish *Paraphoxinus ghetaldi* that inhabit siphons of estavelles in the Popovo Polje (Bosnia Herzegovina) during dry periods of the year and intermittent lakes of the karst poljes during the flood season. After the karst polje is permanently protected from flooding, fishing at the openings of estavelles, for centuries an important traditional food source for the local inhabitants, is no longer practical.

The unique tubeworm *Mariphugia cavatica* (Figure 8.11) that colonises the temporary flooded karst channels are seriously endangered after dewatering temporarily flooded karst poljes. A large colony of this warm was discovered, also in a karst channel beneath, 10 m of alluvium (Popovo Polje, Figure 8.9). After construction of the impervious blanket over the alluvial deposits to prevent seepage from the reservoir bottom the survival of *Mariphugia* is questionable.

To avoid possible conflicts and to provide the sustainable and peaceful utilisation of transboundary waters the designers of TMH undertook a series of long lasting investigations. These were carried out collaboratively with a wide spectrum of scientists and engineers including geologists, hydrologists, chemists, civil engineers, biologists, archeologists and seismologists. An inventory of 120 springs and a monitoring programme began 8 years before the project was operational; a programme of regional water quality monitoring has been organised; the possible influence of TMH on a shell farm at the sea coast was investigated; while some important historical monuments were displaced by the reservoir water (Figure 8.12); submerged water sources were replaced by new capturing structures; and number of different environmental analyses were performed. In most instances the impact is positive and predictable. However, some impacts are negative and unpredictable, such as, for example, karst induced seismicity.

8.6 NEEDS FOR PREVENTION AND REMEDIATION

With the initiation of the construction of large structures the technology of investigating karst phenomena, including many new methods in hydrogeology, hydrology,

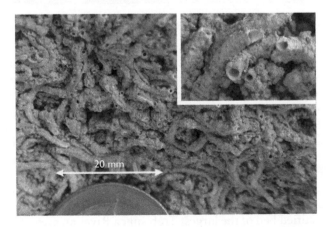

Figure 8.11 Mariphugia cavatica (photo P. Milanović).

Figure 8.12 Monastery displaced from 'Bileća' Reservoir.

Figure 8.13 Trebišnjica river bed blanketed with reinforced shotcrete.

geophysics developed quickly. Specific prevention and remediation methods are required to cope with geotechnically unconventional problems. The most frequent technical difficulties were the presence of caverns along tunnel routes or at dam sites, leakage from reservoirs, groundwater intrusion during underground excavations and induced subsidences in reservoir bottom, which all needed unconventional prevention and remediation methods (Milanović, 2000). To reduce or eliminate these difficulties two approaches were feasible. Surface treatment and sealing underground was dealt with by compaction of surface layer blankets, construction of shotcrete blankets, use of geosynthetics, plugging and grouting, and construction of cylindrical dams, aeration pipes and one-way valves. To secure the impermeability of the highly karstified bed of the largest Trebišnjica River an area of 2.2×10^6 m^2 was blanketed with shotcrete (Figure 8.13). Common underground structures are grout curtains, cut-off walls, concrete plugs and underground dams. A number of these methods has been applied during the construction of the different TMH structures.

During past 50 years the constructed part of the TMH has considerably improved the quality of life.

One of the important lessons learned is that in karst engineering a certain degree of flexibility is necessary during the construction of large structures in order to find the best compromise between different possibilities and solutions, including protection of environment and historical heritage.

Experience and knowledge are important bases for further utilisation of the groundwater potential, as water is presently lost to the sea through a number of permanent, temporary and submarine springs. Successful safe reservoirs and impervious canals, as well as long tunnels, confirm high level of knowledge and high quality of experience in engineering karstology.

8.7 LARGE WATER POTENTIAL – FUTURE OF REGIONAL DEVELOPMENT

The large water potential of this region will be of great importance in the future. The water potential of the Eastern Herzegovina and Boka Kotorska Bay area is not only of local importance but is internationally significant as well. This region would have a role of the Mediterranean 'Water Treasure' (Milanović, 2004). New technologies make possible tapping and control of groundwater flows deep underground.

By abstracting underground water in periods of high precipitation, at elevation between 200 and 300 m, distribution is possible by gravity over large distances. Particularly important is the large, but temporary stored amount of water in the Orjen Mountain (Figure 8.14).

Successful abstraction of only 20% of Orjen water provides 10–15 m³/s throughout the year. The Water Treasure concept requires permanent and perfect protection of surface and underground water over the entire region of Eastern Herzegovina and the wide area of the Boka Kotorska Bay. Eventual problems caused by transboundary aquifers have to be overcome and the entire region should be treated as a unique hydrogeological and hydrological entity.

8.8 CONCLUSIONS

Large sinking rivers, numerous sinkholes, ponors, temporarily flooded karst poljes and deeply concentrated underground flows characterise the Eastern Herzegovina karst region. In spite of abundant rainfall, karstic terrain and uneven distribution of precipitation makes its inhabitants vulnerable to frequent floods and droughts. To provide optimum water management the TMH was initiated as the best feasible solution. The ultimate aim was to improve the livelihoods of the people in the region by regulating water availability all year round for all users. The Reservoir Bileća was the first reservoir with a volume of more than 1 billion cubic meters of water constricted entirely over heavily karstified rock. There was no seepage. The TMH is ongoing and is the most important agent of economic development in Eastern Herzegovina.

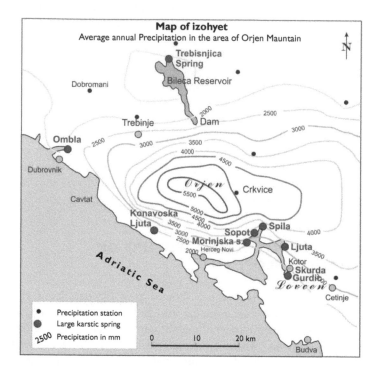

Figure 8.14 Orjen Mountain. Area with largest precipitation in Europe.

However, the environmental properties of complex karstic system and the state borders dissecting the region make integrated management of water resources extremely complex. The need for optimal use of the large water resources still calls for a great effort by all users in the region and for a holistic approach.

REFERENCES

Čučković S. (1978) The question of the survival of well known endemic Proteus Anguinus in the Trebišnjica Power System. Conference on environmental effects of storage reservoirs. Yugoslav Committee for Large Dams. Trebinje.

Milanović P. (2000) *Geological engineering in karst.* Zebra, Publishing Ltd, Beograd.

Milanović P. (2002) The environmental impacts of human activities and engineering constructions in karst regions. *Episodes,* 25(1), 13–27.

Milanović P. (2004) Water without boundary – water resources potential in deep karst of South-Eastern Dinarides. Book of abstracts 32th International Geological Congress, Florence, Italy.

Milanović P. (2006) *Karst of Eastern Herzegovina and Dubrovnik littoral.* ASOS, Belgrade.

Milanović P. (2010) Transboundary aquifers in karst – source of water management and political problems. Case study, S-E Dinarides. ISRAM 2010 International Conference. Transboundary aquifers. UNESCO, Paris.

Chapter 9

Spring hydrograph recession: A brief review focused on karst aquifers

Francesco Fiorillo
Dipartimento di Scienze e Tecnologie, University of Sannio,
Benevento, Italy

ABSTRACT

Spring hydrograph recession analysis is a well-known topic in karst hydrogeology. The different models in the literature can be separated into empirical and physically-based models; in the latter, only analytical models provide the discharge equation during recession. Under constant geometrical and hydraulic aquifer characteristics, the 'exponential form' appears to be the most recurrent theoretical type, at least during the long-term flow recession. During this stage, any deviation from the exponential form, may suggest hydraulic anisotropy of the aquifers, as well as aquifer geometry which has a fundamental role in controlling the shape of spring hydrographs. Karst aquifer hydraulics, under different hydrological conditions, are described along with the relationship to different spring hydrograph segments.

9.1 INTRODUCTION

Spring discharge hydrographs can be used to define the regime of springs, and shape provides a useful tool to investigate some of the karst features of the aquifers. Well developed karstification results in spring hydrographs characterised by pronounced peaks, which follow the main rainfall events, and indicate both a good connection of the conduit networks with the spring and the presence of active recharge pathways. A smoother shape, without sharp peaks and characterised by one or few flood peaks during the hydrological year, indicates a coarse development of karst conduits or that they are poorly connected.

During the recession, the hydrograph is generally the most stable, and it is believed to express some geometrical and hydraulic characteristics of aquifers. This has stimulated numerous studies, and hydrograph recession analysis is a well-known topic in hydrogeology. There are a large number of studies recorded in the literature, and these try to describe the analytical expressions of the recession and their physical meaning (Dewandel *et al.*, 2003; Bailly-Comte *et al.*, 2010; Fiorillo, 2014).

Figure 9.1 is a theoretical example of the spring hydrograph and its recession limb. In this example, an intense pulse is transmitted through the system, which causes the water to rise in spring. The recession limb is considered here as an exponential function, characterised by the response time, t_R, which is the inverse value of the recession coefficient, α ($t_R = 1/\alpha$). The beginning of the recession would occur at the inflection point, Q_0, which splits a convex (left side) and concave (right side) form of the hydrograph. Hydrograph recession can occur for short periods (several days or weeks) between two main rainfall events, but has a longer duration in the dry season.

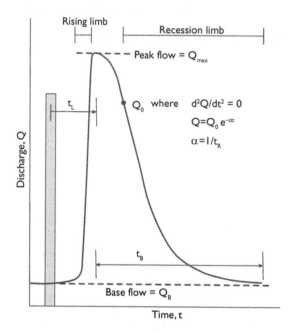

Figure 9.1 Different parts of a spring hydrograph triggered by a storm pulse (modified from White, 1988): t_L, length of time between the storm pulse and the peak in the discharge hydrograph; t_B, time to return to base flow; α, recession coefficient; t_R, response time.

Figure 9.2 shows an example of a spring hydrograph in a Mediterranean area, where several intra-annual recessions occur during the wet season, and a long seasonal recession occurs during the late spring-summer period, extending also into the following hydrological year.

The most simple equations to describe the discharge, $Q(t)$, during the recession phase are the linear:

$$Q(t) = Q_0 - \alpha_T \cdot t \tag{9.1}$$

and the exponential:

$$Q(t) = Q_0 e^{-\alpha t} \tag{9.2}$$

where Q_0 is the initial discharge at time $t = 0$, and the recession coefficient α_T and α depend only on the aquifer hydraulic systems; they are constant during the recession period.

Equation 9.1 describes the linear decrease of discharge with time; the main consequence of this behaviour is the constant slope of the discharge-time plot at any time, independently of the initial value, Q_0. This equation is obtained by draining a cylindrical tank-reservoir, where no energy is lost during the emptying (Table 9.1).

Equation 9.2 was first provided by Maillet (1905) to describe the spring drainage from a porous aquifer, obtained from a cylindrical tank with a porous plug. The physical meaning of the recession coefficient, α, using different hydraulic laws to drain

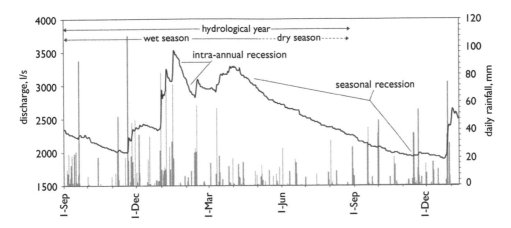

Figure 9.2 Example of karst spring hydrograph recession in a Mediterranean climate: Torano daily spring discharge (Matese Mountain, southern Italy) and daily rainfall recorded in the recharge area (Lago Matese rain gauge, annual mean 1948 mm). Intra-annual recession occurs during the wet season, interrupted by the next rainfall event; during these periods, recharge processes progressively increase the basal water table level of the aquifer. A seasonal recession begins when recharge processes decrease as a consequence of temperature increase and rainfall decrease throughout the (Mediterranean) summer and can extend in the following hydrological year (modified from Fiorillo, 2014).

Table 9.1 Discharge equations found from draining a cylindrical tank-reservoir using different physical laws (from Fiorillo, 2011). Application of Darcy's Law is carried out filling the connected tank-tube (with area A_2 and length L) by sand with hydraulic conductivity K. Application of Poiseuille's Law is carried out considering a small diameter d of the connected tank-tube; μ is the dynamic viscosity of water; ρ is the water density; g is the gravity acceleration.

Tank shape	Physical law	Discharge equation	Recession coeff.	Discharge for t = 0
	Torricelli	$Q(t) = Q_0 - \alpha_T \cdot t$	$\alpha_T = \dfrac{A_2^2}{A_1} g$	$Q_0 = A_2 \sqrt{2gh_0}$
	Darcy	$Q(t) = Q_0 \cdot e^{-\alpha_D \cdot t}$	$\alpha_D = \dfrac{K \cdot A_2}{L \cdot A_1}$	$Q_0 = \dfrac{KA_2 h_0}{L}$
	Poiseuille	$Q(t) = Q_0 \cdot e^{-\alpha_P \cdot t}$	$\alpha_P = n \cdot \dfrac{\pi \cdot d^4 \cdot \rho \cdot g}{128 \cdot L \cdot \mu \cdot A_1}$	$Q_0 = n \cdot \dfrac{\pi d^4 \rho g}{128 \mu} \cdot \dfrac{h_0}{L}$

a cylindrical tank-reservoir is shown in Table 9.1. Equation 9.2 is that of a linear reservoir where the discharge, $Q(t)$, is proportional at any time to the water volume stored, $V(t)$, into reservoir:

$$Q(t) = \alpha \cdot V(t) \tag{9.3}$$

As it gives a straight-line in the semilogarithm plot, the exponential form of the recession has found wide use in hydrogeology. It can describe a large part of the recession.

Real examples of the recession limb of karst spring hydrographs in the semilogarithmic plot have shown a variation in the angle of the straight line. This indicates the presence of more than one recession coefficient during the entire recession period (Forkasiewicz & Paloc, 1967; Milanovic, 1976; Atkinson, 1977).

Other equations describing the recession exist and differ from the linear and exponential form.

9.2 LITERATURE MODELS

The main models analysing the spring hydrograph recession have been discussed in Fiorillo (2014). They vary from pure empirical to physically-based models. In the last ones, only analytical models provide the discharge equation during the recession phase. Most models which analyse the spring hydrograph recession can be considered as lumped (or global) models, as the flow of the total catchment is often focused on one single spring and, therefore, allows for an integral characterisation of the flow behaviour of the whole catchment just by measuring the outflow at a single point (Jeannin & Sauter, 1998). Only a few models are able to consider the spatial heterogeneity of the aquifer.

Most of theoretical models consider the hydrograph recession that is not conditioned by recharge processes. Under this assumption, the only aquifer zone believed to be involved is the saturated zone, and the role of the vadose zone is generally neglected. However, during the recession, some recharge will reach the saturated zone and influences the discharge of springs; this process occurs especially during the first part of the recession and characterises the influenced-stage. Only the last part of the recession, which is the most stable part of the hydrograph, could be considered as a non-influenced stage and it is known as the recession limb.

Empirical models do not consider any geometrical and hydraulic characteristics of the aquifers, as they are generally based on the best fits of experimental recession data; thus, they could not provide any information on the structure of aquifers. However, many empirical models start from very simple reservoirs to obtain discharge equations or try to obtain some hydraulic parameters of aquifers from hydrographs. As these last models evaluate some physical characteristics of the aquifers, they can be considered as semi-empirical models.

Physically-based models require a definition of the geometrical and hydraulic conditions of the aquifers, and also need specific physical laws which describe the drainage. These characteristics are very difficult or impossible to obtain for actual karst aquifers, and for this reason many physical models are based on simple theoretical geometric and hydraulic characteristics of the aquifers, which allow the equation of the discharge to be obtained from analytical models. Most models were developed for porous media, but they have been largely applied to karst media; in any cases no-recharge is assumed in the models.

Forkasiewicz & Paloc (1967) characterised the karst aquifer with different parallel linear reservoirs, all contributing to a spring hydrograph. They reconstructed three segments of the hydrograph in a semilogarithm plot as the different contributions of each reservoir: the conduit network, the intermediate system of well integrated karstified fissures, and the low-permeability network of pores and narrow fissures.

Nutbrown & Downing (1976) stated that different exponential terms can derive exclusively from groundwater flow dynamics. Szilagyi (1999) has highlighted that the changing slope in the recession plot can be simply the consequence of baseflow drainage. Estrela & Sahuquillo (1997) also modelled the recession hydrograph by several decreasing exponentials, without identifying different flow regimes. Eisenlohr et al. (1997) showed that the sum of three exponentials can be fitted on the hydrograph of a system consisting of only two classes of hydraulic conductivities.

Baedke and Frothe (2001) state that the largest slope of the hydrograph is related to the conduit flow system and the smallest slope to the diffuse flow system (and deriving the aquifer diffusivities from each slope following the approach proposed by Rorabaugh, (1964) and Sahuquillo & Gómez-Hernández (2003). They highlighted that the different slopes are never representative of the response of a more or less pervious portion of the aquifer.

Similarly, Kovacs & Perrochet (2008) showed that different exponential components do not correspond to aquifer volume with different hydraulic conductivities as they can be extracted from a 2D analytical solution for diffuse flux from symmetric and asymmetric rectangular blocks.

These examples indicate that it is possible to explain the different slope of the hydrograph in the semilogarithmic plot without requiring different flow regime (in the conduits, minor fracture, etc.). The exact analytical solution provides a shape of the hydrograph which is a sum of different exponentials, as in the model of Rorabaugh (1964), Brutsaert (1994) or Kovacs et al. (2005). Only the latest part of the hydrograph can be approximated to an exponential, as it follows an initial non linear part on the semilogarithmic plot.

Most physically-based models are based on the assumption that the aquifer is characterised by homogeneous and isotropic media and the formation constants Physically-based models generally assume that flow is Darcian.

Table 9.2 shows analytical solutions found draining two different tank-shapes: the exponential function (Equation 9.2) can be obtained using the Torricelli Law, and the linear function (Equation 9.1) can be obtained using the Darcy/Poiseuille Laws, contrary to the cylindrical tank-reservoir of Table 9.1. Table 9.2 show that it is possible to back-calculate the variation of the shape of the reservoir to obtain the desired hydrograph, but it is more difficult to fix the type of flow and the geometrical model for an actual karst aquifer. Analysing an actual spring hydrograph is difficult when Darcian flow occurs, and unclear how the aquifer geometry controls the recession.

Fiorillo (2011) noted that many of the equations for the recession coefficient could be linked to:

$$\alpha = \frac{c}{n_{eff} \cdot A_c} \tag{9.4}$$

where n_{eff} is the effective porosity computed along the water table with an area A_c and c is the hydraulic constant, involving the geometric and hydraulic characteristics of the discharge zone which does not vary during the drainage. This means that, when the exponential form is expected (during the long-term flow recession), any variation of the recession coefficient would depend on the product $n_{eff} \cdot A_c$ (Fiorillo, 2014).

Table 9.2 Discharge equations obtained while draining tank-reservoirs that are characterised by different geometry using different physical laws (extracted from Fiorillo, 2012): *l*, length of the reservoir in the orthogonal direction of *x*; *B*, constant of the reservoir; $C = A_l/l$ for $x = 0$; other symbols are described in the text. Hyperbolic tank shape has a rectangular base area $A_l = x \cdot l$; triangular tank shape has a rectangular base area $A_l = (C - x) \cdot l$.

Tank shape	Physical law	Discharge equation	Recession coefficient
$A_l = x \cdot l$; $h = \frac{B}{x^2}$; h_0; x; L; A_2	Torricelli	$Q(t)=Q_0 \cdot e^{-\alpha t}$	$\alpha_T = \dfrac{A_2}{2 \cdot l}\sqrt{\dfrac{2 \cdot g}{B}}$
$A_l = (C - X) \cdot l$; $\arctan(B)$; h_0; $h(t) = (C - X)B$; x; L; A_2	Darcy/ Poiseuille	$Q(t)=Q_0 - \alpha \cdot t$	$\alpha_D = \left(\dfrac{A_2 \cdot K}{L}\right)^2 \dfrac{B}{l}$ $\alpha_P = \left(\dfrac{n \cdot \pi \cdot d^4 \cdot \rho \cdot g}{128 \cdot \mu \cdot L}\right)^2 \dfrac{B}{l}$

The effective porosity is an important factor controlling the recession coefficient, as it can vary across a wide range due to the heterogeneity of the karst media, and can approximated to storativity for unconfined aquifers (Stevanović *et al.*, 2010). The change in the effective porosity inside the epiphreatic zone could be connected to the different ways of development of the karst system, caves, conduits and voids, according to the main theories on karst development (Ford & Ewers, 1978). Numerical models have shown how the phenomenon of chemical solution and attrition causes the development of conduits and porosity just below the water table of an unconfined aquifer (Gabrovšek & Dreybrodt, 2010; Dreybrodt *et al.*, 2010). Just below the water table there could be a strong variation in void distribution. This means that during the recession, the effective porosity can change along the water table, conditioning the shape of the hydrograph, independently from other hydraulic parameters and the catchment area (Fiorillo, 2011).

Many highly-elevated springs (non-basal springs) dry up during the recession, and remain dry for some time. These springs follow a recession law which is different from the exponential form, as the discharge decreases faster than exponential decay. The simplest explanation of this behaviour is the progressive reduction of the water table area, Ac, as the aquifer is drained.

Commonly, basal karst springs tend to have a progressively decrease in their recession coefficient during the recession phase, whereas other spring types may present a different hydrograph shape. Analysing the spring recession in the semi-logarithmic plot, a deviation from the straight line produced by a simple exponential

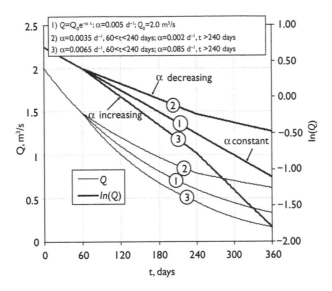

Figure 9.3 Arithmetic and logarithmic plots of constant (curve 1) and non-constant (curve 2 and 3) recession coefficient value α during emptying (modified from Fiorillo *et al.*, 2012).

decay of discharge through time provides information on the actual emptying rate of the aquifer compared to a simple exponential decline (Figure 9.3). If the recession coefficient, α, decreases under drought conditions (curve 2), springs guarantee water during a long dry period and can be considered drought-resistant (Fiorillo *et al.*, 2012). This hydraulic behaviour could be connected to an increase of the effective porosity with depth. If the recession coefficient increases under drought conditions (curve 3), a different hydraulic behaviour occurs, as the aquifer is drained more quickly than expected, and the springs can be considered as drought-vulnerable (Fiorillo *et al.*, 2012). This hydraulic behaviour could be connected to a decrease of the water table area, A_c, or to the fact that the aquifer is also drained by other springs with lower ground-elevation.

The different hydraulic behaviour of springs suggests that the common and diffuse computation of the water stored in a karst aquifer obtained by integrating the Maillet formula (Equation 9.2), over the time interval t_0 to t, should be used with caution in the evaluation of water resources, as it underestimates or overestimates the volume discharged for drought-resistant and drought-vulnerable springs, respectively. Independently from other hydrogeological and hydrological features of karst aquifers, this observation may provide useful tools for water management purpose, especially under droughts.

9.3   KARST AQUIFER HYDRAULIC UNDER DIFFERENT HYDROLOGICAL CONDITIONS

Ford & Williams (2007) provided a discussion on the applicability of the *Darcy*'s law to karst and highlighted how a large part of the flow passes through the conduit system, Thus, according to Mangin (1975), they assumed that the range of condition under which Darcy's Law is valid is very restricted. The restriction has to refer to different parts of the aquifer (spatial meaning), but also to the different recharge conditions of the aquifer, which vary during the hydrological year (temporal meaning) (Fiorillo, 2014). A Darcian flow prevalently occurs during the latest part of the recession, when the mean water flow velocity in the aquifer is the lowest with respect to all previous stages. The Darcian hydraulic behaviour has to be connected to the drainage of the water from the minor fissures, even if the conduit network provides the final water transfer to spring (Fiorillo, 2014).

During the initial part of the recession, due to higher water flow velocites in the conduit network, a different hydraulic flow type could occur. The initial and steepest part of the recession can be modelled by a non-Darcian flow (Bailly-Comte et al., 2010; Fiorillo, 2011; Malík & Vojtková, 2012). The exchange of water between conduits and matrix occurs during floods (Haliban et al., 1998; Martin and Dean, 2001; Bailly-Comte et al., 2010), and it is caused by the higher conduit head in the conduit network than the matrix (Atkinson, 1977; Drogue, 1980; Bailly-Comte et al., 2010; Fiorillo, 2011). This hydraulic behaviour depends on two main hydraulic heads which characterise karst aquifers, and is caused by the two main permeability characteristics of the medium: (i) the hydraulic head of the saturated zone (water table) and (ii) the hydraulic head of the conduit network. Only during the long-term flow recession do the two main different hydraulic heads tend to be the same, and they diverge after recharge events, due to faster increase of the conduit hydraulic head.

Schematic illustrations highlight the main points of the aquifer hydrodynamics, and help to associate the different parts of a hydrograph to specific hydraulic conditions (Figure 9.4). Non-allogenic recharge has been considered here, and a single outlet (spring) drains the aquifer. In each sketch, the hydraulic head into the conduit network is marked by the water level in the shafts; this level can be different from the saturated aquifer zone, which is a function of the hydrologic condition of the aquifer.

Phase A shows the hydrological condition after a non-recharge period, characterised by a decreasing trend in the discharge, and the contribution from the vadose zone can be neglected. In this phase the saturated zone is drained by the conduit network, which has an hydraulic head similar or little lower than the saturated zone.

Phase B occurs after an intense rainfall event and is characterised by several rainy hours or a few days of rainfall. The shafts and sinkholes allow surface runoff to infiltrate and rapidly reach the saturated zone of the aquifer. The amount of rainfall which causes the concentrated infiltration depends on the rainfall intensity and its distribution over time, but is strongly controlled by the hydraulic characteristics of the vadose zone, such as its thickness, the presence of shafts and the morphological characteristics of the ground surface (presence of endorheic areas and swallow holes, slope angle distribution, etc.). During phase B, shafts are temporarily filled with water, and an increase of hydraulic head in the karst conduit network occurs, up to the spring (Figure 9.4). As the concentrated recharge varies according to the catchment features, it provides a different water level for each shaft, which can cause a temporary and locally different water flow direction into the saturated aquifer zone. The rapid response of a spring to an intense rainfall event, with a typical peak in the spring hydrograph, can be associated with the rise and lowering of the water level inside conduits above the saturated zone terminating into the phreatic zone (Drogue, 1980;

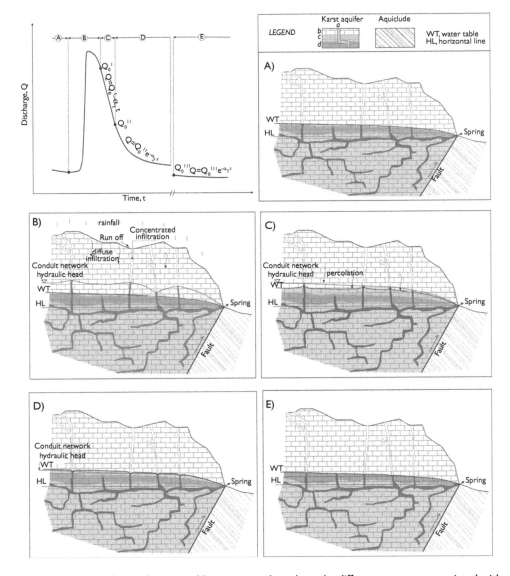

Figure 9.4 Spring hydrograph triggered by a storm pulse, where the different parts are associated with sketches A–E (modified from Fiorillo, 2014), showing the hydrodynamics in a karst aquifer. Recharge occurs by diffuse percolation through the vadose zone, and in a concentrated manner at points such as sinkholes/swallow holes and shafts; a basal spring drains the water table by a conduit network. The hydrodynamics is shown by different sketches where the slope of the water table is exaggerated. A, after a period of no recharge (*a* conduits and shafts; *b* vadose zone; *c* and *d* saturated zone with different effective porosity). B, during a period of strong recharge, concentrated infiltration causes the temporary filling of shafts, later the aquifer water table rises as a consequence also of the diffuse infiltration. C, after a period of strong recharge, water level in the shafts falls, and the water table could still rise. D, lowering of the water table during a non-recharge phase; E, lowering of the water table after a long period without recharge.

Bonacci & Živaljevic, 1993; Bonacci, 1995; Halihan *et al.*, 1998). However, the very high hydraulic conductivity of the conduit network can limit the rise (and lowering) of the water level in conduits.

After the beginning of the increase in flow, a decrease of the chemical hardness of the spring water is observed, which may mark a large part of the peak hydrograph zone. Ashton (1966) provided a simple procedure to evaluate the volume of storage in fully submerged conduits feeding a spring using discharge and the hardness of the spring water. These observations indicate that the first part of the recession (the steepest part) discharges fresh water, which comes from the shafts and conduits, connected to the springs by the conduit network.

The exchange of water between conduits and matrix occurs during floods (Martin & Dean, 2001; Bailly-Comte *et al.*, 2010), and it is caused by the higher conduit head in the conduit network than the matrix (Atkinson, 1977; Drogue, 1980; Bailly-Comte *et al.*, 2010; Fiorillo, 2011). During phase B shafts and the conduit network feed the water table as diffuse infiltration takes place as well. Under such hydraulic conditions, spring discharge is connected to concentrated recharge processes and depends on the hydraulic head in the conduit network. Some karst springs (for example, weakly karstified systems) do not show this hydraulic behaviour, and their hydrograph shape may appear completely smooth; in this case, spring discharge increases or decreases as a function of long wet or dry periods, respectively (Fiorillo, 2011).

In heavily karstified systems, when recharge reduces or ends, a fall of discharge can be observed at each spring. This process reflects rapid drainage of the conduit network which and its temporarily filled shafts (Phase C). Phase C has a linear equation (which represents the nonlinear part in the semilogarithmic plot), but it may have also another analytical form (Malík & Vojtková, 2012). Mangin (1975) explained this initial part of the recession as the influenced stage, characterised by the arrival (at the water table) of the diffuse recharge from the unsaturated zone, which decreases with time.

During phase C the hydraulic head in the conduit network decreases but it is still higher than the water table. It may increase due to percolation from diffuse recharge.

When the water level in the shafts decrease and reach the water table level, the hydraulic head in the conduit network reaches that of the matrix/minor fractures, causing the drainage of the saturated zone by the conduit network, and phase D begins. Many karst springs show an abrupt change of the hydrograph slope from phase C to phase D; the gently sloped part of phase D highlights the higher water volume stored in the saturated zone, which causes a lower decrease of the discharge through time (Fiorillo, 2014). During this stage, the emptying processes cause the lowering of the water table and involves water from minor conduits and the matrix, where fluid viscosity and friction control the hydrograph recession shape. The concave shape of this part of the hydrograph highlights, at least in part, the energy lost by the water flow. Figure 9.4 shows the recession during phase D by a single exponential term, but the first part could still be influenced by the arrival of the diffuse infiltration, and only later the exponential form could appear. Besides, following the theoretical models (Rorabaugh, 1964; Brutsaert, 1994; Kovacs *et al.*, 2005), this first part of Phase D could be characterised by the sum of several exponentials.

Phase E refer to the hydrological condition after a long period without recharge, during which the water table reaches the minimum height, following a different recession coefficient, α_3 (or α_n). If the exponential form is expected, the changing of the recession coefficient, from α_2 (phase D) to α_3 (phase E) may depend on several factors, and reflects the anisotropy of the aquifer or its non constant area during drainage (Fiorillo, 2011).

9.4 CONCLUSION

Interpretation of the recession curve of a spring hydrograph could be complex in any attempt to characterise karst aquifers, without any knowledge of the actual physical characteristics of the aquifers. Generally, the initial part of the recession (in the arithmetic plot) can be approximated linearly and could mean drainage following of non-Darcian type. The last part of the recession, especially for basal springs, presents a concave shape and indicates drainage according to Darcian flow. The concave shape is generally associated with the lowering of the water-table, and highlights the energy lost through water transport. If this concave shape follows the exponential decay (linear reservoir) model, a single value of the recession coefficient is able to explain the entire draining of the aquifer. This condition appears to be rare in nature, especially under drought conditions, but provides an opportunity to investigate the lowest part of the recession limb of spring hydrographs. Under such conditions, hydrographs tend to be similar to curves 2 or 3 in Figure 9.3, indicating a non-constant recession coefficient during drainage. Many karst springs show a typical variation of the recession coefficient during drainage, especially if several continuous years are considered, including droughts (Fiorillo et al., 2012). Fiorillo (2011) explained that the variation in the recession coefficient is due to non-constant geometric or hydraulic characteristics during the emptying process. In particular, the recession coefficient appears to be strongly controlled by the product of the effective porosity along the water table and the area occupied by the water table. Both parameters can vary during the emptying process, and they control the shape of the hydrograph on the semilogarithmic plot.

REFERENCES

Ashton K. (1966) The analysis of flow data from karst drainage systems. *Trans. Cave Res. Group of Great Britain*, 7, 161–203.

Atkinson T.C. (1977) Diffuse flow and conduit flow in limestone terrain in Mendip Hills, Somerset (Great Britain). *Journal of Hydrology* 35, 93–100.

Baedke SJ, Krothe NC (2001). Derivation of effective hydraulic parameters of a karst aquifer from discharge hydrograph analysis. *Water Resources Research* 37(1), 13–19.

Bailly-Comte V., Martin J.B., Jourde H., Screaton E.J., Pistre S., Langston A. (2010) Water exchange and pressure transfer between conduits and matrix and their influence on hydrodynamics of two karst aquifers with sinking streams. *Journal of Hydrology* 386, 55–66.

Bonacci O. (1995) Ground water behaviour in karst: example of the Ombla Spring (Croatia). *Journal of Hydrology* 165, 113–134.

Bonacci O., Zivaljevic R. (1993) Hydrological explanation of the flow in karst: example of the Crnojevica spring. *Journal of Hydrology* 146, 405–419.

Brutsaert W. (1994) The unit response of ground water outflow from a hill-slope – *Water Resources Research* 30(10), 2759–2763.

Dewandel B., Lachassagne P., Bakalowicz M., Weng Ph., Al-Malki A. (2003) Evaluation of aquifer thickness by analysing recession hydrographs. Application to the Oman ophiolite hard-rock aquifer. *Journal of Hydrology* 274, 248–269.

Dreybrodt W., Romanov D., Kaufmann G. (2010) Evolution of caves in porous limestone by mixing corrosion: a model approach. *Geologia Croatica* 63(2), 129–135.

Drogue C. (1980) Essai d'identification d'un type de structure de magasins carbonatés fissurés: application à l'interprétation de certains aspects du fonctionnement hydrogéologique. *Mémoire hors série de la Société Géologique de France* 11, 101–108.

Eisenlohr, L., Kiraly, L., Bouzelboudjen, M., Rossier, I. (1997) A numerical simulation as a tool for checking the interpretation of karst springs hydrographs. *Journal of Hydrology* 193, 306–315.

Estrela T., Sahuquillo A. (1997) Modelling the response of a karstic spring at Arteta aquifer in Spain. *Ground Water* 35(1), 18–24.

Fiorillo F. (2011) Tank-reservoir emptying as a simulation of recession limb of karst spring hydrographs. *Hydrogeology Journal* 19, 1009–1019.

Fiorillo F., Revellino P., Ventafridda G. (2012) Karst aquifer drainage during dry periods. *Journal of Cave and Karst Studies*. 74(2), 148–156.

Fiorillo F. (2014) The recession of spring hydrographs, focused on karst aquifers. *Water Resources Management*, 28, 1781–1805.

Ford D. C., Ewers R. O. (1978). The development of limestone cave systems in the dimensions of length and depth. *Canadian Journal of Earth Science* 15(11), 1783–1798.

Ford D., Williams P. (2007). Karst Hydrogeology and Geomorphology. Wiley, England, 562 pp.

Forkasiewicz, J., Paloc, H. (1967). Le regime de tarissement de la Foux-de-la-Vis [Analysis of the recession period of the Foux-de-la-Vis spring]. Etude preliminaire. *Chronique d'Hydrogeologie*, BRGM 3(10), 61–73.

Gabrovšek F., Dreybrodt W. (2010). Karstification in unconfined limestone aquifers by mixing of phreatic water with surface water from a local input: a model. *Journal of Hydrology* 386(1–4), 130–141.

Halihan T., Wicks C.M., Engeln J.F. (1998). Physical response of a karst drainage basin to flood pulses: example of the Devil's Icebox Cave system (Missouri, USA) – *Journal of Hydrology* 204, 24–36.

Kovács A., Perrochet P. (2008). A quantitative approach to spring hydrograph decomposition. *Journal of Hydrology* 352, 16–29.

Kovács A., Perrochet P., Király L., Jeannin P. (2005). A quantitative method for characterisation of karst aquifers based on the spring hydrograph analysis. *Journal of Hydrology* 303, 152–164.

Jeannin P-Y., Sauter M. (1998). Analysis of karst hydrodynamic behaviour using global approaches: e review. *Bulletin d'Hydrogeologie* 16, 31–48.

Maillet E. (1905). Essais d'Hydraulique souterraine et fluviale [Underground and river hydrology]. Hermann, Paris, 218 pp.

Malík P., Vojtková S. (2012). Use of recession-curve analysis for estimation of karstification degree and its application in assessing overflow/underflow conditions in closely spaced karstic springs. *Enveromental Earth Science* 65(8), 2245–2257.

Martin, J.B., Dean, R.W. (2001) Exchange of water between conduits and matrix in the Floridan aquifer. *Chemical Geology* 179(1–4), 145–165.

Mangin A. (1975). Contribution à l'étude hydrodynamique des aquiféres karstiques [A contribution to the study of karst aquifer hydrodynamics]. 3éme partie, *Annales de Speleologie* 30(1), 21–124.

Milanovic P. (1976). Water regime in deep karst: case study of Ombla spring drainage area. In: V. Yevjevich (ed.) Karst Hydrology and Water resources, vol. 1 Karst Hydrology, *Water resources Publications*, Colorado, 165–191.

Nutbrown D.A., Dowing R.A. (1976) Normal-mode analysis of the structure of base flow recession curves. *Journal of Hydrology* 30, 327–340.

Rorabaugh M.I. (1964) Estimating changes in bank storage and groundwater contribution to streamflow. *International Association of Scientific Hydrology* 63, 432–441.

Sahuquillo A, Gómez-Hernández, J.J. (2003) Comment on 'Derivation of effective hydraulic parameters of a karst aquifer from discharge hydrograph analysis' by SJ Baedke and NC Krothe. *Water Resources Research* 39(6): 1152.

Stevanovic Z., Milanovic S., Ristic V. (2010) Supportive methods for assessing effective porosity and regulating karst aquifers. *Acta Carsologica* 39(2), 301–311.

Szilagyi J. (1999) On the use of semi-logarithmic plots for baseflow separation. *Ground Water* 37(5), 660–662.

White W.B. (1988). Geomorphology and Hydrology of Karst Terrain. Oxford Press University, pp. 464.

Chapter 10

Characterisation of selected karst springs in Slovenia by means of a time series analysis

Gregor Kovačič[1] *& Nataša Ravbar*[2,3]
[1]*University of Primorska, Faculty of Humanities and Science and Research Centre, Koper, Slovenia*
[2]*Karst Research Institute ZRC SAZU, Postojna, Slovenia*
[3]*Urban Planning Institute of the Republic of Slovenia, Ljubljana, Slovenia*

ABSTRACT

The relation between discharge or water level, temperature and rainfall is evaluated for nine karst springs/rivers using a daily time series analysis (autocorrelation and cross-correlation) covering a 30-year period (1984–2013). The study was conducted in order to qualitatively assess the applicability of long-term hydrological data set usage in time series analysis in order to compare different karst springs and their catchments characteristics. The results reveal that the storage capacity of larger systems and systems characterised by a more complex structure is typically greater, although such systems should not necessarily be defined as poorly karstified. Factors influencing the obtained results are determined in the study. Application of longer hydrological data sets is shown to provide valuable information on the hydrological properties of springs, hydrodynamic behaviour and the hydraulic properties of the corresponding aquifers. However, caution should be used when classifying karst systems into groups solely on the basis of the results of a time series analysis.

10.1 INTRODUCTION

Univariate and bivariate time series analysis of hydrological data sets in the time (autocorrelation and cross-correlation) and frequency (single spectral density and cross-spectral density with corresponding amplitude, coherence, gain and phase functions) domains have been widely used in karst hydrology. Jeannin & Sauter (1998) argue that it is inappropriate to use these methods without good knowledge of the investigated area to characterise the hydrogeological structure of karst aquifers, and that the results obtained should be verified by deterministic models and/or direct observations. Despite this, the time series analysis methodology, first applied by Mangin (1984) to study the input-output relationship in karst aquifers, has been further developed, improved and applied by many authors (e.g. Benavente *et al.*, 1985; Padilla & Pulido-Bosch, 1995; Angelini, 1997; Eisenlohr *et al.*, 1997; Larocque *et al.*, 1998; Labat *et al.*, 2000; Samani, 2001; Amraoui *et al.*, 2003; Jukić & Denić-Jukić, 2004; Mathevet *et al.*, 2004; Rahnemaei *et al.*, 2005; Valdes *et al.*, 2005, 2006; Massei *et al.*, 2006; Panagopoulos & Lambrakis, 2006; Novel *et al.*, 2007; Bailly-Comte *et al.*, 2008; Covington *et al.*, 2009; Herman *et al.*, 2009; Jemcov & Petrič, 2009; Fiorillo & Doglioni, 2010; Jemcov & Petrič, 2010; Kovačič, 2010b; Terzić *et al.*, 2012;

Delbart *et al.*, 2014; Mayaud *et al.*, 2014; Katsanou *et al.*, 2015). All these studies of various hydrological data sets (discharge, water level, precipitation, electrical conductivity, temperature, turbidity) have proved the usefulness of time series analysis as a stand-alone or complementary method for the study of the functioning, hydrodynamic behaviour and dynamics of karst springs and the corresponding aquifers.

Time series analysis has mostly been applied to periods covering i) one hydrological year with daily or hourly time series (e.g. Larocque *et al.*, 1998; Massei *et al.*, 2006; Herman *et al.*, 2009; Kovačič, 2010b; Terzić *et al.*, 2012) and ii) several successive hydrological years with daily data (e.g. Benavente *et al.*, 1985; Padilla & Pulido-Bosch,1995; Angelini, 1997; Eisenlohr *et al.*, 1997; Larocque *et al.*, 1998; Labat *et al.*, 2000; Samani, 2001; Amraoui *et al.*, 2003; Mathevet *et al.*, 2004; Rahnemaei *et al.*, 2005; Panagopoulos & Lambrakis, 2006; Novel *et al.*, 2007; Bailly-Comte *et al.*, 2008; Herman *et al.*, 2009; Jemcov & Petrič, 2009; Fiorillo & Doglioni, 2010; Jemcov & Petrič, 2010; Kovačič, 2010b; Delbart *et al.*, 2014; Katsanou *et al.*, 2015). Only a few studies (e.g. Valdes *et al.*, 2005, 2006; Bailly-Comte *et al.*, 2008; Covington *et al.*, 2009; Mayaud *et al.*, 2014) have applied the method to a very short or single-event timescale with hourly or even shorter time series. These studies have demonstrated that single-event analysis provides relevant information on how a karst system reacts to a single recharge event. However, long-term analysis provides information about more average karst aquifer functioning (Panagopoulos & Lambrakis, 2006; Kovačič 2010b).

Herman *et al.* (2009), Delbart *et al.* (2014) and Kovačič (2010b) have shown that the selection and duration of the reference period (one or more hydrological years, parts of an individual hydrological year) within a time series analysis can have strong effects on the results. For example, a comparative autocorrelation and cross-correlation analysis performed for the Unica karst spring (Slovenia) in the five successive hydrological years 1997–2002 revealed important differences in the calculated memory effects ($r_k = 0.2$) and the response times and values of maximum precipitation-discharge cross-correlation coefficients (Kovačič, 2009). The characteristic lags for the Unica spring for an individual hydrological year range from 17 to 86 days, whereas the maximum cross-correlation coefficients vary between 0.21 and 0.33 with lags of 0 to 3 days (Kovačič, 2009). According to Mangin's classification (1984), the Unica spring could, therefore, be classified as a system with a low (well-karstified) or very high memory effect (poorly karstified), which is rather confusing. The study showed that the selection of the hydrological year (the structure of the time series itself) could have significant effects on the results of a time series analysis (Kovačič, 2010b). Additionally, the study confirmed that the calculated memory effects and response times to precipitation events also depend on the selection of the beginning and the end of the hydrological year(s) in an investigated period.

Since in the literature time series analyses are only rarely employed for the characterisation of karst aquifers for periods longer than 10 hydrological years (e.g. Labat *et al.*, 2000; Mathevet *et al.*, 2004; Fiorillo & Doglioni, 2010), this chapter represents a step forward, in that it is a comprehensive study of nine well-investigated karst springs, the sinking rivers recharging these springs and the corresponding karst aquifers in Slovenia. Time series analysis was performed on daily precipitation, discharge, water level and temperature data sets for the 30-year period 1984–2013. A 30-year period was selected because this is a standard long-term reference period used in climatology and hydrology. The objective of this study is to contribute to existing knowledge of the hydrodynamic characteristics and hydraulic properties of selected karst aquifers, previously studied by means of other investigation techniques used in karst hydrology. The results of the time series analysis can, therefore, be properly validated. A comparison of these results allows a qualitative assessment of the applicability of the long-term hydrological data sets used in the time series analysis that provide different information about the hydrogeological characteristics of karst aquifers.

10.2   METHODS AND THEORETICAL BACKGROUND

In karst hydrology, time series analysis of hydrological data requires the mathematical analysis of the response of a karst system to recharge (precipitation, concentrated infiltration via ponors) and indirectly provides information regarding the structure and functioning of karst aquifers (Box & Jenkins, 1970; Mangin, 1984; Box et al., 1994). Univariate analysis reveals the structure of an individual time series, in terms of either the time domain (autocorrelation) or the frequency domain (spectral density). The autocorrelation function, presented in a correlogram, quantifies the memory effect of the system, which is computed on the basis of the decorrelation lag time, defined as the time at which the autocorrelation function attains a predetermined value, usually 0.2 (Mangin, 1984; Benavente et al., 1985). The value of $r_k = 0.2$ was also used in this study. The memory effect reflects the inertia of the system. The slope of the correlogram is determined by the response of the system to an event and gives indirect information on the storage capacity of the system and its degree of karstification. Generally, a high memory effect indicates that the system is influenced by an event for a long time, which often relates to a large storage capacity (Mangin, 1984). The correlogram of such karst aquifers will decrease slowly. Such systems are characterised by a network of smaller fissures through which the water flows at much lower velocities. In contrast, according to the literature (e.g. Larocque et al., 1998; Panagopoulos & Lambrakis, 2006), a well-developed karst aquifer with larger conduits and without a significant water storage capacity should correspond to a low memory system, which, however, is a rather problematic statement. In general, a correlogram of a well-developed karst system decreases steeply and quickly. Kovačič (2010b) proved that large and well-developed karst aquifers such as the Malenščica and Unica karst springs systems can also have a large storage capacity, even though the water flows through the system of unobstructed conduits at relatively high velocities (several tens of m/h; Gabrovšek et al., 2010), and cannot be considered as poorly karstified. In this regard caution is needed when interpreting the results on the basis of the expected values. A spectral density function quantifies the regulation time, which defines the duration of the influence of the input signal and gives an indication of the length of the impulse response of the system (Larocque et al., 1998).

Cross-correlation and cross-spectral density functions imply the transformation of input signals to output signals. In karst hydrology these are mostly discharge-discharge, precipitation-discharge or water level-discharge relationships. The shape of the cross-correlogram indicates the degree of karstification of a karst system. The delay, which is the time lag between lag 0 and the lag of the maximum value of the cross-correlation coefficient $(r_{xy}(k))$, gives an estimation of the pressure pulse transfer time through the aquifer (Panagopoulos & Lambrakis, 2006). Gently sloping cross-correlograms indicate a significant storage capacity and a low degree of karstification. However, a well-developed karst aquifer is characterised by much shorter response times (steep cross-correlogram), usually showing no delay (Padilla & Pulido-Bosch, 1995). Cross-spectral density function allows a calculation of amplitude, phase, coherence and gain functions. In the present study the values of mean coherence functions, periods of mean delays and cut-off frequencies between variables were also calculated. The coherence function expresses the linearity of karst systems. Higher values correspond to highly karstified aquifers and vice versa (Panagopoulos & Lambrakis, 2006). Cut-off frequency defines the duration of precipitation events that can be efficiently filtered by the karst system and is determined by the period when the amplitude function drops below 1 (Mathevet et al., 2004). A mean delay is defined on the basis of a phase function and shows an average delay between precipitation and discharge, or discharge and discharge in the case of the concentrated recharge of karst aquifers (Padilla & Pulido-Bosch, 1995; Rahnemaei et al., 2005). It is important to mention that the mean delay and cut-off frequency parameters are not often used in karst hydrology, owing to the many uncertainties regarding the calculation and interpretation of their values.

The study deals with nine karst springs/rivers in Slovenia: the Malenščica, the Unica, the Stržen, the Pivka, the Ljubljanica, the Bistrica, the Vipava, the Hubelj and the Rižana; and five precipitation stations: Pokojišče, Podgrad, Postojna, Juršče and Podkraj. Daily data, used in the time series analysis (discharge, water level, precipitation and temperature), were obtained from the Environmental Agency of the Republic of Slovenia (EARS, 2014a, 2014b). The selection of the reference precipitation stations for some of the karst springs (Rižana, Bistrica and Ljubljanica) was difficult, since precipitation data for the observed 30-year period were not available for the stations in their catchments. Nevertheless, the analysis conducted by Kovačič (2010b) showed that the precipitation regime within the broader regions in Slovenia is more or less homogeneous at a daily time step (almost simultaneously) in terms of rainfall events occurrence, which means that some of the remote precipitation stations (e.g. Podgrad, Pokojišče) could also be considered as relevant for the individual cross-correlation analysis. Not all of the hydrological data sets included in the study were complete, a fact that has been taken into account when performing the analysis. All the analyses were performed using *Statistica* software. The autocorrelation time series analysis for the 30-year period 1984–2013 (N = 10,958) was carried out at a daily time step and at a 130-day cutting point for autocorrelation and at a 100-day cutting point for cross-correlation analysis. All the time series were previously automatically detrended by a software module to avoid the influence of long-term changes in the time series on the final results.

10.3 STUDY AREA

The catchment areas (relating to karst aquifers) of the karst springs (rivers) included in this study are mostly developed in deep sequences of Mesozoic limestones and dolomites, with the exception of the Pivka and Rižana springs, where Palaeocene limestones also occur (Figure 10.1). The Rižana, Malenščica, Hubelj, Vipava and Bistrica karst springs are tapped for drinking water supply.

10.3.1 The Ljubljanica river basin

The Stržen, Pivka, Malenščica and Unica karst springs (rivers) with their catchments are part of the common Ljubljanica karst river basin (Figure 10.2, Table 10.1). The major outflow from the binary karst system (1,100 km²), which has a complex structure, is the Ljubljanica karst spring (several outflows) (Žibrik *et al.*, 1976). The mean discharge of the Ljubljanica spring at the Vrhnika gauging station (the Ljubija and Bistra karst springs lie downstream of the station) in the reference period 1984–2013 is 22.68 m³/s, with a minimum of 0.95 and a maximum of 120.48 m³/s (EARS, 2014a). The ratio between minimum, mean and maximum discharge is 1: 24.8: 126.3. Long-term mean annual precipitation is estimated at 1800 mm (Žibrik *et al.*, 1976).

The catchment of the Ljubljanica springs is characterised by (i) autogenic recharge from the high Dinaric plateaus (Snežnik, Javorniki, Menišija) and (ii) numerous sinking streams on higher-lying poljes in the Notranjska region, the Pivka river basin (the Pivka) and the Bloke plateau (the Bloščica) (Figure 10.2). Waters from Babno Polje, the Snežnik platcau and the Racna Gora plateau emerge on Loško Polje as the Obrh sinking stream. To the north east is Cerknica Polje, the biggest of the poljes in the area, which is famous for its intermittent Lake Cerknica and is an important confluence of

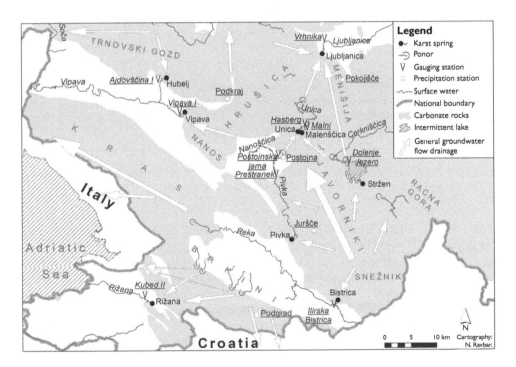

Figure 10.1 Location of investigated karst springs (rivers), precipitation and gauging stations.

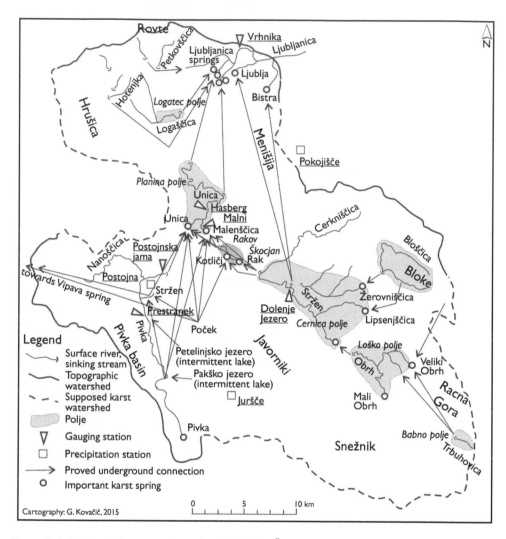

Figure 10.2 Ljubljanica karst river basin (modified after Žibrik *et al.*, 1976; Kovačič, 2010a). Precipitation and gauging stations included in the study are underlined.

karst waters from the Javorniki plateau, Loško Polje and the Bloke plateau. The major sinking stream flowing into the polje is the Stržen, which collects karst waters emerging from numerous karst springs of varying abundance on the polje's eastern and southern sides and partly on the western side at the foot of the Javorniki and Bloke plateaus (Habič, 1976; Kovačič, 2010a). The right tributary of the Stržen is the Cerkniščica, a surface river with a catchment of about 45 km² (EARS, 2014a). The sum total of inflows to the polje is 210–240 m³/s; maximum runoff can surpass 90 m³/s (Kranjc, 1986; Kovačič, 2010a). As well as disappearing underground in the principal ponor caves on the polje's north western margin (underground flow towards Rakov Škocjan), water from the polje sinks in numerous swallow holes. The water sinking through

Table 10.1 Key characteristics of the karst springs and watercourses in the study area (daily data for the period 1984–2013) with characteristic lag times (days), where the autocorrelation coefficient exceeds the 0.2 value, and calculated regulation times. The locations of the measuring sites are shown in Figures 10.1 and 10.2.

Spring or river	Mean	Min	Max	q5	q95	$Q_{min}:Q_{mean}:Q_{max}$	SD	CV (%)	Characteristic lag (days) ($r_k = 0.2$)	Treg
Malenščica										
Q (m³/s)	6.33	1.10	11.24	2.20	9.90	1:5.80:10.21	2.39	37.44	56	32
Unica										
Q (m³/s)	20.25	1.08	89.22	2.29	62.10	1:18.72:82.46	19.96	98.57	47	39
T (°C)	8.96	0.8	18.4	3.60	14.40		3.23	36.14	72	107
Stržen										
h (cm)	225.88	0	654	33	437		120.96	47.27	51	32
Pivka (Prestranek)										
Q (m³/s)	2.49	0	43.35	0	11.10		3.95	158.61	40	13
Pivka (Postojnska jama)										
h (cm)	123.55	7	829	34	220		60.24	48.76	39	17
Ljubljanica										
Q (m³/s)	22.68	0.95	120.48	2.14	75.40	1:23.77:126.29	23.80	104.95	37	39
T (°C)	9.88	3.7	14.8	6.1	13.2		2.30	23.25	78	58
Bistrica										
Q (m³/s)	1.33	0.04	24.90	0.13	4.16	1:30.11:565.93	1.38	104.20	45	25
Vipava										
Q (m³/s)	6.25	0.73	81.42	1.11	21.90	1:8.59:112.00	7.75	124.09	7	7
T (°C)	10.35	8.0	15.0	8.8	13.0		1.28	12.38	$r(999) = 0.385$	
Hubelj										
Q (m³/s)	2.76	0.18	51.26	0.34	9.38	1:15.77:229.94	3.67	132.79	12	25
T (°C)	8.53	5.0	13.3	7.3	10.3		0.92	10.78	72	
Rižana										
Q (m³/s)	3.37	0.03	63.16	0.13	15.50	1:112.3:2105.27	5.48	162.60	15	22

these swallow holes re-emerges in the Ljubljanica springs (Behrens *et al.*, 1976). Within the Rakov Škocjan karst depression, waters from Cerknica Polje and, in part, from the Javorniki plateau flow into the Rak, a surface stream. When the Rak finally disappears underground, it flows towards the Unica and Malenščica springs.

To the north west is Planina Polje (the outflow of the Malenščica and Unica springs), an important confluence of karst waters from three contributing sub-catchments of the Ljubljanica basin: Cerknica Polje, the Pivka basin, and the Snežnik and Javorniki karst plateaus. The Malenščica karst spring is a tributary of the more abundant Unica river. Their catchment areas almost completely overlap, with the exception of the Lower Pivka basin area (the Pivka with the Nanoščica), which is a part of the Unica's catchment but not part of the Malenščica's catchment. In comparison to the Unica spring, the portion of autogenic recharge in the Malenščica spring is greater. The Unica river sinks on the eastern and northern rim of Planina Polje and flows directly towards the Ljubljanica karst springs. Surface drainage occurs in the Lower Pivka basin (Eocene flysch) and in the Cerkniščica and Bloščica drainage basins (Triassic clastic formations). Alluvial deposits from the Quaternary period are found along the Cerkniščica, Bloščica and Pivka rivers and at the bottom of the higher-lying poljes.

According to the results of the water balance analysis for the 30-year observation period (1971–2000), the Malenščica spring has a catchment area of 200 km² (Kovačič, 2010b). However, tracer test results conducted in the past show that its catchment is much larger and extends over 726 km² (EARS, 2009). The catchment area of the Unica is estimated at 800 km² (Žibrik *et al.*, 1976). The mean discharge of the Malenščica and Unica springs in the reference period 1984–2013 is 6.38 and 20.25 m³/s respectively (EARS, 2014a). The ratio between minimum, mean and maximum discharge of the Malenščica spring is 1: 5.8: 10.2, with a minimum discharge of 1.10 and a maximum of 11.24 m³/s. Minimum discharge of the Unica spring in the reference period is 1.08 while the maximum is 89.22 m³/s (EARS, 2014a). The ratio between minimum, mean and maximum discharge is 1: 18.7: 82.5. When water levels are low, the Malenščica contributes almost all of the water in the Unica. Mean annual precipitation in the common catchment area over the long-term period is estimated at 1780 mm; mean annual evapotranspiration is 720 mm and mean annual run-off is 1060 mm, with a runoff coefficient of around 60% (Kovačič, 2010b).

The Pivka river basin has a dual hydrogeological structure. The upper part of the basin (Prestranek gauging station), which also drains the western parts of the surrounding Javorniki and Snežnik karst plateaus, is in limestones, while the lower part is in almost impermeable Eocene flysch. The Pivka sinks in the world-famous Postojna Cave and flows towards Planina Polje as the Pivka branch of Planina Cave. The cave is a spring of the Unica river, where two subsurface river channels (the Rak and Pivka branches) meet. The Rak branch drains the waters from the Rakov Škocjan karst depression and the Javorniki plateau. The catchment area of the Pivka river at its ponor is estimated at 300 km² (Žibrik *et al.*, 1976). In the reference period 1984–2013 the discharges of the Pivka (Prestranek gauging station) ranged between 0 and 43.35 m³/s, with a mean of 2.49 m³/s (EARS, 2014a). When water levels are low, the Pivka dries up completely in the upper part of the basin.

North of Planina Polje is Logatec Polje with several small sinking streams (Figure 10.2). Together with the Logatec and Rovte plateaus, this part of the Ljubljanica catchment covers about 100 km² (Habič, 1976). The area is mainly built of Triassic dolomites where surface drainage occurs. Streams sink underground in numerous ponors at the point of contact with Cretaceous limestones and flow towards the Ljubljanica springs.

10.3.2   The Vipava and Hubelj catchments

The catchment area of the Vipava karst spring (several outflows) is estimated at 149 km² (Kranjc, 1997; Jemcov & Petrič, 2009). It encompasses the Nanos karst plateau, which is mostly built of Cretaceous limestone. Only a small part of the catchment (an area of 9 km²) is developed on very low permeable Eocene flysch rocks of the lower Pivka basin, where several surface streams sink into the karst aquifer and flow towards the Vipava springs (Kranjc, 1997). Though the results of the tracer test proved an underground connection of the Vipava springs with the Stržen (tributary of the Pivka) (Habič, 1989) and the Poček area of the western part of the Javorniki karst plateau (Kogovšek, 1999; Kogovšek et al., 1999), neither area is included in the estimated area of the spring's catchment. The mean discharge of the Vipava spring in the reference period 1984–2013 is 6.25 m³/s, with a minimum of 0.73 and a maximum of 81.42 m³/s (EARS, 2014a). The ratio between minimum, mean and maximum discharge is 1: 8.6: 112.0. Mean annual precipitation (1984–2013) in the catchment (Podkraj precipitation station) is 2068 mm (EARS, 2014b).

The Hubelj spring drains the high karst plateau of Trnovski Gozd in South West Slovenia, with elevations ranging from 800 to 1200 m asl. The plateau is built of highly karstified carbonate rocks, mainly Jurassic limestones and dolomites, which are bounded by Triassic dolomites in the north and Eocene flysch in the south and east (Janež et al., 1997), acting as a partial hydrogeological barrier. Mean annual precipitation on the plateau ranges from 2000 to 3000 mm. The Hubelj spring represents an outflow of karst waters stored in the south-eastern part of the plateau. Its catchment is estimated at 50–90 km² (Kranjc, 1997; Turk et al., 2013). The spring discharges at 240 m asl when the water level is low and 40 m higher when the water level is high. The mean discharge of the Hubelj spring in the reference period 1984–2013 is 2.76 m³/s, with a minimum of 0.18 and a maximum of 51.26 m³/s (EARS, 2014a). The ratio between minimum, mean and maximum discharge is 1: 15.77: 229.94.

10.3.3   The Bistrica and Rižana catchments

The Bistrica karst springs emerge at the western foot of the Snežnik karst plateau (700 to 1796 m asl), at the contact with the highly impermeable Eocene flysch of the Reka Valley. The Ilirska Bistrica gauging station does not include all the water emerging at the contact. Cretaceous and Jurassic limestones, dolomites and dolomite-limestone breccias predominate on the deeply karstified plateau. The unary recharge area of the Bistrica springs only stretches over a small part of the plateau and is estimated at 50 km² (Kovačič, 2003). The mean discharge of the Bistrica spring in the period 1984–2013 is 1.33 m³/s, with a minimum of 0.04 and a maximum of 24.90 m³/s (EARS, 2014a). The ratio between minimum, mean and maximum discharge is 1: 30.1: 565.9. If in the calculation the amount of water abstracted for water supply is included (100 l/s), the ratio changes to 1: 9.2: 172.9. Mean annual precipitation in the catchment is estimated at 1800 mm (Kovačič, 2003).

The mean discharge of the Rižana spring in the period 1984–2013 is 3.37 m³/s, with a minimum of 0.03 and a maximum of 36.16 m³/s (EARS, 2014a). The ratio between minimum, mean and maximum discharge is 1: 112.3: 2105.3. If in the calculation the amount of water abstracted for water supply is included (200 l/s), the ratio changes to 1: 15.5: 275.5. Mean annual precipitation (1984–2013) in the catchment (Podgrad precipitation station) is 1465 mm (EARS, 2014b). The recharge area of the Rižana spring covers an area of 245 km². The spring discharges at an elevation of 70 m asl, while the recharge area is at 500 to 1100 m asl. It is predominantly composed of carbonate rocks (Cretaceous and Palaeocene limestones). Several small surface streams from the Eocene flysch of the Brkini hills sink underground and additionally recharge the karst aquifer (Krivic et al., 1987; Krivic et al., 1989).

10.4   RESULTS AND DISCUSSION

10.4.1   Univariate time series analysis

The autocorrelation functions of daily data from precipitation stations diminish rapidly and reach the $r_k = 0.2$ value immediately, reaching maximum values of between 0.204 (Pokojišče) and 0.255 (Postojna) after 1 day and the $r_k = 0$ value after 18 to 19 days. This means that no memory effect can be detected in precipitation on a daily scale and that signals (events) appear to be rather random and homogeneous within a daily period in the investigated area.

The spectral density functions of the daily discharges (water levels) of the springs (rivers) and those of daily precipitation show peaks at the low frequencies of 0.0027 (364 days) and 0.0055 (182 days), which confirm the presence of important annual and seasonal (half-year) cycles. The latter is most likely associated with the sub-Mediterranean precipitation regime with a precipitation maximum in autumn. The seasonal cycle is less evident in the spectral density functions of the Pivka and almost invisible in those of the Rižana spring. However, the spectral density functions of daily precipitation also show a periodicity at a frequency of 0.0059 (169 days), which is not so clearly expressed in the spectral density functions of discharge (water level).

Autocorrelation functions for the daily discharge series show that the Malenščica spring has the largest storage capacity (Table 10.1, Figure 10.3) of all the springs. The autocorrelation coefficient of the spring reaches the characteristic lag after 56 days, revealing that the memory effect of the system is high according to the theoretical values (Mangin, 1984). The results are in accordance with the calculated characteristic lags (16 to 80 days), calculated for the individual hydrological years of the period 1997–2002 (Kovačič, 2010b). The regulation time of the spring (32 days), however, indicates the medium storage capacity of the system.

The absence of a fast recharge component is the reason why the density function of the Stržen stream water level has a relatively high regulation time of 32 days, which is comparable to the high memory effect (51 days) of the stream. The Stržen stream floods all the watercourses emerging on or flowing on the surface into Cerknica Polje. Once it has reached its flood, the water level of the stream shows long-term persistence and small variability (CV = 47%). The rank correlation coefficient (Spearman's rho) between the water level of the Stržen and the discharge of the Malenščica is high (0.92). In a study conducted by Kovačič (2009), the calculated memory effects of the

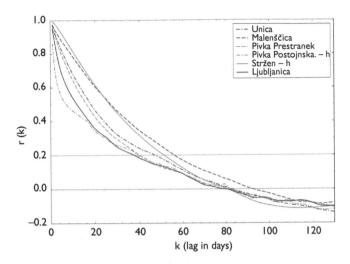

Figure 10.3 Autocorrelation functions of daily discharges for the Unica, Malenščica, Ljubljanica, Stržen (water level) and Pivka (water levels in its middle course and at its ponor).

water levels of the Stržen and the Malenščica in one hydrological year were identical; both autocorrelograms reached the $r_k = 0.2$ value after 32 days.

Only the Pivka spring has two gauging stations that show a good correlation (Spearman's rho = 0.86). The upper and lower courses of the Pivka show slightly different results in the time series analysis. The upper part of the basin is characterised by a shallow karst aquifer with relatively fast recharge at high water levels (floods), but is also subject to fast subsurface drainage towards the Planina Polje springs when the water level is low. Unlike the lower course, the upper course of the Pivka at the Prestranek gauging station usually dries up completely. The autocorrelation of the Pivka (Prestranek) discharges exceeds the characteristic lag after 40 days and has a regulation time of 13 days, showing that the storage capacity of the upper part of the Pivka basin is small to medium. Data for the levels of the Pivka at the Postojna Cave ponor show similar values: the characteristic lag is reached after 39 days and the regulation time is 17 days – which, surprisingly, is four days longer than at the Prestranek gauging station. In its lower course, the Pivka is highly influenced by the surface-flowing Nanoščica, which ought to considerably decrease the memory effect of the Pivka at its ponor (Postojnska jama) and shorten the regulation time. This, however, is not the case, owing to: i) floods that occur outside the Pivka ponor when the amount of incoming water exceeds the swallow capacity of the ponor, which results in a prolongation of the water level time series persistence and ii) the use of a water level hydrological time series in the analysis, which generally show smaller variability (CV = 49%; Table 10.1) than discharges. The influence of the Nanoščica is apparent in a CV for the temperature of the Pivka that is higher than that for any of the other springs in the study (63%). The temperature characteristic lag at the Pivka ponor is 77 days. The highest value of r_{xy} between the temperatures of the Pivka ponor and the Unica is reached at $k = -14$ ($r_{xy} = 0.74$), indicating that the temperatures of the Unica react faster.

The most abundant of the karst springs in the Ljubljanica river basin is the Unica. Its memory effect is 47 days (Figure 10.3), which corresponds to a medium to high storage capacity. Surprisingly, and for an unexplained reason, the regulation time of the spring (39 days) exceeds that of the Malenščica by seven days and indicates a longer impulse response, which is exactly the opposite of the results calculated in the study conducted by Kovačič (2009). The daily temperature autocorrelation function of the spring shows a memory effect of 72 days, meaning that the impulse response of the system to temperature is long. This has also been observed in the study area by means of short-term time series analysis (Kogovšek & Petrič, 2010). Temperature commonly shows a longer memory effect than other hydrological time series (discharge, water level, electrical conductivity, turbidity), which corresponds to the low variability of temperatures (Table 10.1). At the same time, a temperature time series provides less information about the structure of the karst system, since it is not a conservative tracer (Larocque et al., 1998; Kovačič, 2009).

As we have already mentioned, the characteristics of the discharges of the Malenščica (Malni gauging station) are reflected in those of the Unica (Hasberg gauging station); the Spearman's rho between the stations is 0.96. A minor part of the recharge of the Unica spring comes from the Pivka. The value of the same correlation coefficient between water levels at the Pivka ponor and discharges of the Unica is 0.93. The Spearman's rho correlation between the Unica and the Ljubljanica (0.94) also confirms the significant contribution of the Unica to the recharge of the Ljubljanica springs; their mean discharge exceeds the Unica's by just 12%. The Ljubljanica springs have the same regulation time of 39 days as the Unica, showing that the length of the impulse response of the system is medium. The memory effect of the springs is, however, slightly lower (37 days), which still indicates the medium storage capacity of the system, although the latter value is the lowest among all the gauging stations studied in the Ljubljanica river basin. One of the reasons for this could be the relatively high outflow capacity of the Ljubljanica permanent and intermittent karst springs, which can exceed 110 m³/s, meaning that the system could be subject to rapid drainage. The Ljubljanica karst springs (CV = 105%) are also characterised by greater discharge variability in comparison to the Unica (CV = 99%). Another reason could be the influence of the Bela surface stream during intense precipitation. The latter collects water from partly karstified dolomite and flows into the Ljubljanica upstream of the gauging station (Habič, 1976). The temperature memory effect of the Ljubljanica spring (78 days) is similar to that of the Unica. The cross-correlation temperature function between the Unica and Ljubljanica shows a clear asymmetry towards positive k values, reaching the maximum value at $k = +10$ (0.872) and once more indicating the significant influence of the Unica on the Ljubljanica springs.

The autocorrelation function of the Bistrica spring, recharged only by autogenous precipitation from the Snežnik plateau, shows that the memory effect of the spring is medium (45 days), while the regulation time of 25 days indicates the relatively short duration of the influence of the input signals (Figure 10.4). It is uncertain whether the memory effect also reflects the medium storage capacity of the spring or can be attributed at least to some extent to the small total outflow capacity of the springs. Unlike in the Ljubljanica river basin, where some large spring caves are present, there are no known caves in the outflow zone of the Bistrica springs (Cave Registry, 2014).

The Hubelj karst spring drains the deeply karstified Trnovski Gozd plateau. The low memory effect of the spring (12 days) corresponds to the small storage capacity of the

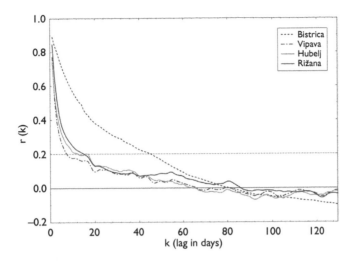

Figure 10.4 Autocorrelation functions of daily discharges for the Bistrica, Vipava, Hubelj and Rižana karst springs.

system and indicates the very rapid flow of infiltrated precipitation water through the system of a hydraulically connected network of fissures and conduits in its vadose and phreatic zones towards the spring. Over a shorter period, Kranjc (1997) calculated the memory effect of the spring to be 10 days. The high value of the spring's discharge CV (133%) and groundwater level fluctuations of several tens of metres show that the out-flow capacity of the spring is not limited and the spring can drain large amounts of water in a short period of time. Surprisingly, the regulation time (25 days) is not completely in accordance with the calculated memory effect of the spring, but again it implies the small storage capacity and short impulse response of the spring. In terms of the temperature autocorrelation function, the spring shows similar values to the Unica and Ljubljanica springs (72 days). It also has the smallest variability (CV = 11%) of the temperature time series in this study. The regulation time of the daily temperatures could not be calculated.

The highest discharge variability among all the spring studied is that of the Rižana karst spring (CV = 163%). The memory effect of the spring is low (15 days) which corresponds to the small storage capacity of the aquifer, while the regulation time (22 days) indicates that the length of the impulse response of the system is short.

Among all the springs studied, the Vipava spring has the lowest memory effect (7 days) which indicates a well-drained karst aquifer with larger and well-connected conduits that prevent significant water storage. For a much shorter period, Jemcov & Petrič (2009) calculated a memory effect of 6 days. Although the Vipava springs are also recharged by sinking streams, the importance of the latter is insignificant when compared to autogenic recharge by precipitation water. The hydraulic conductivity of the system of active fissures and conduits within the vadose zone of the Vipava aquifer must be quite high to immediately transport the pressure pulse of infiltrated precipi-tation water from the Nanos plateau towards the springs. The regulation time of the spring is 7 days, compared to 14 days (Jemcov & Petrič, 2009) and 12 days (Kranjc, 1997). The temperature autocorrelation function decreases slowly and shows values above $r_k = 0.2$ at lags exceeding 999 days.

The memory effects ($r_k = 0.2$) of the various springs show no significant (p < 0.05) correlation (Spearman's rho = 0.32) with regulation times, while they do show a weak positive (though not statistically significant) correlation with mean discharges (Spearman's rho = 0.50). Generally, this means that more abundant karst springs with larger catchments show higher memory effects, although these aquifers cannot as a rule be considered poorly karstified. Interestingly, the values of the springs' coefficient of variation show a relatively high and statistically significant negative correlation with the memory effects of the springs (Spearman's rho = −0.82), meaning that the greater the variability of the springs' discharges, the shorter the memory effects of the springs.

10.4.2 Bivariate time series analysis

Delays between precipitation events and the reaction of the discharges of springs vary between 0 and 7 days and the values of maximum correlation coefficients vary between 0.19 and 0.68 (Table 10.2). The lowest are the cross-correlation functions between the Juršče precipitation station and the Malenščica spring ($r_{xy}(1) = 0.19$) and the water levels of the Stržen stream ($r_{xy}(7) = 0.19$) (Figure 10.5). In comparison to the Vipava, Hubelj and Rižana springs, the precipitation-discharge (water level) cross-correlation functions in the Ljubljanica basin (Unica, Malenščica, Pivka, Ljubljanica, Stržen) and those of the Bistrica spring decrease relatively slowly with gentle slopes (maximum r_{xy} ranging between 0.19 and 0.44), which indicate their higher filtration capacity of rainfall signals. The highest cross-correlation coefficient is that of the Pivka at its ponor (water levels) with the Postojna precipitation station ($r_{xy}(0) = 0.44$), although the shape of the function in distant lags does not differ considerably from the rest of the functions (Figure 10.5). The response of the Pivka at its ponor to precipitation events is immediate, which is partly due to the influence of the surface drainage in the lower Pivka basin, while at the Prestranek gauging station the Pivka shows a delay of two days. Its discharge does not start to increase until the shallow aquifer in its upper course fills up and the river appears on the surface. The mean values of the precipitation-discharge coherence function in the Ljubljanica basin vary between 0.48 and 0.64 (Figure 10.6), which is comparable to the results from the literature (Larocque et al., 1998; Panagopoulos & Lambrakis, 2006) and does not indicate highly karstified aquifers.

According to the precipitation-discharge cross-correlation analysis, the Vipava, Hubelj and Rižana springs show an apparently smaller filtration capacity of precipitation events in their catchments, with higher maximum values of cross-correlation coefficients (0.54 to 0.68) and shorter response times (Table 10.2, Figure 10.7). The reaction to precipitation events is immediate and more intense, which indicates that the systems can be characterised as being well-karstified, something that is already shown by the results of the autocorrelation analysis. All three systems have a more linear response to precipitation (coherence functions vary between 0.63 and 0.72). The Vipava spring shows the greatest cross-correlation and highest coherence function coefficient of all the springs studied.

The mean delay between precipitation and discharges varies between 0.47 days (Pivka, Prestranek gauging station) and 5.98 days (Bistrica); in general the results correspond to the calculated values of the cross-correlation function. Cut-off frequencies

Table 10.2 Summary results of precipitation-discharge and discharge-discharge cross-correlation analysis with maximum values of cross-correlation coefficients (r_{xy}) with corresponding lags in days (k), average coherence function (CO_{xy}), mean delay (MD) and cut-off frequency (COFF) of a cross-spectral analysis in the study area (daily data for the period 1984–2013). The locations of the measuring sites are shown in Figures 10.1 and 10.2.

Spring or river	Juršče (P)	Postojna (P)	Pokojišče (P)	Podgrad (P)	Podkraj (P)	Stržen (h)	Pivka (river) (Q)	Pivka (ponor) (h)	Unica (Q)
Malenščica									
Max r_{xy}(k)	0.194 (1)					0.907 (–1)	0.692 (0)		
CO_{xy}	0.519					0.508	0.417		
MD	5.92					0.72	0.93		
CUFF	16.89					9.24	17.17		
Unica									
Max r_{xy}(k)	0.266 (1)	0.293 (1)				0.806 (–2)	0.906 (0)	0.756 (0)	
CO_{xy}	0.577	0.627				0.529	0.534	0.559	
MD	1.30	0.48				6.60	1.50	5.50	
CUFF	5.25	3.18				5.60	7.95	3.32	
Stržen (h)									
Max r_{xy}(k)	0.191 (7)								
CO_{xy}	0.538								
MD	1.53								
CUFF	3.44								
Pivka (river)									
Max r_{xy}(k)	0.249 (2)								
CO_{xy}	0.482								
MD	0.47								
CUFF	11.46								
Pivka (ponor) (h)									
Max r_{xy}(k)	0.384 (1)	0.436 (0)					0.756 (0)		
CO_{xy}	0.587	0.631					0.569		
MD	1.85	3.64					2.82		
CUFF	2.32	3.16					5.55		

(Continued)

Table 10.2 (Continued).

Spring or river	Juršče (P)	Postojna (P)	Pokojišče (P)	Podgrad (P)	Podkraj (P)	Stržen (h)	Pivka (river) (Q)	Pivka (ponor) (h)	Unica (Q)
Ljubljanica									
Max $r_{xy}(k)$		0.392 (1)	0.355 (1)						0.934 (0)
CO_{xy}		0.631	0.636						0.703
MD_{xy}		2.74	1.32						0.66
CUFF		2.91	3.00						3.27
Bistrica									
Max $r_{xy}(k)$				0.278 (2)					
CO_{xy}				0.552					
MD_{xy}				5.98					
CUFF				17.05					
Vipava									
Max $r_{xy}(k)$					0.679 (0)				
CO_{xy}					0.722				
MD_{xy}					0.49				
CUFF					3.02				
Hubelj									
Max $r_{xy}(k)$					0.634 (0)				
CO_{xy}					0.634				
MD_{xy}					0.61				
CUFF					9.51				
Rižana									
Max $r_{xy}(k)$				0.538 (0)					
CO_{xy}				0.645					
MD_{xy}				1.16					
CUFF				4.76					

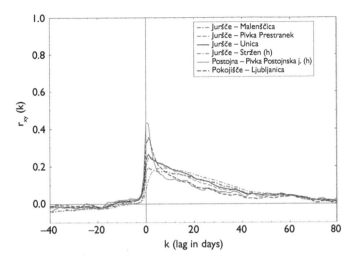

Figure 10.5 Cross-correlation functions of discharges and daily precipitation series as inputs for the Unica, Malenščica, Ljubljanica, Stržen (water level) and Pivka (water levels in its middle course and at its ponor).

range from 2.32 days (Pivka, Postojnska jama) and 17.05 days (Bistrica), which points to evident differences in the abilities of the springs in question to filter precipitation events of varying duration.

Discharge (water level)-discharge (water level) cross-correlation coefficients show higher values then precipitation-discharge cross-correlation values, and range from 0.69 to 0.93 with delays of between −2 and 0 days (Figures 10.8 and 10.9). The negative time lags (Stržen-Malenščica and Stržen-Unica) correspond to the faster response to recharge of the gauging station positioned downstream in comparison to the upstream station. The highest cross-correlation coefficient is calculated for the Unica-Ljubljanica springs (r_{xy} = 0.93). The cross-correlogram shows a slight asymmetry to the negative values, but an instant response to the recharge of both springs. The value of the average coherence function (0.70), the mean delay (0.66 days) and the cut-off frequency (3.27 days) all imply well-developed underground connections and a similar response of the springs to the more or less homogeneous recharge of karst aquifers by precipitation in the catchments. The second highest cross-correlation coefficient is calculated for the Stržen-Malenščica pair of variables (r_{xy} = 0.91). The high correlation value of the almost symmetrical function reflects the fact that both time series have a similar structure. The discharge variability is small, whereas the water levels of the Stržen show a slow increase once the water in the river channel starts to overflow and induces the appearance of Lake Cerknica (Kovačič, 2010b).

The Pivka (Prestranek gauging station) is not directly connected to the Malenščica and Unica springs, but the cross-correlation coefficient shows how the springs react to recharge from the upper part of the Pivka basin. Due to the complexity of its aquifer, the Malenščica spring shows much lower cross-correlation coefficient and coherence function values than the Unica. However, the cross-correlation coefficient between the

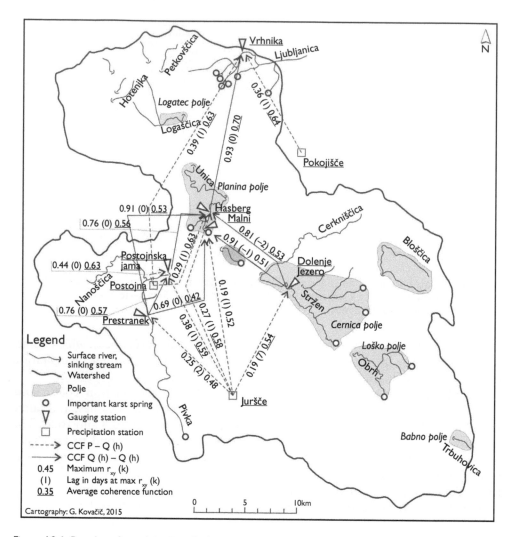

Figure 10.6 Results of precipitation-discharge and discharge-discharge cross-correlation and coherence function of daily time series in the period 1984–2013 in the Ljubljanica river basin.

water levels of the Pivka at its ponor and the discharges of the Unica is not high either (0.76), which means that the input signal of the Pivka ponor is significantly transformed on its way to the Unica spring. The influence of the Nanoščica on the Pivka at its ponor is reflected in the low value of the cross-correlation coefficient (0.76) between the Prestranek and Postojnska jama gauging stations.

In comparison to the calculated precipitation-spring mean delays are much shorter (ranging from 0.66 to 6.6 days) between sinking streams and springs, while the values of the cut-off frequencies are very similar (Table 10.2).

Calculations of Spearman's rho show that a statistically significant ($p < 0.05$) negative correlation exists between the calculated memory effects of the springs (rivers)

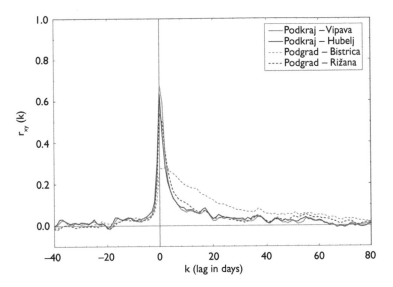

Figure 10.7 Cross-correlation functions of discharges and daily precipitation series as inputs for the Bistrica, Vipava, Hubelj and Rižana karst springs.

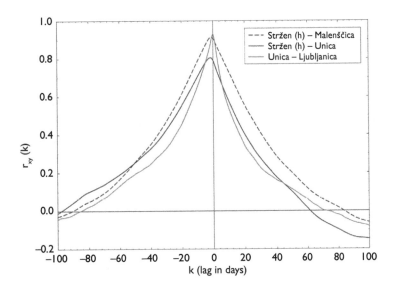

Figure 10.8 Cross-correlation functions of discharges (water levels) as inputs and discharges as outputs for the Unica, Malenščica and Ljubljanica karst springs.

and the values of precipitation-discharge (water level) average coherence functions (–0.65), meaning that the higher the coherence function the lower the memory effect of the spring. Furthermore, a statistically negative correlation also exists between the average coherence function and the cut-off frequency (–0.64). Both values are consistent with expectations. The cut-off frequencies also show a significant positive

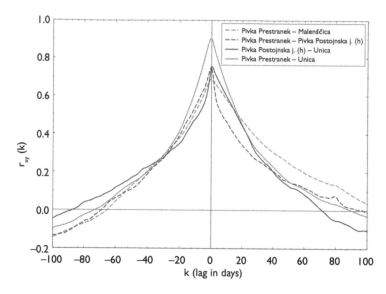

Figure 10.9 Cross-correlation functions of discharges (water levels) as inputs and discharges (water levels) as outputs for the Unica and Malenščica karst springs and for the Pivka river at its ponor.

correlation with the mean discharges (0.66), meaning that more abundant karst springs filter longer-lasting precipitation events.

10.4.3 Significance of the results and outlook

The results of the time series analysis for daily data (period 1984–2013) of nine selected karst springs show that springs and their corresponding aquifers in Slovenia can be characterised as poorly karstified (high memory effect), medium karstified or well-karstified systems (low memory effect). A generalised comparison is shown in Figure 10.10.

For the Vipava, Hubelj and Rižana springs, it can be concluded that their systems are characterised by rapid drainage of the infiltrated precipitation water through a hydraulically well-connected network of conduits and fissures, which corresponds to the calculated low to very low storage capacity. On the other hand, the karst springs of the Ljubljanica river basin (Malenščica, Stržen, Unica, Pivka and Ljubljanica) and the Bistrica karst spring show higher memory effects, which indicates the larger storage capacity of these aquifers.

The results of the auto- and cross-correlation analysis in the present study show that well-developed karst aquifers of complex structure (autogenic plus allogenic recharge) can also have a relatively large storage capacity, although water flows through the system of unobstructed conduits (caves) at high velocities (Gabrovšek *et al.*, 2010) of up to several hundreds of m/h. One would expect these systems to have lower memory effects, as is the case of the Malenščica and Unica springs. These aquifers cannot, however, be considered systems with low hydraulic conductivity in absolute terms (Mangin, 1984). According to these findings, not all karst systems

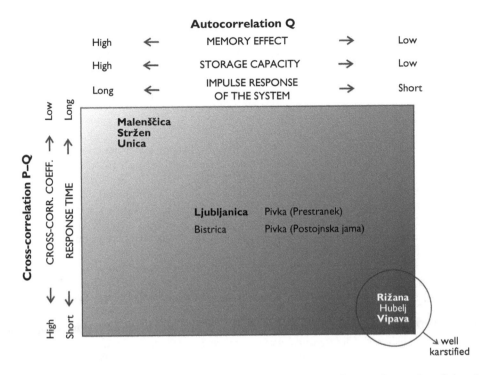

Figure 10.10 Classification of the studied springs into groups according to the results of the time series analysis. Binary karst systems are marked in bold.

characterised by a higher memory effect should be considered to be poorly karstified (i.e. the Malenščica, Unica, Stržen (Figure 10.10). The higher memory effect of these springs can also be attributed to the limited discharge capacity of their aquifers, which is reflected in the low values of coefficients of variation (e.g. the Malenščica spring). The last parameter shows an inverse correlation with the calculated memory effects of the studied springs, meaning that the greater the variability of the springs' discharges, the shorter the memory effects of the springs (see the Vipava and the Malenščica as examples). For the same reason, in cases where only water level data sets are available, as opposed to discharges (meaning a lower variability of the variable), caution is needed, since the results of the autocorrelation will as a rule show higher memory effects.

An important conclusion of the study is that there is no link between the type of karst aquifer in terms of its recharge characteristics (binary or unary system) and the calculated memory effects. For example, the Bistrica and Hubelj springs (both unary karst systems) show completely different values for the calculated memory effects and the corresponding storage capacities of their aquifers.

In view of the above, caution needs to be used when classifying karst systems into groups on the basis of the results of a time series analysis. Furthermore, in order to obtain a more general picture of the hydrogeological structure and functioning of a karst system in absolute terms, it is important to consider longer hydrological data sets (more successive hydrological years), since they include more information and reflect more average conditions. Finally, the results of the time series analysis should be interpreted together with the results of other methods (e.g. speleological investigations, tracer tests, modelling, etc.) used in karst hydrology in order to avoid bias in conclusions.

REFERENCES

Amraoui F., Razack M., Bouchaou L. (2003) Turbidity dynamics in karstic systems. Example of Ribaa and Bittit springs in the Middle Atlas (Morocco). *Hydrological Sciences Journal-Journal Des Sciences Hydrologiques* 48(6), 971–984.

Angelini P. (1997) Correlation and spectral analysis of two hydrogeological systems in Central Italy. *Hydrological Science - Journal-des Sciences Hydrologiques* 42(3), 425–438.

Bailly-Comte, V., Jourde, H., Roesch, A., Pistre, S., Batiot-Guilhe, C. (2008) Time series analyses for karst/river interactions assessment: case of the Coulazou River (southern France). *Journal of Hydrology* 349(1–2), 98–114.

Benavente J., Pulido-Bosch A., Mangin A. (1985) Application of correlation and spectral procedures to the study of the discharge in a karstic system (Eastern Spain). In: Gunay, G., Johnson A.I. (editors) *Karst water resources – Proceedings of the Ankara-Antalya Symposium*, July 1985, 67–75.

Behrens H., Zupan M., Zupan M. (1976) Tracing with fluorescent tracers. In: Gospodarič R., Habič P. (editors) *Underground water tracing: Investigations in Slovenia 1972–1975.* Third International Symposium of Underground Water Tracing (3. SUWT) Ljubljana-Bled, 139–164.

Box G.E.P., Jenkins G.M. (1970) *Time series analysis: Forcasting and control.* Holden Day, San Francisco.

Box G.E.P., Jenkins G.M., Reinsel C. (1994) *Time series analysis: forecasting and control, 3rd ed.* Prentice Hall, New Jersey.

Cave Registry (2014) Speleological Association of Slovenia and Karst Research Institute ZRC SAZU.

Covington, M.D., Wicks, C.M., Saar, M.O. (2009) A dimensionless number describing the effects of recharge and geometry on discharge from simple karstic aquifers. *Water Resources* 45(11), W11410.

Delbart C., Valdes D., Barbecot F., Tognelli F., Richon P., Cauchoux L. (2014) Temporal variability of karst aquifer response time established by sliding-windows cross-correlation method. *Journal of Hydrology* 511, 580–588.

Eisenlohr L., Bouzelboudjen M., Kiraly L., Rossier Y. (1997) Numerical versus statistical modelling of natural response of a karst hydrogeological system. *Journal of Hydrology* 202, 244–262.

Environmental agency of the Republic of Slovenia (EARS), Digital data on water (2009) URL: http://gis.arso.gov.si; last visited 1/7/2009.

Environmental agency of the Republic of Slovenia (EARS) (2014a) Daily discharges and water levels at the Malni (the Malenščica), Hasberg (the Unica), Postojna Cave (the Pivka ponor), Prestranek (the Pivka), Dolenje Jezero (the Stržen), Vipava I (the Vipava), Ajdovščina I (the Hubelj), Kubed II (the Rižana), Ilirska Bistrica (the Bistrica) and Vrhnika (the Ljubljanica) gauging stations for the period 1984–2013. The archive of the Environmental agency of the Republic of Slovenia, Ljubljana.

Environmental agency of the Republic of Slovenia (EARS) (2014b) Daily precipitation amounts at the Podkraj, Podgrad, Juršče, Postojna and Pokojišče rain gauge stations for the period 1984–2013. The archive of the Environmental agency of the Republic of Slovenia, Ljubljana.

Fiorillo F., Doglioni A. (2010) The relation between karst spring discharge and rainfall by cross-correlation analysis (Campania, southern Italy). *Hydrogeology Journal* 18, 1881–1895.

Gabrovšek F., Kogovšek J., Kovačič G., Petrič M., Ravbar N., Turk J. (2010) Recent results of tracer tests in the catchment of the Unica River (SW Slovenia). *Acta Carsologica* 39(1), 27–37.

Habič P. (1976) Geomorphologic and hydrographic characteristics. Investigations in Ljubljanica river basin, Description of investigated area. In: Gospodarič R., Habič P. (editors) *Underground water tracing: Investigations in Slovenia 1972–1975.* Third International Symposium of Underground Water Tracing (3. SUWT) Ljubljana-Bled, 12–27.

Habič P. (1989) Kraška bifurkacija na jadransko črnomorskem razvodju. *Acta carsologica 18*, 235–264.

Herman E.K., Toran L., White W.B. (2009) Quantifying the place of karst aquifers in the groundwater to surface water continuum: a time series analysis study of storm behaviour in Pennsylvania water resources. *Journal of Hydrology* 376, 307–317.

Jemcov I., Petrič M. (2009) Measured precipitation vs. effective infiltration and their influence on the assessment of karst systems based on results of the time series analysis. *Journal of Hydrology* 379(3–4), 304–314.

Jemcov I., Petrič M. (2010) Time series analysis, modelling and assessment of optimal exploitation of the Nemanja karst springs, Serbia. *Acta Carsologica* 39(2), 187–200.

Janež J., čar J., Habič P., Podobnik R. (1997) *Vodno bogastvo Visokega krasa.* Geologija d.o.o., Idrija.

Jeannin P. Y., Sauter M. (1998) Analysis of karst hydrodynamic behaviour using global approach: a review. In: Jeannin, P. Y., Sauter M. (editors) *Modelling in karst systems.* Bulletin d'Hydrogeologie 16, 31–48.

Jukić D., Denić-Jukić V. (2004) A frequency domain approach to groundwater recharge estimation in karst. *Journal of Hydrology* 289, 95–110.

Katsanou K., Lambrakis N., Tayfur G., Baba A. (2015) Describing the Karst Evolution by the Exploitation of Hydrologic Time-Series Data. *Water Resources Management* 29, 3131–3147.

Kogovšek J. (1999) Nova spoznanja o podzemnem pretakanju vode v severnem delu Javornikov (Visoki kras). *Acta Carsologica* 28(1), 161–200.

Kogovšek J., Knez M., Mihevc A., Petrič M., Slabe T., Šebela S. (1999) Military training area in Kras (Slovenia). *Environmental Geology* 38(1), 69–76.

Kogovšek J., Petrič M. (2010) Water temperature as a natural tracer – a case study of the Malenščica karst spring (SW Slovenia). *Geologia Croatica* 63(2), 171–177.

Kovačič G. (2003): Kraški izviri Bistrice (JZ Slovenija). *Annales: Series historia naturalis* 13(1), 111–120.

Kovačič G. (2009) Hydrology of the Malenščica karst spring and its catchment. PhD thesis, University of Primorska, Koper.

Kovačič G. (2010a) An attempt towards an assessment of the Cerknica Polje water balance. *Acta Carsologica* 39(1), 39–50.

Kovačič G. (2010b) Hydrogeological study of the Malenščica karst spring (SW Slovenia) by means of a time series analysis. *Acta Carsologica* 39(2), 201–215.

Kranjc A. (1986) Cerkniško jezero in njegove poplave. *Geografski zbornik* 25, 73–123.

Kranjc A. (editor) (1997) Karst hydrogeological investigations in south-western Slovenia. *Acta Carsologica 26*(1), 1–388.

Krivic P., Bricelj M., Trišič N., Zupan M. (1987) Sledenje podzemnih vod v zaledju Rižane. *Acta Carsologica* 16, 83–104.

Krivic P., Bricelj M., Zupan M. (1989) Podzemne vodne zveze na področju Čičarije in osrednjega dela Istre (Slovenija, Hrvatska, NW Jugoslavija). *Acta Carsologica* 18, 265–295.

Labat D., Ababou R., Mangin A. (2000) Rainfall-runoff relations for karstic springs. Part I: convolution and spectral analyses. *Journal of Hydrology* 238, 123–148.

Larocque M., Mangin A., Razack M., Banton O. (1998) Contribution of correlation and spectral analyses to the regional study of a large karst aquifer (Charente, France). *Journal of Hydrology* 205, 217–231.

Mangin A. (1984) Pour une meilleure connaissance des systèmes hydrologiques à partir des analyses corrélatoire et spectrale. *Journal of Hydrology* 67, 25–43.

Massei N., Dupont J.P., Mahler B.J., Laignel B., Fournier M., Valdes D., Ogier S. (2006) Investigating transport properties and turbidity dynamics of a karst aquifer using correlation, spectral, and wavelet analyses. *Journal of Hydrology* 329, 244–257.

Mathevet T., Lepiller M., Mangin A. (2004) Application of time-series analyses to the hydro-logical functioning of an Alpine karstic system: the case of Bange-L'Eau-Morte. *Hydrology and Earth System Sciences* 8, 1051–1064.

Mayaud C., Wagner t., Benischke R., Birk S. (2014) Single event time series analysis in a binary karst catchment evaluated using a groundwater model (Lurbach system, Austria). *Journal of Hydrology* 511, 628–639.

Novel, J.P., Dimadi A., Zervopoulou A., Bakalowicz M. (2007) The Aggitis karst system, Eastern Macedonia, Greece: Hydrologic functioning and development of the karst structure. *Journal of Hydrology* 334, 477–492.

Padilla A., Pulido-Bosch A. (1995) Study of hydrographs of karstic aquifers by means of cor-relation and cross-spectral analysis (France, Spain). *Journal of Hydrology* 168, 73–89.

Panagopoulos G., Lambrakis N. (2006) The contribution of time series analysis to the study of the hydrodynamic characteristics of the karst systems: Application on two typical karst aquifers of Greece (Trifilia, Almyros Crete). *Journal of Hydrology* 329, 368–376.

Rahnemaei M., Zare M., Nematollahi A.R., Sedhi H. (2005) Application of spectral analysis of daily water level and spring discharge hydrographs data for comparing physical characteris-tics of karst aquifers. *Journal of Hydrology* 311, 106–116.

Samani N. (2001) Response of karst aquifers to rainfall and evaporation, Maharlu basin, Iran. *Journal of Cave and Karst Studies* 63(1), 33–40.

Terzić J., Stroj A., Frangen T. (2012) Hydrogeological investigation of karst system properties by common use of diverse methods: a case study of Lička Jesenica springs in Dinaric karst of Croatia. *Hydrological Processes* 26(21), 3302–3311.

Turk J., Malard A., Jeannin P-Y., Vouillamoz J., Masini J., Petrič M., Gabrovšek F., Ravbar N., Slabe T. (2013) Interpretation of hydrogeological functioning of a high karst plateau using the KARSYS approach: the case of Trnovsko-Banjška planota (Slovenia). *Acta Carsologica* 42(1), 61–74.

Valdes, D., Dupont, J.P., Massei, N., Laignel, B., Rodet, J. (2005) Analysis of karst hydrody-namics through comparison of dissolved and suspended solids' transport. *C. R. Geoscience* 337, 1365–1374.

Valdes, D., Dupont, J.P., Massei, N., Laignel, B., Rodet, J. (2006) Investigation of karst hydro-dynamics and organization using autocorrelation and T-curves. *Journal of Hydrology* 329, 432–443.

Žibrik K., Lewicki F., Pičinin A. (1976): Hydrologic investigations. Investigations in Ljubljanica river basin, Investigations before the tracing test 1972–1975. In: Gospodarič R., Habič P. (editors) *Underground water tracing: Investigations in Slovenia 1972–1975*. Third International Symposium of Underground Water Tracing (3. SUWT) Ljubljana-Bled, 1–56.

Chapter 11

Characterisation of the influence of evaporite rocks on the hydrochemistry of carbonate aquifers: The Grazalema Mountain Range (Southern Spain)

Damián Sánchez & Bartolomé Andreo
Department of Geology and Centre of Hydrogeology of the University of Málaga (CEHIUMA). Faculty of Science, Málaga (Spain)

ABSTRACT

Alpine and Mediterranean karst aquifers are frequently associated with Upper Triassic (Keuper) deposits which generally constitute the substratum or impervious base of the system given their predominantly clayey nature. When the groundwater stored in the carbonate aquifer comes into direct contact with the Triassic formation, the result is a modification of the chemical composition of the water due to the dissolution of high solubility evaporite rocks interspersed with the Triassic clays. A hydrochemical and temporal characterisation of the effect of evaporites on the chemical composition of two carbonate aquifers located in Sierra de Grazalema Mountain Range (Southern Spain) has been undertaken. Results show differences in the intensity of the evaporite signal as well as in the mineral species responsible for the hydrochemical modification: from sulphate-related minerals (gypsum/anhydrite) to rock salt.

11.1 INTRODUCTION

Numerous carbonate aquifers located in Alpine regions are made up of Jurassic limestones and dolostones the base of which comprises rocks of Upper Triassic age. These rocks largely consist of fine terrigenous deposits of continental or shallow-marine origin (Martín-Algarra & Vera, 2004) which do not allow water to circulate through them.

The origin of these Triassic deposits is related to the first stages of the Alpine cycle, when large sedimentary basins were created after the rifting of the Pangea during the Upper Permian-Lower Triassic. The basins were filled with thick clayey-sandstone and evaporite sequences locally accompanied by basic volcanism (Martín-Algarra & Vera, 2004). This sedimentation took place over large areas located to the south and south east of the Iberian Massif as well as over other areas of central Europe. Subsequent tectonic movements during the Alpine orogeny led to the current chaotic mixture of clays, sandstones, carbonate and subvolcanic rocks, and evaporites. The extensive areas occupied by these deposits became shallow carbonate platforms during the Lower Jurassic.

The Triassic rocks have traditionally been considered as impervious units given their primary clayey nature, constituting both the lower and lateral hydrogeological boundaries of many alpine carbonate aquifers. Nevertheless, the high solubility of the evaporites contained in this unit leads to a process of 'cross formational flow', or karstification processes that resulted in an increase in permeability, at least on a local basis. The presence of karst features such as dolines, sinkholes, groundwater-dependent wetlands, pumping wells and saline springs reflect both the karstification processes and groundwater flow (Carrasco et al., 2005; Sánchez et al., 2009).

The presence of evaporite rocks in carbonate sedimentary sequences is a characteristic inherent to the alpine aquifers in southern Spain (López-Chicano et al., 2001; Moral et al., 2008; Mudarra & Andreo, 2011). Alpine aquifers are generally found in Jurassic limestones and dolostones overlaying Upper Triassic deposits. Similar geological configurations can be found in other European regions such as the Northern Appennines in Italy (Capaccioni et al., 2001), south western Switzerland (Kilchmann et al., 2004) and Southern France (Aquilina et al., 2002).

The most common mineral species in the Triassic evaporite facies are sulphates and halides: gypsum ($CaSO_4 \cdot 2H_2O$), its dehydrated equivalent anhydrite ($CaSO_4$) and halite or rock salt ($NaCl$). They all originate from precipitation in over-saturated water under arid climatic conditions (Goldscheider & Andreo, 2007).

The effect of evaporites on the chemical composition of groundwater stored in carbonate aquifers is mainly reflected in the concentrations of the ions that constitute the mineral composition of evaporite rocks: SO_4^{2-}, Ca^{2+}, Na^+ and Cl^-. This effect can range from slight increases of some or all of these components to high contents which determine the chemical facies of the water (Plummer et al., 1990; Capaccioni et al., 2001; Ma et al., 2011). Since Ca^{2+} can also come from the dissolution of carbonate minerals including calcite, aragonite and dolomite, SO_4^{2-}, Na^+ and Cl^- are the best elements to assess the contribution of evaporite rocks to the final composition of groundwater draining from carbonate aquifers. Water that was in contact with evaporites has sometimes also a higher concentration of trace elements (Hunkeler & Mudry, 2007).

This chapter aims to characterise the effect of the evaporite rocks on the chemical composition of the water drained by karst springs as well as their temporal evolution and their relationship with the hydrodynamic response. The study area is the westernmost part of the Sierra de Grazalema Mountain karst system in Southern Spain.

11.2 THE GRAZALEMA MOUNTAIN RANGE

The area of interest corresponds to the El Bosque and La Silla carbonate aquifers, which are located in the western sector of the Grazalema Mountain Range in the province of Cádiz, Southern Spain (Figure 11.1).

The relief is rugged with altitudes ranging from less than 300 m asl in the westernmost part of the study area to almost 1000 m asl in the central sector of the El Bosque aquifer. The latter is separated from La Silla aquifer by a clayey valley (Figure 11.1). The prevailing climate is semi-continental Mediterranean. In general, the annual distribution of precipitation presents a marked seasonal pattern. The first rainfall normally takes place in the autumn, at the beginning of the hydrological year. This is often in the form of intense storms from the Atlantic Ocean, featuring heavy precipitation. Winter rainfall and, to a lesser extent, spring rains are commonly associated

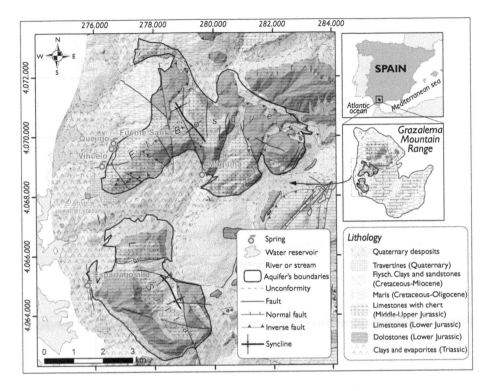

Figure 11.1 Location and geological map of the study area.

with humid winds, which account for most of the annual precipitation. During the rest of the year, rainfall is scarce to non-existent, particularly in summer. The historic mean annual precipitation for the period 1965–2006 ranges from 900 to 1000 mm/year, with a positive gradient towards the east where values exceeding 1000 mm/year are recorded. The mean annual temperature in the region is about 16°C (Gálvez-Maestre, 2005) and both precipitation and temperature vary with the altitude.

Geologically, the two aquifers consist of rocks belonging to the External Zone of the Betic Cordillera. The stratigraphic sequence begins with clays and evaporites from the Upper Triassic age, continues with a thick Jurassic dolostones and limestones series and finishes with Middle-Upper Jurassic limestones with chert and Lower Cretaceous–Oligocene marly limestones and marls at the top (Peyre, 1974; Martín-Algarra, 1987) (Figure 11.1). Flysch-type clays and sandstones outcrop in the central and eastern sector of the area. The structure of both aquifers consists of north west to south east (El Bosque) and north north west to south south east (La Silla) synclines the cores of which are occupied by Jurassic-Cretaceous marly-limestones. The entire folded structure has been affected by more recent fractures, predominantly with north east to south west and north west to south east directions.

El Bosque and La Silla aquifers are made up of fractured and karstified Jurassic dolostones and limestones. Adjacent marly-limestones behave as an aquitard. Impervious lateral boundaries are the Triassic clays (mainly the western and southern limits), flysch clays to the east and Cretaceous marls (only in El Bosque aquifer). The base of La Silla aquifer and probably most of El Bosque aquifer consists of Triassic clays with evaporites.

Recharge to the aquifers takes place exclusively by direct infiltration of rainwater into dolostones, limestones and, to a lesser extent, limestones with chert, while discharge is produced through springs located at the edges of the mountains, coinciding with the contact between the permeable carbonate rocks and the aquitards, as well as by pumping intended for drinking water supply (only in El Bosque aquifer). Table 11.1 summarises the main characteristics of the discharge points that were analysed. Springs are located at altitudes ranging from 290 to 345 m asl. El Bosque aquifer has a total area of 18 km² from which 16 km² correspond to permeable outcrops, while La Silla has an area of 11.5 km², from which 10 km² are permeable.

Karst surface landforms are not well developed in the area. Karrenfield is rare and the presence of exokarstic forms such as dolines is scarce. Travertine formations are recognised in the borders of both aquifers. On the south west edge of El Bosque unit there is an inactive travertine outcrop indicating the existence of a former drainage point at this place. The western limit of La Silla aquifer has clusters of small travertine outcrops, one of them is associated to the discharge of Esparragosilla spring and is still active.

Vegetation is mostly Mediterranean shrub. No known sources of pollution exist on the recharge areas apart from some extensive livestock farming.

11.3   METHODOLOGY

Field measurements and spring sampling were undertaken between November 2012 to November 2013. Records were kept of the discharge rate and the chemical composition of the water drained by selected springs. At the same time as samples were taken, water temperature, pH and electrical conductivity were measured in situ, using portable equipment (WTW Cond 3310, HACH HQ40d), with an accuracy of ±0.1°C, 0.01 and ±1 µS/cm respectively. Spring discharge was also measured using a propeller flowmeter model OTT C31 mostly during low and medium flow conditions, and a salt-dilution flowmeter model SalinoMADD was used in high flow conditions. On average, flow measurements and water sampling were performed every 2–3 weeks, although during flood events monitoring was conducted more frequently; sampling was reduced to fortnightly and monthly during low flow periods.

Table 11.1 General description of the selected springs.

Spring	Aquifer	Discharge elevation (m asl)	Annual discharge (hm³)[a]
Fuente Santa	El Bosque	310	0,5
Quejigo	El Bosque	312	1,0
Vihuelo	El Bosque	296	1,4
Máquina	El Bosque	345	0,8
Esparragosilla	La Silla	290	3,0

[a]: for the period Nov'12–Nov'13

Chemical analyses were carried out at the Centre of Hydrogeology of the University of Málaga. Alkalinity (Alk) was measured by volumetric titration using 0.02 N H_2SO_4 to pH 4.45. The chemical analyses of the major components (Ca^{2+}, Mg^{2+}, Na^+, Cl^-, SO_4^{2-}, NO_3^-) were carried out using high pressure liquid chromatography (Metrohm 791 Basic IC model) with ±0.1 mg/l of accuracy. Samples were filtered before being introduced in the system (filter in line as well as a precolumn-filter).

11.4   RESULTS

11.4.1   Hydrochemical characterisation

Statistical parameters for the chemical composition of the groundwater drained by the monitored springs are shown in Table 11.2. The mean values of electrical conductivity (EC) cover a wide range. The water samples taken at Vihuelo spring are the least mineralised, with a mean value for the monitored period of 551 µS/cm. Esparragosilla drains the most mineralised water with EC values over 2000 µS/cm. Coefficients of variation of EC values are relatively low in all cases (2.3–7.5%). Mean water temperatures are all over 17°C, with a maximum of 17.5°C in Esparragosilla. The spring water temperatures are slightly higher than the average air temperature for the area (16°C; Gálvez-Maestre, 2005). This average temperature is from El Bosque weather station, which is located at a lower altitude (285 m asl) with respect to the recharge areas of both aquifers (230 to 970 m asl), where mean annual temperatures should be slightly colder.

The water drained by the springs can be classified into three groups according to their chemical facies: calcium sulphate (Esparragosilla), calcium bicarbonate

Table 11.2 Main statistical descriptors of a selection of physico-chemical parameters of all water samples analysed (period November 2012–November 2013). (n) number of samples, (cv) coefficient of variation, EC (electrical conductivity), T (temperature), Alk (alkalinity).

Spring		EC (µS/cm)	T(°C)	Alk (mg/l)	SO_4^{2-} (mg/l)	Ca^{2+} (mg/l)	Mg^{2+} (mg/l)	Cl^- (mg/l)	Na^+ (mg/l)
Fuente Santa	n	11	11	11	11	11	11	11	11
	Mean	853	17.1	318	72	102	24	121	77
	cv (%)	5.1%	1.5%	6.6%	5.5%	8.2%	1.3%	12.1%	9.8%
Quejigo	n	24	24	24	24	24	24	24	24
	Mean	631	17.1	325	62	96	25	43	30
	cv (%)	7.5%	1.1%	3.6%	8.4%	6.5%	2.5%	34.3%	31.9%
Vihuelo	n	24	24	24	24	24	24	24	24
	Mean	551	17.4	320	57	94	26	17	11
	cv (%)	4.2%	1.1%	3.8%	5.4%	9.6%	1.9%	7.7%	4.2%
Máquina	n	22	22	22	22	22	22	22	22
	Mean	564	17.4	310	89	95	32	11	8
	cv (%)	2.3%	2.6%	1.9%	5.2%	4.2%	1.5%	7.4%	2.7%
Esparragosilla	n	21	21	21	21	21	21	21	21
	Mean	2,243	17.5	278	831	324	68	322	193
	cv (%)	5.2%	1.2%	12.4%	12.4%	8.2%	9.1%	6.7%	5.9%

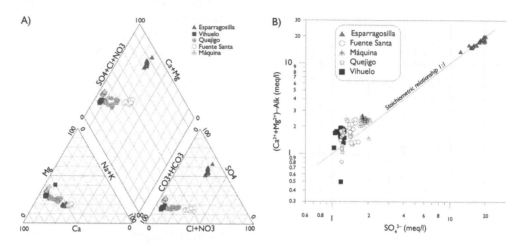

Figure 11.2 (A) Piper diagram. (B) Binary diagram showing the relationship between SO$_4$$^{2-}$ and non-alkalinity Ca^{2+}+Mg^{2+}.

(Vihuelo, Quejigo and Máquina) and calcium-sodium bicarbonate-chloride facies (Fuente Santa) (Figure 11.2A).

In Figure 11.2B the SO$_4$$^{2-}$ concentration has been plotted versus the non-alkalinity Ca^{2+}+Mg^{2+} content, the latter being calculated as the sum of the concentrations of these two components minus the alkalinity, in meq/l. The stoichiometric line 1:1 is shown in the graph. Water samples from Esparragosilla spring show the highest contents of SO$_4$$^{2-}$ and non-alkalinity Ca^{2+}+Mg^{2+}, with values exceeding 10 meq/l for both parameters, whereas the rest of springs display much lower values ranging from 0.5 to 2.5 meq/l for non-alkalinity Ca^{2+}+Mg^{2+} and from 1 to 2 meq/l for SO$_4$$^{2-}$. Water samples taken at Esparragosilla spring plot on the stoichiometric line 1:1, indicating that the Ca^{2+} and Mg^{2+} are not related to the dissolution of carbonate species (calcite, dolomite). The increases/decreases in SO$_4$$^{2-}$ concentrations registered during the monitoring period are accompanied by similar stoichiometric variations of non-alkalinity Ca^{2+}+Mg^{2+}, suggesting a similar origin for the three components which should be related to the dissolution of evaporite rocks. The calcium sulphate facies of the water drained by this spring (Figure 11.2A) suggests that the chemical composition is mainly controlled by the dissolution of sulphate mineral species (gypsum and anhydrite). Nevertheless, some of the water samples show slight increases in Na+K concentration, which might be indicative of halite dissolution also contributing to the final chemical composition of the groundwater drained by Esparragosilla spring.

Water samples from the rest of springs show a different pattern. They do not plot on the 1:1 stoichiometric line. Most of them are above this line, suggesting an excess of dissolved Ca^{2+} and Mg^{2+} in the water drained by these springs.

In Figure 11.3A the SO$_4$$^{2-}$/Cl$^-$ ratio (mg/l) has been plotted against Na$^+$ concentration for each water sample analysed, whilst the Cl$^-$ concentration is represented as the diameter of each circle. The line representing the SO$_4$$^{2-}$/Cl$^-$ ratio=1 has been highlighted since it separates samples with higher SO$_4$$^{2-}$ (ratio>1) or Cl$^-$ (ratio<1) concentrations with respect to the other ion. The value of this ratio can be related to the

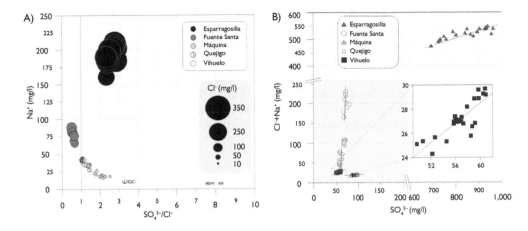

Figure 11.3 (A) Diagram showing the SO_4^{2-}/Cl^- ratio versus Na^+ concentration and the Cl^- content (circles size). (B) Binary diagram showing SO_4^{2-} against (Cl^-+Na^+) concentrations.

relative contribution of gypsum/anhydrite dissolution ($CaSO_4$) with respect to halite dissolution (NaCl) in each water sample.

Water samples collected at Máquina spring show the largest SO_4^{2-}/Cl^- ratio values, ranging from 7.5 to 8.3 (Figure 11.3A). Máquina is, after Esparragosilla, the sampling point with the highest dissolved SO_4^{2-} contents, with a mean value of 89 mg/l (Table 11.2). During the 12-month monitoring period the water drained by this spring displayed small variations in both Cl^- and Na^+ contents (circles of the same diameter and no vertical deviations).

The greatest Na^+, Cl^- and SO_4^{2-} concentrations were found at Esparragosilla spring (Table 11.2). All samples show SO_4^{2-}/Cl^- ratios higher than 2.3, indicating that SO_4^{2-} concentrations are at least more than double those of Cl^-. The only water samples with Cl^- being more abundant than SO_4^{2-} are those drained by Fuente Santa spring (SO_4^{2-}/Cl^- <1, Figure 11.3A).

Quejigo spring is characterised by lower concentrations of Cl^- and Na^+ and greater SO_4^{2-}/Cl^- ratios compared to those coming from Fuente Santa spring. This suggests a more important contribution of gypsum/anhydrite dissolution to the hydrochemistry of Quejigo waters as opposed to halite dissolution. Finally, the water drained by Vihuelo spring has low Cl^- and Na^+ concentrations and intermediate SO_4^{2-}/Cl^- ratios.

Figure 11.3B shows the relationship between SO_4^{2-} and (Cl^-+Na^+) concentrations. The SO_4^{2-} concentrations range from 51 mg/l in Vihuelo to 963 mg/l in Esparragosilla, and those of (Cl^-+Na^+) from 18 mg/l in Máquina to 549 mg/l in Esparragosilla.

Water samples from Fuente Santa and Quejigo springs plot along an almost vertical pattern. This indicates that during the 12 month period of monitoring the water drained by these outlets has undergone large variations with respect to Na^+ and Cl^+ and small changes with respect to SO_4^{2-}. The difference between the maximum and minimum values for (Cl^-+Na^+) concentrations is 65 mg/l for Fuente Santa and 67 mg/l for Quejigo, while SO_4^{2-} variations for both springs are 12.5 and 15.1 mg/l respectively.

Points representing water samples from Máquina spring show a nearly horizontal trajectory. This pattern is the result of larger SO_4^{2-} variations throughout the monitoring period than those of Cl^- and Na^+, which remained quite stable (Figure 11.3B).

Water samples collected at Esparragosilla and Vihuelo springs have inclined trajectories. This reflects increases in dissolved SO_4^{2-} in these springs accompanied by increases in Cl^- and Na^+ concentrations, and vice versa. However, the ranges of variation in both springs are not the same. The difference between the maximum and minimum values measured in Esparragosilla has been 377 mg/l (SO_4^{2-}) and 138 mg/l (Cl^-+Na^+), in contrast to 11.2 mg/l for SO_4^{2-} and 5.4 mg/l for (Cl^-+Na^+) in Vihuelo.

11.4.2 Temporal evolutions

The temporal characterisation of the evaporite signal on selected springs has been carried out by means of temporal evolutions of discharge, water temperature and chemical components of the water. The springs are those in which the signal of evaporites on their chemical compositions is the most intense: Esparragosilla, Fuente Santa and Quejigo (Figure 11.4).

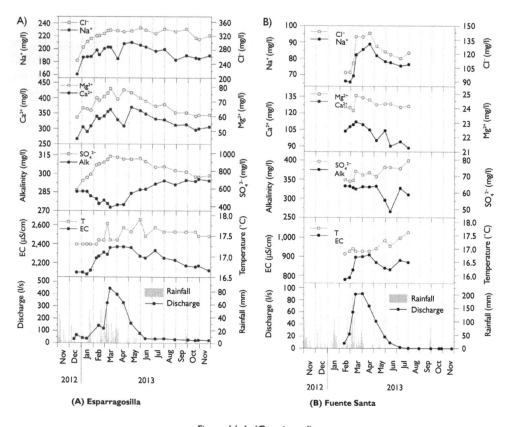

Figure 11.4 (Continued).

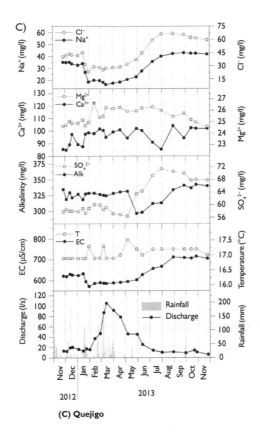

(C) Quejigo

Figure 11.4 Temporal evolutions of discharge, water temperature and the principal chemical components of the water drained by Esparragosilla (A), Fuente Santa (B) and Quejigo (C) springs.

The hydrograph of Esparragosilla spring shows a relatively gradual increase of flow in response to rainfall (Figure 11.4A). No intermediate peaks are identified during the flood period, even after intense precipitation events, and the discharge peak is followed by a gentle recession limb. The effect of recharge on Esparragosilla spring is a progressive increase of water mineralisation. The highest EC values occur during the peak of flow discharge, whereas in low water conditions the EC decreases. Water mineralisation is mainly controlled by SO_4^{2-}, Ca^{2+}, Mg^{2+}, Cl^- and Na^+ contents, which are representative of the saturated zone of the aquifer and the evaporite substratum. The temporal evolution of alkalinity is opposite to that of EC and evaporite-related ions, showing minimum values during flooding and a rising trend in low water conditions.

Fuente Santa is a seasonal spring in which water usually starts flowing several months after the beginning of the recharge season and dries up in summer. During the 12 months of monitoring, Fuente Santa was active from the middle of February until August 2013 (Figure 11.4B). The hydrograph shows a gradual increase in discharge until April, when the maximum value is achieved, and a subsequent flow rate decrease, which ends in August when the spring dried up. The water drained by Fuente Santa was more mineralised during flooding, when the discharge rate was maximum. Hence, the highest electrical conductivity values (over 900 μS/cm) were measured in March and April. During the sampling period the mineralisation of water was mainly controlled by Cl^- and Na^+ contents; their temporal evolutions and that of EC are almost similar (Figure 11.4B). This dependency is especially evident for the three first months of sampling, when alkalinity, SO_4^{2-} and Ca^{2+} concentrations stayed almost constant. This indicates that the changes in electrical conductivity observed during that period (>120 μS/cm) were neither caused by the dissolution of carbonate nor sulphate mineral species.

The hydrograph of Quejigo spring also shows a gradual evolution in which sudden discharge variations do not occur even after the major rainfall events (Figure 11.4C). Unlike Esparragosilla and Fuente Santa springs, in which recharge coincides with the drainage of the more mineralised water, at Quejigo spring the recharge tends to dilute the mineralised water. This dilution, nevertheless, is lagged behind the start of the rainfall season.

The mineralisation of the water from Quejigo spring is principally controlled by Cl^- and Na^+ concentrations. The dilution recorded in January 2013 was produced by decreased concentrations of both ions, while other chemical components such as SO_4^{2-} or Mg^{2+} showed little variation. In low water conditions the EC increased and coincided with an increase in both Cl^- and Na^+ concentrations (their concentrations doubled in flood conditions) and with a slight increase in SO_4^{2-} concentration. This suggests the arrival at the spring of less mineralised water (mainly due to lower Cl^- and Na^+ concentrations) during flooding conditions. Thus recharge promotes the drainage of water from the unsaturated zone of the aquifer which has not been in contact with the Triassic evaporite rocks located at deeper zones. The steep decrease in alkalinity during May seems to be the consequence of a common ion effect caused by a higher rate of gypsum/anhydrite dissolution, which produces an increase in Ca^{2+} and SO_4^{2-} concentrations and the removal of $CaCO_3$ by precipitation.

11.5   DISCUSSION

The mean annual temperature of the water drained by all springs was stable during the monitoring period (coefficients of variation ranging from 1.1 to 2.6%, Table 11.2) and slightly higher than the average air temperature at the study area (16°C). These increased spring water temperatures suggest relatively deep groundwater flow through the saturated zone of the aquifers, where water temperatures are higher than in the shallower – recharge – parts of the system.

Groundwater stored in the aquifer drained by Esparragosilla spring is in direct contact with the Upper Triassic deposits which constitute the lower and lateral hydrogeological limits of La Silla aquifer (Figure 11.1). Thus, dissolution of evaporites – mainly gypsum/anhydrite and halite – occurs and, as a consequence, water becomes more mineralised principally because of large increases in the concentrations of SO_4^{2-}, Ca^{2+}, Na^+ and Cl^-. The water from this spring is the most mineralised and has the greatest SO_4^{2-}, Na^+ and Cl^- concentrations (Table 11.2, Figure 11.3A). This suggests the existence of groundwater flowpaths through the more permeable parts of the Upper Triassic body, or longer residence times of groundwater in contact with the evaporites.

The hydrochemistry of Esparragosilla spring water is influenced by both gypsum and halite dissolution (Figure 11.3B), although their contribution is not similar. Mean SO_4^{2-} concentrations are considerably higher than those of Cl^- and Na^+ (Table 11.2), and all samples are of a calcium sulphate facies (Figure 11.2A). These data indicate that the chemical composition of the groundwater drained by this spring is principally controlled by the dissolution of gypsum/anhydrite.

The temporal evolution of the chemical composition of Esparragosilla spring water is mainly controlled by a 'piston-flow' effect in the saturated zone of the aquifer (Figure 11.4A). Recharge produces the mobilisation of water from deeper zones of the aquifer where groundwater is characterised by longer residence times in contact with the evaporites and higher mineralisation, temperature and SO_4^{2-}, Ca^{2+} and Mg^{2+} concentrations. The evaporites located at the bottom of the aquifer drained by this spring contain both gypsum and halite. While the effect of the dissolution of these minerals on the hydrochemistry of the water is evident, the role played by carbonate-related minerals in the temporal evolution of the water chemistry of this spring is much more attenuated. The hydrodynamic and hydrochemical evolution of Esparragosilla is characteristic of aquifers with a low degree of karstification whose output signal (flow discharge) tends to smooth the input signal (rainfall).

The effect of the Triassic substratum on the chemical composition of the water drained by Fuente Santa and Quejigo springs (El Bosque aquifer) is mainly the result of dissolution of halite. The water samples collected at these discharge points show the highest Cl^- and Na^+ contents with respect to SO_4^{2-} concentrations (Figure 11.3A). In the case of Fuente Santa the mean concentrations of Cl^- and Na^+ during the monitoring period have exceeded that of SO_4^{2-} (SO_4^{2-}/Cl^- ratio <1 in Figure 11.3A). The contribution of halite to the hydrochemistry of Quejigo spring water is less evident since Cl^- and Na^+ concentrations are lower and the SO_4^{2-}/Cl^- ratio is higher (Figure 11.3A). In both springs the annual variations of Cl^- and Na^+ concentrations are larger than that of SO_4^{2-} (Figure 11.3B), suggesting a limited and not variable contribution of the latter (i.e. gypsum/anhydrite) to the final chemical composition of the water flowing out of these two springs.

In Fuente Santa spring recharge produces the drainage of more mineralised water stored in deeper zones of the aquifer, where it has been in contact with Triassic evaporites for long periods (Figure 11.4B). The relatively high mean temperature of the water (17.1°C) also suggests the existence of groundwater flow paths through deeper parts of the aquifer.

Recharge produces the dilution of the water drained by Quejigo spring (Figure 11.4C). This is consequence of the mixing of groundwater stored in the saturated zone of the aquifer with more recent and less mineralised infiltrating water coming from the unsaturated zone of the aquifer. The chemical composition of the recently infiltrated water is characterised by lower Cl^- and Na^+ concentrations (Figure 11.4C). The EC of the spring water is principally dependent upon the concentrations of these two components. The final chemical composition of the water is controlled by the dissolution of halite as well as by other hydrochemical processes such as the common ion effect, which can provoke relatively sudden reductions in alkalinity.

The contribution of evaporites at Máquina spring is the dissolution of sulphate minerals since it is only reflected by increased SO_4^{2-} concentrations, while Cl^- and Na^+ concentrations remained low and stable during the monitoring period (Figure 11.3B); this indicates little halite dissolution.

The chemical composition of Vihuelo spring water is the least influenced by the Triassic evaporites. Mean annual concentrations of Cl^- and Na^+ are similar to those from springs draining aquifers with similar characteristics elsewhere (López-Chicano et al., 2001; Barberá et al., 2014; Mudarra & Andreo, 2011). Nevertheless, dissolution of sulphate minerals contribute to a mean annual SO_4^{2-} concentration higher than the average in springs draining karst aquifers (Table 11.2 and Figure 11.3A).

11.6   CONCLUSIONS

Groundwater stored in carbonate aquifers can be in direct contact with underlying evaporites which constitute the base of the systems. The consequence is a change in the chemical composition of the groundwater mainly reflected in SO_4^{2-}, Ca^{2+}, Cl^- and/or Na^+ concentration increases. The strength of this hydrochemical modification and the ions involved in it depend to a large extent on the characteristics of each individual karst system. Thus, the groundwater may only show slight concentration increases of these ions, in keeping with the original calcium bicarbonate facies derived from the carbonate aquifer. In other cases the dissolution of evaporites is much more intense and can lead to a change of the chemical facies of the water towards calcium sulphate or sodium chloride.

The most mineralised and warmest waters, coming out from La Silla aquifer, have their origin in deeper flow paths through the Triassic substratum and/or longer residence times in contact with the evaporites.

The hydrochemical characterisation of the water flowing out of the five springs has permitted two different patterns regarding the dominant mineral species in the evaporite body to be identified. The chemical composition of the water drained by some springs is mainly controlled by the dissolution of gypsum/anhydrite, although halite dissolution can also exist. The chemical facies of other springs is principally dependent upon the dissolution of halite, with sulphate-related minerals playing a secondary role.

The analysis of the temporal variations of discharge, water temperature and the main chemical components of the water has led to the identification of different hydrogeological patterns in the groundwater drained by the springs. Some springs have shown an increase in water mineralisation during flood periods, coinciding with the highest discharge rates, and a subsequent and progressive reduction during low flow conditions. In these cases recharge water produces the mobilisation of groundwater stored in deeper zones of the aquifer ('piston-flow' effect), which is characterised by longer residence times and higher mineralisation. The contribution of carbonate-related minerals to the final composition of the water drained by these outlets is limited. In other springs, recharge tends to dilute the minerals in solution due to the mixing of recently infiltrated water with more mineralised flows coming from the saturated zone of the system. During low flow conditions, when recharge is minimum or inexistent, water only comes from the saturated zone of the aquifer and, consequently, is more mineralised.

ACKNOWLEDGEMENTS

This work is a contribution to the projects RNM-02161, CGL-2012-32590 of DGICYT and IGCP 598 of UNESCO, and to the Research Group RNM-308 of the Junta de Andalucía.

REFERENCES

Aquilina L., Ladouche B., Doerfliger N., Seidel J.L., Bakalowicz M., Dupuy C., Le Strat P. (2002) Origin, evolution and residence time of saline thermal fluids (Balaruc springs, southern France): implications for fluid transfer across the continental shelf. *Chemical Geology* 192, 1–21.

Barberá J.A., Andreo B., Almeida C. (2014) Using non-conservative tracers to characterize karstification processes in the Merinos-Colorado-Carrasco carbonate aquifer system (southern Spain). *Environmental Earth Science* 71, 585–599.

Capaccioni B., Didero M., Paletta C., Salvadori P. (2001) Hydrogeochemistry of groundwaters from carbonate formations with basal gypsiferous layers: an example from the Mt Catria Mt Nerone ridge (Northern Apennines, Italy). *Journal of Hydrology* 253, 14–26.

Carrasco F., Andreo B., Linares L., Sánchez D., Rendón M., Cobos A., Ortega F., Vadillo I., Pérez I. (2005) Contexto hidrogeológico de humedales del norte de la provincia de Málaga [Hydrogeological context of wetlands located in northern Málaga province]. In: López-Geta J.A., Rubio-Campos J.C., Martín-Machuca M. (editors) VI Simposio del Agua en Andalucía, Sevilla, España. Serie Hidrogeología y Aguas Subterráneas, vol. 14. Publicaciones del IGME, 605–618.

Gálvez-Maestre M.J. (2005) Climatología [Climatology]. In: López-Geta J.A. (editor) *Atlas Hidrogeológico de la Provincia de Cádiz*. Instituto Geológico y Minero de España y Diputación de Cádiz. Madrid, Spain, 53–58.

Goldscheider N., Andreo B. (2007) The geological and geomorphological framework. In: Goldscheider N., Drew D.P. (editors) *Methods in Karst Hydrogeology*. Taylor & Francis, London, United Kingdom, 9–24.

Hunkeler D., Mudry J. (2007) Hydrochemical methods. In: Goldscheider N., Drew D.P. (editors) *Methods in Karst Hydrogeology*. Taylor & Francis, London, United Kingdom, 93–121.

Kilchmann S., Waber H.N., Parriaux A., Bensimon M. (2004) Natural tracers in recent groundwaters from different Alpine aquifers. *Hydrogeology Journal* 12, 643–661.

López-Chicano M., Bouamama B., Vallejos A., Pulido-Bosch A. (2001) Factors which determine the hydrochemical behaviour of karstic springs: a case study from the Betic Cordilleras, Spain. *Applied Geochemistry* 16, 1179–1192.

Ma R., Wang Y., Sun Z., Zheng C., Ma T., Prommer H. (2011) Geochemical evolution of groundwater in carbonate aquifers in Taiyuan, northern China. *Applied Geochemistry* 26, 884–897.

Martín-Algarra M. (1987) Evolución geológica alpina del contacto entre las Zonas Internas y Externas de la Cordillera Bética [Alpine geological evolution of the contact between the Internal and External Zones of the Betic Cordillera]. PhD thesis, University of Granada. 1171 p.

Martín-Algarra M., Vera J.A. (2004) Evolución de la Cordillera Bética [Evolution of the Betic Cordillera]. In: Vera J.A. (editor) *Geología de España*. Sociedad Geológica de España e Instituto Geológico y Minero de España, 437–444.

Moral F., Cruz-Sanjulián J., Olías M. (2008) Geochemical evolution of groundwater in the carbonate aquifers of Sierra de Segura (Betic Cordillera, southern Spain). *Journal of Hydrology* 360, 281–296.

Mudarra M., Andreo B. (2011) Relative importance of the saturated and the unsaturated zones in the hydrogeological functioning of karst aquifers: the case of Alta Cadena (Southern Spain). *Journal of Hydrology* 397, 263–280.

Peyre Y. (1974) Géologie d'Antequera et de sa région (Cordillères Bétiques, Espagne) [Geology of Antequera and its region (Betic Cordillera, Spain)]. PhD thesis, Institute National Agronomique, Paris. 528 p.

Plummer L.N., Busby J., Lee R., Hanshaw B. (1990) Geochemical modelling of the Madison aquifer in parts of Montana, Wyoming, and south Dakota. *Water Resources Research* 26, 1981–2014.

Sánchez D., Carrasco F., Andreo B. (2009) Proposed methodology to delineate bodies of groundwater according to the European water framework directive. Application in a pilot Mediterranean river basin (Málaga, Spain). *Journal of Environmental Management* 90, 1523–1533.

Chapter 12

3D Spatial modelling of karst channels – The Beljanica karst massif

Saša Milanović

University of Belgrade – Faculty of Mining and Geology, Centre for Karst Hydrogeology of the Department of Hydrogeology, Belgrade, Serbia

ABSTRACT

It is very difficult to define hydrogeological parameters in an anisotropic aquifer such as carbonate matrix. Even the application of many different methods cannot provide a robust case for the karst aquifer. However, in the past decade a whole new branch of karst investigation, 3D modeling of karst aquifers and karst conduits, has become widely used in the earth sciences. The main topic of this chapter is 3D reconstruction and modelling of the Beljanica karst aquifer. The input parameters of the model are represented by the 69 caves, 15 sinkholes and 1682 dolines, 7 major karst springs, around 70 minor springs, and all available geological, morphological and hydrogeological data. As output from the model, more than 6000 nodes (registered in the database) were calculated. There are in some cases 60 different numbers for each node. The 3D model of the karst conduit network, enables the reconstruction of a complex karst aquifer, including its spatial, temporal, quantitative and qualitative characteristics.

12.1 INTRODUCTION

Creating 3D (physical) models of karst conduits as well as 3D models of a karst aquifer has been the interest of several researchers for many years (Fish, 1996; Gogu *et al.*, 2001; Ohms and Reece, 2002; Kovacs, 2003; Strassberg, 2005; Kincaid, 2006; Butscher and Huggenberger, 2007; Jeannin *et al.*, 2007; Filipponi and Jeannin, 2008, Filipponi, 2009; Borghi *et al.*, 2010). All of these authors, and many others, have practiced a method of forming a 3D model of the karst system based on large scale parameters that are collected first by the ground survey and further by special speleological or cave diving survey. The modelling technique used here is based on a multiparameter approach with 3D shape of conduits as output, connected to the data base with all collected spatial data. Generally, the field of physical or 3D modelling of the karst is a work in progress, as creating a 3D model depends on the development both of software and underground karst survey techniques.

The model of the Beljanica karst aquifer was developed using ArcGIS software and its 3D Analyst, Spatial Analyst and Network Analyst extensions, as well as COMPASS software and with some special programs made especially for reconstruction of conduit distribution. All spatial data, such as geological maps and cross sections, caves, sinkholes, springs and karst channels were converted into digital form, and each spatial unit was defined by its x, y and z coordinates. The compilation of all elements in the 3D environment produced a real, spatially oriented network of potential karst-conduit pathways. The case study of the Beljanica aquifer generally shows the procedure and results of the 3D modelling methodology of karst conduits, and further analysis for the purpose of watershed delineation, water pathway determination and assessment of karst reservoir storativity and consequently static and dynamic reserves.

12.2　OVERVIEW AND METHODOLOGY

Analysing the geometry of the main karst conduits in the saturated zone and the connection with conduits in the aeration zone, provides insight into karst channels and has enabled the creation of a 3D model. Analysis of various parameters obtained by quantitative and qualitative monitoring of groundwater characteristics and their analysis through the established physical model provides data on the relationship between the water recharge and discharge zones. Such a model can be further used for an analysis of the speleogenesis and hydrogeology of the area.

Generally there are three zones of research in karst aquifer spatial modelling. The first is the recharge zone and all underground forms that are related to this area such as dolines, sinkholes, caves and pits. The third zone is the phreatic zone with main forms such as karst springs and their submerged karst channels that often lead to the deeper parts of karst aquifers. The most problematic is the second zone between these two sections, but with the support of different software for 3D modelling and knowledge of karst development and intensity, this part can be reconstructed and the pathways (conduits) connections from inlet to outlet can be made.

The basic problem in the determination of the methodology was how to perform a quality analysis, or how to state the problem whose goal was to develop a model which would be an analysis of the geometry of karst conduits, integrating hydrogeological laws and geological characteristics.

The methodology is illustrated with the case study of the Beljanica karst aquifer (area of around 300 km^2). This study shows the procedure and results of the 3D modelling methodology for karst conduits, and further analysis is made for the purpose of watershed delineation, water pathways and the definition of karst reservoir storativity.

Data entry is the starting point for the establishment of a karst aquifer model and includes the conversion of available data into a digital format (graphic and alphanumeric). Such a model-designed system includes four main elements:

1. Preparation and data entry
2. Logical grouping and data connections
3. Data analysis and
4. Visualization and interpretation

Data preparation includes:

1. Ensuring appropriate graphic layers are in place – topographic, hydrographic, geological, hydrogeological, speleological and geomorphological.
2. Geocoding of available layers, or 'merging' in the appropriate map projection.

Main layers collected (as raster and vector entities) from detailed field studies are:

1. Topographic layer

    1. DEM (Digital Elevation Model)
        1. DEM of Surface
        2. DEM of Base of Karstification

2. Geological map
3. Hydrogeological map
4. Map of sinkholes and dolines
5. Map of groundwater levels
6. Map of forests, small plants and bare karst
7. Map of cave channels distribution
8. Tectonic map
9. Map of base of karstification
10. Map of the hypsometric location of the sinkholes, caves and karst springs.

12.3 ESTABLISHING OF PHYSICAL MODEL AND ITS RELATION BASE

The model with all the layers can generally be shown through the relationship of the spatial entities as a single model (Figure 12.1).

Components of the model include all of the classes and entities which, through a series of computational network and geostatistical analyses, form the 3D output of the Beljanica karst aquifer. The result of the output data is the connection of surface spatial-oriented data with the defined position of karst channels in the karst systems that are in connection between the implemented database and the spatial 2D and 3D entities (Figure 12.1).

Network spatial data 2D and 3D entities represent their field of numerical clusters that are 2D data transformed into a 3D object through a simulation model. 3D objects belong to the 'multipatch feature class' and contain three-dimensional information.

12.4 CREATING THE 3D KARST CONDUIT

The creation of a 3D model of karst channels is the conversion of known nodes of groundwater and conduit directions from a 2D to a 3D model. It is built through the known factors that could be used to create a 3D mesh. One of the basic parameters of the formation of a 3D model are faults and other tectonic features and their

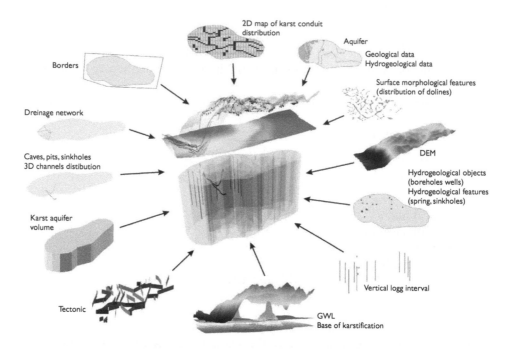

Figure 12.1 Scheme of 3D karst model forming.

inter-sections, together with dolines, sinkholes and speleological features. These parameters convert 2D to 3D points and determine the z coordinates and possible corrections of the x, y coordinates to a new position (Milanović, 2015).

The interactive interface of well-known 2D and 3D partially defined parameters can produce the three-dimensional character. The resulting surface potential distribution of karst channels is the basis for the definition of the orientation of a channel of karst aquifer features and follows the deterministic law that can be established with only small errors.

The basic steps of physical modelling of karst conduits formed from an initial 2D model or map of the distribution of karst channels to form the final 3D model of karst channel network is presented in Figure 12.2.

12.5   CASE STUDY – BELJANICA KARST AQUIFER

The Beljanica karst massif is located in the eastern part of Serbia (Figure 12.3), and the mountain range is an anticline (Kučaj-Beljanica structure) composed of Jurassic and Cretaceous carbonate rocks generally inclined from the central to peripheral parts of the massif (Stevanović, 1991). The carbonate rock complex of the Beljanica massif is the result of the Alpine orogenic phase. Tectonic events resulted in a complex system of faults and fractures which are preferred subterranean water flow pathways. The main forms are dolines, caves, pits, sinkholes and karstic springs and their

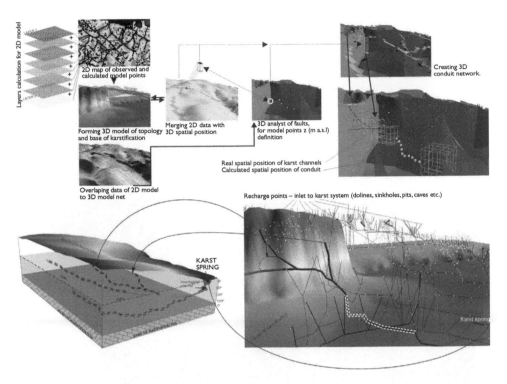

Figure 12.2 Schematic procedure of creating the 3D physical model (Milanovic, 2015).

classification and exploration was of great importance in the process of creating the Beljanica karst aquifer physical model. The main hydrogeological characteristic of the aquifer is its deep circulation. Cave divers explored channels from the Krupaja Spring to a depth of 133 m and Mlava Spring to a depth of over 73 m (Figure 12.4), the latter channels go deeper (survey not yet completed) (Milanović, 2007).

This case study includes an analysis of the 3D ArcGIS physical model of the karst system. The input data for the model are represented by the 69 caves, 15 sink-holes and 1682 dolines, 7 major karst springs and around 70 minor springs that were also explored plus the geological, morphological and hydrogeological data connected with the area of investigation (Figure 12.5A). As 2D output from the model, more than 6000 data points (registered in the database) were calculated. There are 60 different data for each point (coordinates, groundwater level, type of channel, dimensions of channels, next point of conduit, connection with (conduit), orientation of channel, hydrogeological function etc.). Eventually, the model is determined by more than 360 000 data inputs (Milanović *et al.*, 2010). The 2D model mesh is shown on Figure 12.5B.

The total length of the karst channel network, which is calculated using this model and presented in a 3D environment, is 647.3 km. Detail for the model are shown on Figure 12.6.

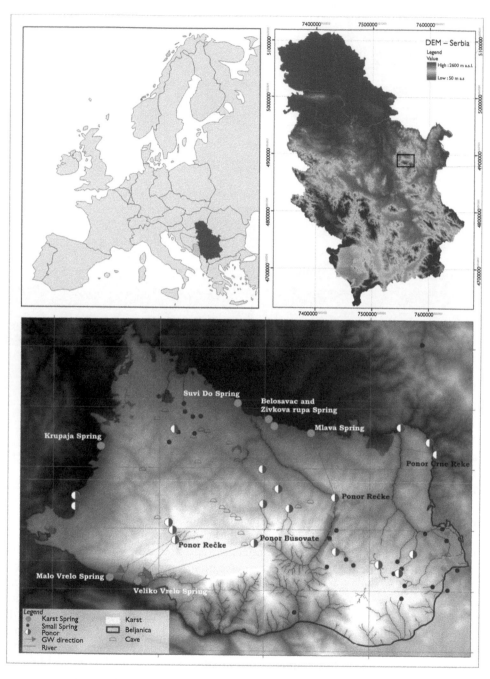

Figure 12.3 Geographic position of Beljanica massif with main ponors and karstic springs.

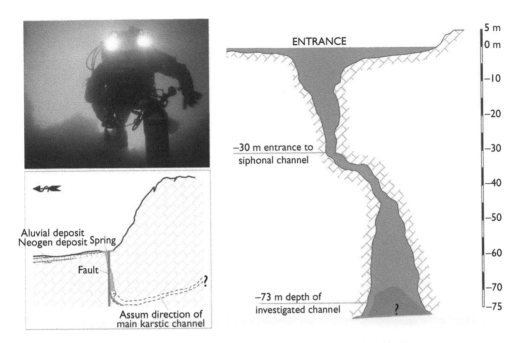

Figure 12.4 Result (cross section) of Mlava spring cave diving investigation.

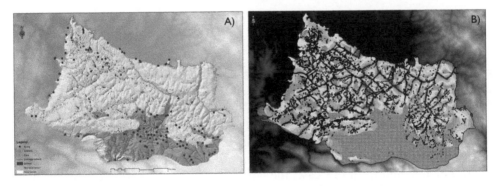

Figure 12.5 Map of Beljanica karst massif including position of hydrogeological, geological, geomorphological and speleological features – A, 2D model mesh according Beljanica physical model data – B.

12.6 DISCUSSION AND CONCLUSION

A spatial-oriented network, the potential karst channels providing a base for the assessment of storativity, is a crucial component for groundwater resources analysis. The underground karst development along with the saturation status were important elements from the 3D model to be correlated with data collected from simultaneous groundwater quantity (discharge) observations (Milanović *et al.*, 2013).

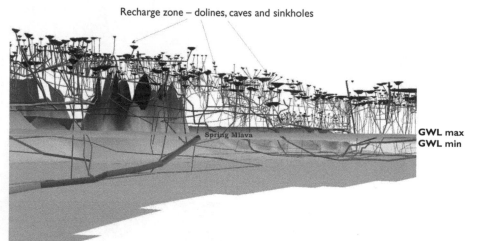

Figure 12.6 3D model of conduit system of the Beljanica karst aquifer with Mlava spring physical model.

Despite inconsistencies the complex 3D model enables an approximation of the hydrogeological watersheds to be determined. As watersheds have been proclaimed 'realistic' they represent a scenario for periods of medium flow (neither extreme maximum nor minimum). In this case, the largest is the Mlava Spring catchment which comprises some 124 km². The Belosavac Spring watershed covers 27 km², while 85 km² belongs to the Krupajsko Spring. MaloVrelo Spring is the smallest catchment in Beljanica covering some 7 km², while VelikoVrelo Spring comprises some 24 km² (Figure 12.7). The spatial position of watersheds is shown on Figure 12.7 (Milanović, 2010).

Finally, we should emphasize that the complex field research and its results, along with continuous monitoring and the correlation of those results with the newly created 3D model of the karst conduit network, enable the reconstruction of the complex karst aquifer, including a forecast of its spatial, temporal, quantitative and qualitative characteristics. The results favour wide application of such an approach as a basis for sustainable water management and appropriate utilisation of aquifers elsewhere with large groundwater reserves.

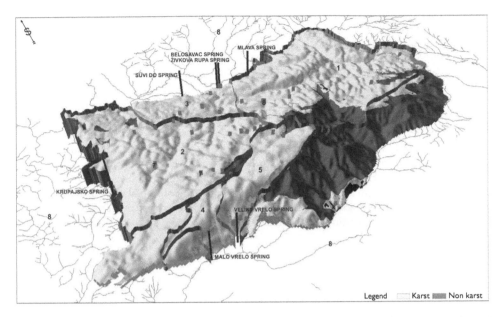

Figure 12.7 Delineation of watershed of Beljanica karst aquifer through analysis of karst channels distribution (3D analyses).

REFERENCES

Borghi A., Renard P., Jenni S. (2010) How to model realistic 3D karst reservoirs using a pseudo-genetic methodology – Example of two case studies, *Advance in Research in Karst Media*, Springer.

Butscher C., Huggenberger P. (2007) Implications for karst hydrology from 3D geological modeling using the aquifer base gradient approach, *Journal of Hydrology* 342, 184–198.

Fish L. 1996. Compass, http://members.iex.net/lfish/compass.html.

Filipponi M., Jeannin P.Y. (2008) Possibilities and Limits to predict the 3D Geometry of Karst Systems within the Inception Horizon Hypothesis, Geophysical Research Abstracts, Vol. 10, EGU General Assembly 2008, EGU2008-A-02825.

Filipponi M. (2009) Spatial Analysis of Karst Conduit Networks and Determination of Parameters Controlling the Speleogenesis along Preferential Lithostratigraphic, Horizons, thèse no 4376, Suiss.

Gogu R. C., Carabin G., Hallet V., Peters V., Dassargues A. (2001) GIS-based hydrogeological databases and groundwater modelling, *Hydrogeology Journal* 9, 555–569.

Jeannin P.Y., Groves C., Hauselmann P. (2007) Speleological investigations, Methods in karst hydrogeology, IAH Book Series, Taylor & Francis, London.

Kincaid T.R. (2006) A method for producing 3-D geometric and parameter models of saturated cave systems with a discussion of applications, *Groundwater Flow and Contaminant Transport in Carbonate Aquifer*, Taylor & Francis, 2000, pp. 169–190.

Kovacs A. (2003) Geometry and hydraulic parameters of karst aquifers: a hydrodynamic modeling approach, Doctoral theses, University in Neuchatel.

Milanović S. (2007) Hydrogeological characteristics of some deep siphonal springs in Serbia and Montenegro karst. *Environmental Geology*. 51(5), 755–759.

Milanović S. (2010) Creation of physical model of karstic aquifer on example of Beljanica Mt. (eastern Serbia), Doc. dissert, FMG, University of Belgrade, Belgrade.

Milanović S., Stevanović Z., Vasić Lj. (2010) Development of karst system model as a result of Beljanica aquifer monitoring. *Vodoprivreda* Vol. 0350-0519, 42 (2010) pp. 209–222.

Milanović S., Stevanović Z., Vasić Lj., Ristić-Vakanjac V. (2013) 3D Modeling and monitoring of karst system as a base for its evaluation and utilization – A case study from eastern Serbia, *Environmental Earth Science*, 71(2), 525–532.

Milanović S. (2015) Physical modeling of karst environment karst aquifers. In: Karst Aquifers – Characterisation and Engineering. (Ed. Z. Stevanović), Springer, Professional Practice in Earth Sciences, pp. 267–281.

Ohms R., Reece M. (2002) Using gis to manage two large cave systems, wind and jewel caves, south dakota, *Journal of Cave and Karst Studies*, April 2002.

Stevanović Z. (1991) Hydrogeology of Carpathian-Balkan karst of eastern Serbia and water supply opportunities (in Serbian). Spec. ed. Fac. Min. & Geol., pp. 1–245, Belgrade.

Strassberg G. (2005) A geographic data model for groundwater systems, Doctoral thesis, The University of Texas at Austin December, 2005.

Chapter 13

A field work oriented approach for complex karst aquifer characterisation

Philipp Stadler[1,2], *Hermann Häusler*[3], *Magdalena Rogger*[1,4], *Domenico Savio*[1,5] *& Hermann Stadler*[6]

[1]*TU Wien, Centre for Water Resource Systems, Vienna, Austria*
[2]*TU Wien, Institute for Water Quality, Resources and Waste Management, Vienna, Austria*
[3]*University of Vienna, Department of Environmental Geosciences, Vienna, Austria*
[4]*TU Wien, Institute of Hydraulic Engineering and Water Resources Management, Vienna, Austria*
[5]*TU Wien, Institute of Chemical Engineering, Research Group Environmental Microbiology and Molecular Ecology, Vienna, Austria*
[6]*Joanneum Research, Institute for Water, Energy and Sustainability, Department for Water Resources and Environmental Analyses, Graz, Austria*

ABSTRACT

Hydrogeological and geological field surveys are the indispensable basis for purposeful monitoring campaigns and catchment characterisation. In order to acquire, within a given timeframe, comprehensive information about complex hydrogeological settings or under-studied catchments it is not only advisable, but essential to follow a step-by-step protocol that is based on goal-oriented field work. Information from such an approach applied to the delineation of catchment characteristics enables the conduct of process-oriented monitoring at strategic locations and a more meaningful interpretation of resulting data. The approach described in this chapter becomes particularly important regarding the complexity of karstified aquifers and overlying catchments. Implementation of the methodology was applied to investigations in a karst area within the Outer Dinarides (Gorski Kotar, Western Croatia). In this study, information about the complex local hydrogeology was obtained through geological mapping, hydrogeological field surveys, and hydrological monitoring using stable isotopes.

13.1 INTRODUCTION

13.1.1 Objectives

The aim of the study is to describe, how comprehensive information about the complex karst system around Zeleni Vir spring was gained, particularly regarding discharge and storage dynamics, as well as catchment- and aquifer characteristics. These were obtained by applying a process-orientated scheme of combined hydrogeological field methods, such as mapping springs, hydrogeological quality rating of rocks, geological investigations on local lithologies, and stable isotope hydrology.

13.1.2   Research area

The hydrogeology of the upper Curak Valley north east of Delnice in western Croatia gives an insight to an interlaced geological setting which represents and affects the characteristic karst hydrology of the Gorski Kotar, a south east trending green karst mountain range with elevations between 1000 and 1200 m asl (above sea level) and well known for big karst springs such as Kupa, Kupica and Zeleni Vir.

The sedimentary rocks comprise Permian sandstone, Triassic- and Jurassic dolomite and limestone (Figure 13.1). During the Dinaric Orogeny these formations were folded and locally overthrusted (Herak, 1980). In terms of structural geology the area east of Delnice around Zeleni Vir appears as a tectonic window, where karst-ified Jurassic rocks were overthrusted by confining Permian formations (Figure 13.1). From the hydrogeological point of view, the karstified Jurassic formation acts as a basal karst aquifer (*Karst floor 1*) overlain by an aquiclude consisting of Permian rocks. In the Skrad-mountains (1043 m asl) the Permian rocks are overlain by Triassic carbonates forming a second karst floor *Karst floor 2*.

Karst floor 1 can be described as *Hidden Karst* with deep groundwater flow underneath the confining hanging wall formation. This is represented by Zeleni Vir spring which discharges from *Karst floor 1* with a maximum volumetric rate (Q-max) of 75 m³/s (Biondić *et al.*, 2006). Zeleni Vir's spring water is collected in a pressure pipe line for a power plant and discharged into the Curak river downstream. The hydrogeological catchment of Zeleni Vir spring can be located 10 km to the south where highly karstified Triassic rocks crop out at Ravna Gora at an altitude of 620 m asl. Points of interest for local hydrogeology are springs discharging close to the con-tact of Permian rocks and their Triassic cover (*Karst floor 2*). Such springs are located near the village of Skrad (Figure 13.1).

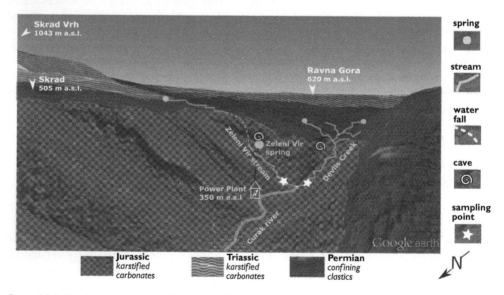

Figure 13.1 South view into the Curak valley. Map showing lithology, springs, streams and sampling locations.

13.2 METHODS

The methodological concept follows the scheme of prior assessing local geology and hydrogeology by field surveys, to gain an overview about the catchment and then to determine the most useful locations for monitoring and sampling (Figure 13.2). This approach will support an emerging insight to the complex hydrogeology of the research area, allow a focused data interpretation and consolidate results from field survey and lab analyses in a comprehensive way.

13.2.1 Local geology

Geological information was gathered from existing maps (e.g. Savić & Dozet, 1985; HGI, 2011) followed by a detailed geological mapping and studies of fault and fold structures in the research area. Some 21 outcrops (Figure 13.3) in the Jurassic, Triassic and Permian formations were surveyed. The striking and dipping of bedding, thrust planes and faults were measured, the lithology determined and assigned to stratigraphy.

13.2.2 Hydrogeology

A hydrogeological quality rating of rocks in the catchment was conducted to obtain an overview of the hydrogeological characteristics of local lithologies. This method is focused on the capability of rocks to act as a potential aquifer or aquiclude. Therefore, peculiarities of rocks such as grade of karstification, open joints and infiltration capacity are important, rather than petrographic or stratigraphic aspects. Seven representative outcrops in the catchment (Triassic) and spring (Jurassic) area were assessed (Figure 13.3). Bedding thickness, joint frequency, number of water-transmitting joints and bedding planes, unfilled joints and karstification grade were determined for each outcrop every 5 m.

Geological and hydrogeological information, including discharge (Q), electrical conductivity (EC) and temperature, was used to identify springs and streams of interest for subsequent monitoring and sampling.

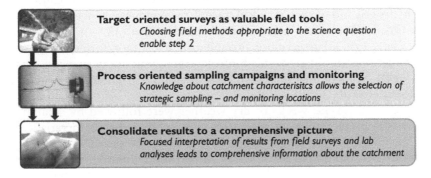

Figure 13.2 Proposed step-by-step work scheme.

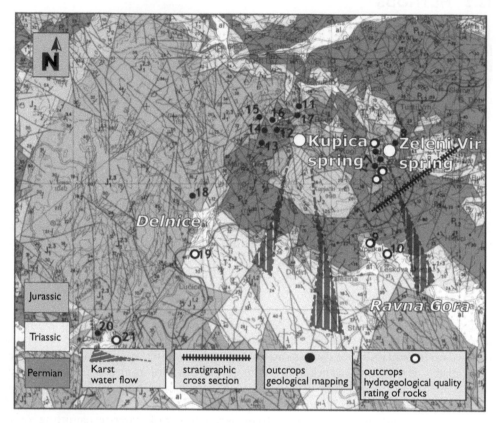

Figure 13.3 Geological map (modified after Savić & Dozet, 1985) of the research area with recorded outcrops, showing the Zeleni Vir window and indicating the Karst water flow underneath the confining Permian formation. Dashed line marks the cross section shown in Figure 13.10.

13.2.3   Event monitoring

Event monitoring by means of environmental isotopes was chosen, because stable isotopes of hydrogen and oxygen are ideal tracers, the conduction involves a manageable effort in time and staff and is a well-established method of isotopic investigations (Clark & Fritz, 1997). Furthermore, precipitation-event monitoring and sampling are indispensable tools to obtain detailed information about aquifer response characteristics, as well as vulnerability to contamination, storage dynamics, and runoff characteristics (Stadler *et al.*, 2008).

In June 2010 several weeks of dry weather were followed by a rainfall event. This allowed a measurement of the baseflow as well as the characteristic signal of coastal Adriatic summer rain (enriched $\delta^{18}O$ values, compared to the signal of the baseflow) in two monitored streams. To obtain representative samples during the event, a strategic position was chosen that allowed a parallel sampling of the two streams, each draining geologically diverse sub-catchments. The selection of monitoring locations was based on the previous hydrogeological investigations. One sampling location was located at Zeleni Vir stream (ZV), which drains what was assumed to be a karstic aquifer, discharging at the contact between *Karst floor 2* and Permian formations. During the monitored precipitation event the entire discharge of Zeleni Vir spring was captured in the power plant's pressure pipeline, the overflow was not active and, therefore, only water originating from the springs close to Skrad was monitored at the sampling location Zeleni Vir stream. The second sampling location was chosen at Devil's Creek (DC), which is dominated by surface runoff from a nearby gorge, discharging from the capping Permian clastic rocks. Some 50 m downstream of the sampling locations both streams flow together, forming the Curak river (Figure 13.1).

Samples were taken manually with an interval from one hour to three hours, covering the whole dynamics of the event. Water temperature and conductivity were measured in hourly intervals. The precipitation event consisted of a short-duration thunderstorm followed by steady rain. Rainfall data were collected continuously for the duration of the event (48 h), during which a total 29 mm of rainfall was measured.

A gauge was installed at the Devil's Creek sampling location and runoff was measured applying the tracer dilution method (Hubbard *et al.*, 1982; Benischke & Harum, 1984) during critical moments of the event. Following this, a discharge stage relation was calculated using six measurements.

A second gauge was installed at the Curak river where runoff was calculated applying the Manning-Strickler formula as described in Strobl & Zunic (2006).

$$Q = k_{st} \times r_H^{2/3} \times S^{1/2} \times A \qquad (13.1)$$

Where Q is the calculated discharge, k_{st} is an empirical parameter (Manning-Strickler coefficient) describing the roughness of the riverbed, r_H is the proportion of wetted area to wetted contour of the profile and S is the base slope of the river bed.

The riverbed was classified as a 'semi-natural' state with respect to the presence of debris and unevenness. Therefore a k_{st} of 30 was selected. For the base slope of the river bed 0.3 m difference in height per 100 m stream length was measured, yielding a S-value of 0.003.

The discharge of the Zeleni Vir stream was calculated as the difference between the total runoff (measured at Curak river) and the measured discharge of the Devil's Creek. Therefore, a discharge stage relation is given for both sampling locations. Combined with the results of environmental isotope analysis this enables an interpretation of the hydraulic dynamics of both aquifers during the event. To determine the isotopic input signal rain samples were taken during the event.

For further characterisation of the discharged aquifers, in addition to the samples for stable isotope analyses, three samples for water chemistry analyses were taken at both monitoring locations during different phases of the event.

13.2.4 Long termed sampling campaign

No existing times series data on isotopic composition of spring water was available for Zeleni Vir spring. Therefore, long-term sampling for stable isotopes compositions was carried out to gain information about seasonal discharge and storage dynamics, and to determine the impact of the alternating influence of Adriatic and continental precipitation on the recharge of Zeleni Vir's karstic reservoir.

The sampling campaign started in February 2011 and lasted nine month. Thirty water samples of Zeleni Vir spring were taken on a regular basis and four precipitation samples during diverse seasons were gathered.

The isotopic composition of the water samples was measured in the laboratory by using cavity ring-down spectroscopy (Berden *et al.*, 2001) with a WS-CRDS (Wavelength-Scanned Cavity Ring-Down Spectroscopy) instrument of Picarro, Inc. coupled to a CTC HTC-Pal liquid autosampler (LEAP Technologies, Carrboro, NC, USA) for automated measurements of liquid water samples. CRDS is a direct absorption technique (Berden *et al.*, 2001) that offers results for pure water samples highly comparable in precision with classical mass spectroscopy (Brand *et al.*, 2009). The instrumental setup is comparable to the system described by Gupta *et al.* (2009).

13.3 RESULTS

13.3.1 Local geology

Jurassic, Outcrop No.: 1–8, 11, 15, 16, 17 (Figure 13.3):

Jurassic formations are composed of north west to south east striking and south west to west dipping (Figure 13.4) limestones and dolomites. Large faults with distinct slickensides and fracture traces can be found, particularly in Devil's Creek gorge. The Devil's Creek gorge is mainly aligned with two major fault systems (K1: 120/85,

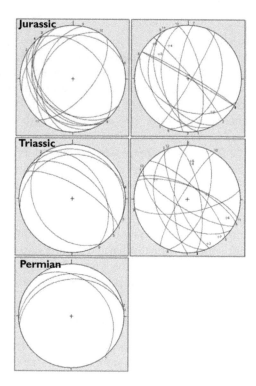

Figure 13.4 Stereo plots showing bedding- (1st plot each) and fault planes (2nd plot each).

K2: 030/90). Caves, like the Muzeva Hiza cave and the Zeleni Vir Cave are linked to strained and notably karstified zones and bound to the major fold systems of the Gorski Kotar.

Triassic, Outcrop No.: 9, 10, 19, 20, 21 (Figure 13.3):

Triassic carbonates are mainly north west to south east striking and north to north east dipping (Figure 13.4). At outcrops, distinct, partly conjugated fault systems can be found. Slickensides and fracture traces are present. Karstification is strongly associated with tectonically strained zones, but also cleavages are considerably karstified. The fractured Triassic limestones around Ravna Gora are, due to the high density of faults and joints with significant karstification, a catchment area with high infiltration capacity.

Permian, Outcrop No.: 12, 13, 14 (Figure 13.3):

Outcrops in the Permian formation are rare. Nevertheless the distribution of Paleozoic clastic rocks can be tracked because of its smooth morphology that differs explicit from that of karst. Bedding planes dip around 30° north (Figure 13.4). The tectonic contact between the overthrusted and confining Permian rocks with the Jurassic carbonates of the footwall can be observed at an outcrop near Dolnje Tihovo (outcrop No. 14, Figure 13.3).

13.3.2 Hydrogeological quality rating of rocks

The field survey determined rocks with different types of infiltration (high, medium, low), aquifer properties potential (fractures, unfilled joints, karstified cleavages), and aquicludes. For the field site, five distinguished categories are listed in Table 13.1 and in Figure 13.5. This rating does not differentiate between rocks in the spring and catchment areas. Jurassic rocks around Zeleni Vir spring were assigned to the same type, e.g., Triassic rocks in the catchment area. Such rocks are characterised

Table 13.1 Types of hydrogeological rock quality, determined by rating of representative outcrops.

Type	Characteristics	Joint frequency [n per 5 m]	Unfilled joints [%]	Water transmitting joints [%]	Comment
1	Limestone Jurassic Bedding (2–3 m)	0	0	0	Karstified bedding planes
2	Limestone Jurassic Fractured	16	25%	16%	Slicken sides, Water bearing joint (120/80)
3	Limestone Jurassic Bedding (cm–dm)	15	0	0	
4	Limestone Jurassic Fractured, Karstified	10	60%	10%	
4	Limestone Triassic Bedding (dm) Fractuered, Karstified	>100	70%	0	Joints filled with red earth, Bedding (225/20)
5	Dolomite Jurassic Bedded and fractured	>100	0	0	
5	Dolomite Triassic Bedded and fractured	>100	0	0	Joints filled with calcite

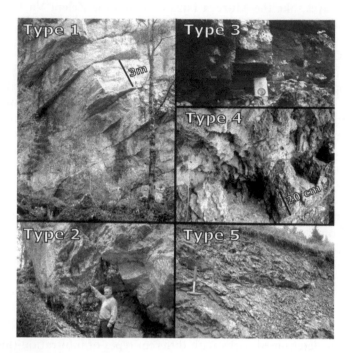

Figure 13.5 Example pictures of the five types of hydrogeological rock quality, determined in the research area.

by comparable properties regarding aquifer capabilities and degrees of karstification (Table 13.1). Whether a rock acts as infiltration lithology or as an aquifer is determined by its topographical position.

Type 1: Limestone with intact bedding, bedding-planes are unfilled and karstified.

This potential aquifer is characterized by intact limestone with a bedding thickness of several meters. The open and unfilled bedding planes are characterised by slight dipping and show distinct signs of karstification. Due to intact and gently dipping beddings, Type 1 rocks are assumed to have minimal potential for direct infiltration of precipitation water and act predominantly as an aquifer allowing lateral water flow. It can be assumed the aquifer of Zeleni Vir spring exhibits similar properties, at least in part.

Type 2: Highly fissured carbonates with discordant thrust faults and water-transmitting joints.

Highly fissured and strained rocks with water-transmitting joints and discordant fault systems. These rocks allow an interchanging flow between different karst water levels, due to discordant faults.

Type 3: Fissured carbonates with closed or filled joints.

Dolomite rocks and limestones with closed joints are assumed to act in a manner similar that of aquiculdes. Much of the flow occurs in the Type 1 and Type 2 rocks.

Type 4: Fissured limestones and highly developed karstification.

The genesis of spelean formations; cavities and caves are mainly linked to these highly fissured and karstified limestones. Type 4-rocks can be found in the spring area e.g. in the surrounds of the cave Muzeva Hiza, located in the upper part of Devil's Creek gorge. In the catchment area, Type 4 rocks are assumed to enable dominant parts of infiltration. Filling of joints and cavities are often palaeosols that contain iron-rich earth. The geomorphology typically involves karstification characterised by frequent occurrences of dolines and ponors.

Type 5: Highly fissured Dolomite with closed or filled joints.

The appearance of dolomite is limited to small scaled areas. The infiltration capability is assumed to be low and the associated landscape does not show the asperity that often characterises karstified limestones.

13.3.3   Event monitoring

Isotopic composition ($\delta^{18}O$ relative to Vienna standard mean ocean water) of baseflow (June 2010) was determined for Devil's Creek −10.32 ‰ and −10.43 ‰ for Zeleni Vir stream (Table 13.2). Electrical conductivity of Devil's Creek baseflow was 200 µS/cm and 307 µS/cm for Zeleni Vir stream (Table 13.2). Precipitation had a $\delta^{18}O$ value of −5.5 ‰ and a deuterium excess of 12.0 (average of three samples taken during the event). Both phases of the event, the thundershower of 15 June 2010 as well as the subsequent steady rain of 16 June affected the isotopic signals in the two monitored streams (Table 13.2, Table 13.3, Figure 13.6). The influence of coastal Adriatic summer precipitation, which has an enriched isotope signal compared to the streams baseflow, increased $\delta^{18}O$ values significantly in both streams. The response of Devil's Creek to the precipitation event was more distinct compared with that exhibited by the Zeleni Vir Stream (Figure 13.6). Electrical conductivity in stream water decreased due to the influence of rain water in the rivulets (Table 13.2, Table 13.3).

For data interpretation a LMWL ($\delta^2H = (7.6 \pm 0.4)\ \delta^{18}O + (10.5 \pm 4.0)$) calculated for Zavažan at Mount Velebit was taken from Vreca et al., (2006) and in addition an estimated LMWL for Ravna Gora was derived from the nine month sampling campaign (Table 13.4) at Zeleni Vir spring ($\delta^2H = 7.9\ \delta^{18}O + 13.0$) (Figure 13.7). The Ravna Gora LMWL lies within the range of the LMWL reported by Vreca et al. (2006) for Mount Velebit. The isotope ratio measured during the monitoring shows a shift towards the LMWL and higher enriched summer values at the maximum of the event, indicating the peak of event water arriving at the sampling locations (Figure 13.7).

Applying a two components isotopic separation the proportion of event water was calculated for both streams. The following equation was used.

$$Q_i = Q_G \times \frac{C_G - C_B}{C_i - C_B} \tag{13.2}$$

Where Q_i is the amount of rainwater in the stream [l/sec], Q_G is the streams discharge [l/sec], C_G is the isotope signal of the stream [$\delta^{18}O$ in ‰], C_B is the isotope signal of the stream baseflow [$\delta^{18}O$ in ‰] and C_i is the isotope signal of rainwater [$\delta^{18}O$ in ‰].

Table 13.2 Environmental isotope- and hydrologic data from event monitoring at sampling location Zeleni Vir Stream.

				ZELENI VIR STREAM			
Date	Sample ID	EC [μS/cm]	Water temp. [°C]	Q [l/s]	δ O18 ‰	δ D ‰	d ‰
15.06.2010 16:30	ZV1	307	10.3	210.9	−10.19	−68.2	14.7
15.06.2010 16:50	ZV2	301	10.7	266.9	−10.13	−67.7	14.6
15.06.2010 17:30	ZV3	305	10.5	258.7	−10.34	−68.8	15.3
15.06.2010 18:20	ZV4	308	10.4	353.6	−10.32	−69.1	14.8
15.06.2010 19:40	ZV5	309	10.5	284.1	−10.27	−68.8	14.7
15.06.2010 21:15	ZV6	313	10.5	249.9	−10.21	−68.5	14.5
15.06.2010 22:15	ZV7	311	11	249.9	−10.23	−68.5	14.7
16.06.2010 02:50	ZV1A	313	11.4	213.9	−10.24	−68.5	14.8
16.06.2010 04:00	ZV2A	311	11.8	249.9	−10.22	−68.1	15.0
16.06.2010 05:00	ZV3A	310	11.7	272.1	−10.26	−68.3	15.1
16.06.2010 06:30	ZV4A	307	10.8	284.2	−10.17	−67.2	15.5
16.06.2010 07:15		306	10.5	284.2			
16.06.2010 07:40	ZV5A	306	10.5	284.2	−10.09	−66.6	15.4
16.06.2010 08:50	ZV6A	301	10.8	341.1	−9.98	−66.0	15.2
16.06.2010 09:20		303	10.8	341.1			
16.06.2010 10:15	ZV7A	301	10.9	419.0	−10.05	−66.2	15.5
16.06.2010 11:00		301	10.9	341.1			
16.06.2010 13:00	ZV8A	300	11.2	314.9	−10.07	−66.0	15.8
16.06.2010 14:00		299	11.3	328.8			
16.06.2010 15:00	ZV9A	303	10.9	272.1	−10.06	−66.3	15.5
16.06.2010 17:00	ZV10A	306	10.7	284.1	−10.16	−67.0	15.6
16.06.2010 18:00		307	10.6	284.1			
16.06.2010 19:00		308	10.3	230.1			
16.06.2010 20:30	ZV11A	310	10.1	230.1	−10.15	−67.5	15.0
16.06.2010 21:45		311	10.3	240.4			
16.06.2010 23:00	ZV12A	311	10.3	249.9	−10.35	−69.1	15.0
17.06.2010 06:00	ZV13A	312	9.8	198.3	−10.17	−69.5	13.2
17.06.2010 11:00	ZV14A	309	10.5	213.9	−10.32	−69.6	14.3
17.06.2010 14:00	ZV15A	309	10.7	220.7	−10.19	−69.4	13.5
17.06.2010 16:00	ZV16A	311	10.4	220.7	−10.21	−69.8	13.2

For Devil's Creek precipitation water (Q_i) contributed a maximum of 21% to total discharge, whereas in the Zeleni Vir stream a maximum of 9% precipitation water could be detected (Figure 13.8). The maximum event water participation in the Devil's Creek was determined 1.5 hours later than in the Zeleni Vir stream (Figure 13.8).

Following the two components isotopic separation method the interflow was calculated from the difference between total discharge and event water in the stream. This again shows clearly a time shift between the different components. At the Devil's Creek sampling location the interflow reached its peak 3 hours before the event water component, showing a time shift between hydraulic reaction and maximum rainwater amount of the spring system in the Permian aquifer (Figure 13.9). After the surface flow passage in the Devil's Creek, this signal is dispersed.

Analyses of water samples reveal distinct differences in hydrochemical parameters of both monitored streams (Table 13.5). Water from Devil's Creek (Table 13.5,

Table 13.3 Environmental isotope- and hydrologic data from event monitoring at sampling location Devil's Creek.

			DEVIL'S CREEK				
Date	Sample ID	EC [uS/cm]	Water temp. [°C]	Q [l/s]	δ O18 ‰	δ D ‰	d ‰
15.06.2010 16:30	K1	200	14.4	68.6	−10.21	−70.1	12.9
15.06.2010 16:50	K2	190	14.7	73.2	−9.96	−67.8	13.2
15.06.2010 17:30	K3	193	14.6	81.3	−9.75	−67.3	12.0
15.06.2010 18:20	K4	194	14.5	121.0	−9.91	−68.1	12.5
15.06.2010 19:40	K5	187	14.5	121.0	−9.84	−68.1	11.9
15.06.2010 21:15	K6	187	14.7	90.1	−9.90	−68.4	12.1
15.06.2010 22:15	K7	192	14.9	90.1	−9.95	−68.4	12.5
16.06.2010 02:50	K1A	200	14.9	65.7	−10.04	−69.1	12.5
16.06.2010 04:00	K2A	198	14.9	90.1	−9.89	−67.9	12.5
16.06.2010 05:00	K3A	195	15.1	133.0	−9.91	−65.9	14.6
16.06.2010 06:30	K4A	179	15.3	190.4	−9.77	−61.8	17.6
16.06.2010 07:15		176	13.9	190.4			
16.06.2010 07:40	K5A	174	13.9	190.4	−9.68	−62.0	16.7
16.06.2010 08:50	K6A	170	13.9	207.4	−9.49	−60.6	16.5
16.06.2010 09:20		168	13.9	207.4			
16.06.2010 10:15	K7A	159	13.8	207.4	−9.29	−59.7	15.9
16.06.2010 11:00		158	13.9	207.4			
16.06.2010 13:00	K8A	168	14.5	159.7	−9.44	−60.7	16.0
16.06.2010 14:00		173	14.7	145.9			
16.06.2010 15:00	K9A	176	14.8	133.0	−9.62	−62.2	16.0
16.06.2010 17:00	K10A	183	14.7	121.0	−9.67	−64.0	14.6
16.06.2010 18:00		185	14.7	121.0			
16.06.2010 19:00		186	14.7	109.9			
16.06.2010 20:30	K11A	189	14.6	109.9	−9.84	−65.6	14.4
16.06.2010 21:45		190	14.5	99.6			
16.06.2010 23:00	K12A	192	14.2	90.1	−9.80	−66.6	13.1
17.06.2010 06:00	K13A	196	13.3	81.3	−9.93	−67.7	13.0
17.06.2010 11:00	K14A	197	13.7	65.7	−10.17	−69.8	12.9
17.06.2010 14:00	K15A	198	14.6	58.8	−10.04	−69.1	12.5
17.06.2010 16:00	K16A	198	14.6	58.8	−10.08	−69.5	12.5

DC), discharging the Permian clastics has a lower mineral content (A-sum, K-sum) than does the Zeleni Vir stream (Table 13.5, ZV), discharging an carbonatic aquifer. This validates the intended selection of sampling locations and also confirms on-site electrical conductivity measurements during baseflow conditions for both streams (DC: 200 μS/cm, ZV: 307 μS/cm). The Mg, Ca and HCO_3 concentrations and especially the Ca/Mg ratio of 3.4 for Zeleni Vir stream water determines its hydrochemical facies to carbonate rocks, most likely Dolomite (Pavuza & Traindl, 1983).

13.3.4   Zeleni Vir spring

$\delta^{18}O$ values of 30 water samples taken at Zeleni Vir spring had a range from −10.24 ‰ to −9.4 ‰, with a mean of −9.9 ‰ (Table 13.4, Figure 13.2). The most depleted isotope signal was measured in samples taken in July 2011 und the least depleted $\delta^{18}O$

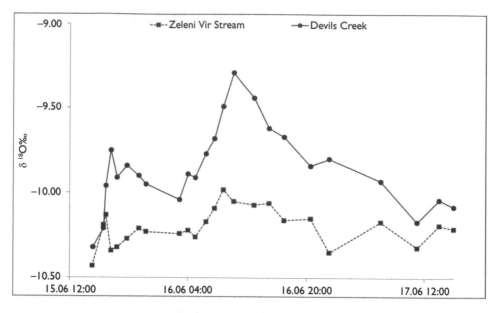

Figure 13.6 $\delta^{18}O$ values of Devil's Creek and Zeleni Vir stream during the event monitoring.

signal was determined for samples taken in September 2011 (Table 13.4). The deuterium excess ranges from 14.0 in July 2011 to 17.1 in February 2011, with a mean value of 15.7. Complementary to the spring water, rain water was sampled (n = 4) in Zeleni Vir. The precipitation showed high $\delta^{18}O$ values during the early summer month (samples from June and July 2011), when a signal of −5.7 ‰ to −5.3 ‰ was measured. Significantly depleted $\delta^{18}O$ values of −11.5 ‰ were determined from samples taken in September 2011. The deuterium excess in precipitation water ranges from 12.0 in June 2011 to 17.2 in September 2011, with a mean value of 14.

Although no precipitation samples from the catchment area of Zeleni Vir spring were available, a Local Meteoric Water Line (LMWL) can be estimated as follows: assuming that the spring water of Zeleni Vir is a mixture of precipitation infiltrated, without major fractionation in the hydrogeological catchment around Ravna Gora and the measured values of the isotope ratio are scattered around a LMWL for Ravna Gora, a substituted LMWL can be derived:

$$\delta^2H = 7.9\ \delta^{18}O + 13.0 \tag{13.3}$$

Evaluation of isotope data from spring water samples taken at Zeleni Vir spring revealed a range of $\delta^{18}O$ values significantly lower (B = 0.8 ‰) than those determined for monitored summer precipitation (A = 6.3 ‰). This may reflect a high reservoir capacity of the aquifer and low dynamics. Partly this may be caused by the relatively small amounts of summer precipitation compared with the annual mean. Accordingly the recharge of Zeleni Vir spring is mainly driven by precipitation during the autumn and winter months. As described by Clark & Fritz (1997) the strong correlation

Table 13.4 Environmental isotope data from long term sampling campaign at Zeleni Vir Spring.

	ZELENI VIR SPRING			
Date	Sample ID	δ O18 ‰	δ D ‰	d ‰
01.02.2011	ZV Spring 1	−9.86	−63.3	16.9
04.02.2011	ZV Spring 2	−9.80	−63.4	16.3
14.02.2011	ZV Spring 3	−9.95	−64.0	16.9
21.02.2011	ZV Spring 4	−9.96	−64.3	16.7
25.02.2011	ZV Spring 5	−9.88	−64.2	16.1
28.02.2011	ZV Spring 6	−9.98	−64.0	17.1
03.03.2011	ZV Spring 7	−9.95	−64.3	16.6
09.03.2011	ZV Spring 8	−9.96	−64.1	16.9
16.03.2011	ZV Spring 9	−9.97	−64.6	16.5
18.03.2011	ZV Spring 10	−9.78	−63.0	16.5
28.03.2011	ZV Spring 11	−9.82	−64.2	15.6
06.04.2011	ZV Spring 12	−9.77	−64.4	15.1
08.04.2011	ZV Spring 13	−9.72	−63.9	15.1
13.04.2011	ZV Spring 14	−9.76	−63.9	15.4
06.07.2011	ZV Spring 15	−10.17	−68.6	14.0
09.07.2011	ZV Spring 16	−10.24	−68.4	14.9
15.07.2011	ZV Spring 17	−10.02	−66.1	15.4
19.07.2011	ZV Spring 18	−9.93	−66.1	14.6
23.07.2011	ZV Spring 19	−9.98	−66.0	15.1
26.07.2011	ZV Spring 20	−9.88	−66.1	14.2
28.07.2011	ZV Spring 21	−9.93	−65.8	15.0
04.08.2011	ZV Spring 22	−9.87	−65.4	14.9
07.08.2011	ZV Spring 23	−9.73	−63.9	15.3
25.08.2011	ZV Spring 24	−10.00	−65.6	15.6
30.08.2011	ZV Spring 25	−9.99	−65.6	15.6
05.09.2011	ZV Spring 26	−9.93	−64.9	15.8
14.09.2011	ZV Spring 27	−9.98	−64.7	16.5
20.09.2011	ZV Spring 28	−9.41	−61.7	14.8
24.09.2011	ZV Spring 29	−9.65	−63.0	15.5
04.10.2011	ZV Spring 30	−9.97	−65.7	15.3

between temperature and stable isotopes in meteoric water provides a seasonal signal that can be used to date ground waters. McGuire & McDonnell (2006) described the following analytical solution for the mean transit time (MTT) parameter for an exponential model by using the damping of the isotope signal:

$$\tau_m = c^{-1}\sqrt{f^{-2} - 1} \tag{13.4}$$

Where τ_m is the calculated mean transit time in days, c is the angular frequency constant $(2\pi/365)$ and f is the damping coefficient $(f = B/A)$.

Although a higher number of spring water and precipitation samples are preferable, this analytical solution still enables an estimate of the mean transit time. Making use of the available data, a MTT for Zeleni Vir spring water of 1.2 years can be calculated.

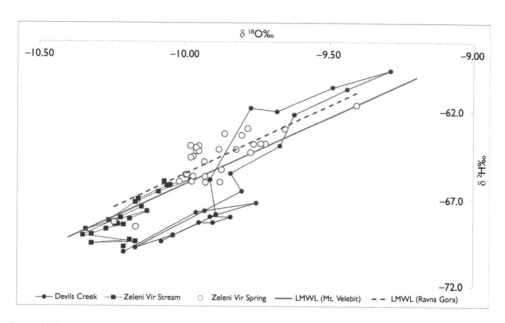

Figure 13.7 Isotope ratio of deuterium and ¹⁸O at both sampling locations during the event monitoring, isotope ratio of Zeleni Vir spring (long termed sampling campaign), LMWL calculated for Ravna Gora and LMWL for Mount Velebit (Vreca *et al.*, 2006).

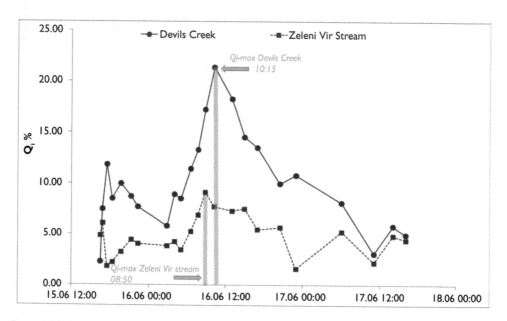

Figure 13.8 Percentage of event water contributing to the discharge of both streams during the event monitoring.

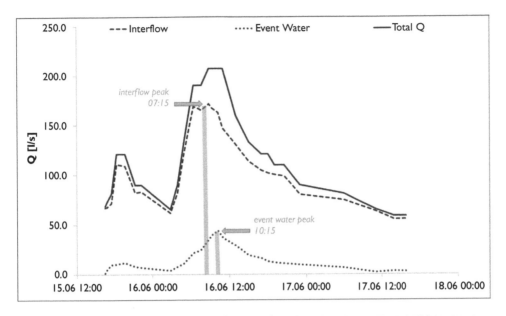

Figure 13.9 Component separation for the Devil's Creek showing a distinct time shift between interflow and event water.

13.4   DISCUSSION AND CONCLUSIONS

The study followed the proposed approach based on selected field methods, suitable to gather information about the complex hydrological and hydrogeological setting of the Zeleni Vir karst area (Figures 13.2 and 13.3). Geological mapping provided a fundamental understanding of local stratigraphy and tectonics (Figures 13.1 and 13.10). Hydrogeological investigations, such as the hydrogeological quality rating of rocks enabled to determine their specific characteristics, governing their overall function as either an aquifer or an aquiclude (Table 13.1, Figure 13.5). Mapping of springs completed this task. This target-oriented survey led to an overview of the local karst hydrology and enabled the selection of strategic sampling locations for subsequent monitoring programmes (Figure 13.1). Due to the merging of results that were based on field work with those from laboratory analyses consolidating information about (i) Zeleni Vir spring, (ii) Devil's Creek and (iii) Zeleni Vir stream was gained, provided information on the local karst hydrogeology.

(i)   The estimated LMWL for Ravna Gora lies within the range of the LMWL calculated for Mount Velebit (Vreca *et al.*, 2006) (Figure 13.7). This confirms the hypothesis that the precipitation in coastal mountain ranges of the Dinarides, like Gorski Kotar and Mount.Velebit emerges mainly from air masses having a comparable origin. Enriched isotopic signatures for rain water at Zeleni Vir during the summer months can be ascribed to precipitation that originates from Adriatic air masses, whereas decisively lower enrichments are caused by the increased

Table 13.5 Results from hydrochemical analyses of samples taken at Devil's Creek and Zeleni Vir stream during the event monitoring.

Sample ID	Sampling date	Na [mg/l]	K [mg/l]	Mg [mg/l]	Ca [mg/l]	Cl [mg/l]	NO₃ [mg/l]	SO₄ [mg/l]	HCO₃ [mg/l]	K-sum [mval/l]	A-sum [mval/l]	Diff [mval/l]	%-Dev.
DC4	16.06.2010 06:30	4.72	0.94	6.98	19.41	4.15	1.05	8.82	92.1	1.77	1.83	−0.06	3.11
DC12	16.06.2010 23:00	5.29	0.95	7.61	21.00	4.76	1.00	9.10	99.5	1.93	1.97	−0.04	2.12
DC16	17.06.2010 16:00	5.44	0.88	7.98	21.78	4.87	0.98	9.41	104.3	2.00	2.06	−0.06	2.80
ZV4	16.06.2010 06:30	3.14	0.42	12.59	42.75	4.77	2.74	6.25	183.7	3.32	3.32	0.00	0.09
ZV12	16.06.2010 23:00	3.28	0.50	13.00	43.34	4.78	2.83	6.51	186.7	3.39	3.38	0.01	−0.32
ZV16	17.06.2010 16:00	3.24	0.63	12.96	43.33	4.90	2.81	6.61	186.1	3.38	3.37	0.01	−0.40

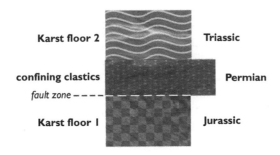

Karst floor 2	Triassic
confining clastics	Permian
fault zone ‒ ‒ ‒ ‒	
Karst floor 1	Jurassic

Figure 13.10 Schematic stratigraphy of the Zeleni Vir window. Compare cross section with Figure 13.3.

influence of continental air masses in autumn and winter. Comparable results were reported by Vreca *et al.*, (2006). The low dynamic of spring water isotopic signature shows also that summer rain has only a marginal influence on the isotope signal of the spring water, as the summer rain fraction is mixed up in the karst reservoir with the dominant fraction of stored water that has depleted $\delta^{18}O$ values. Furthermore the derivation of this LMWL supported the interpretation of data gained from the event monitoring. For calculating the mean transit time (MTT) of water in the aquifer of Zeleni Vir spring a higher number of spring water and precipitation samples are preferable. Nevertheless, by making use of the available data a MTT of 1.2 years could be calculated. Although this is a first estimate, it still provides useful information for assessing the rough dimension and characteristics of the reservoir.

(ii) Hydrographs and stable isotopes data for both streams during the 48 hours of event monitoring revealed their particular storage- and discharge dynamics (Table 13.2, Table 13.3). Environmental isotope analyses and hydrograph evaluation shows, that the Devil's Creek runoff is not only composed of surface runoff but also of spring water (Figure 13.9). The mentioned time shift between the interflow and the event water (Figure 13.9) describes the influence of springs during rainfall events, leading to a difference between hydraulic reaction and arrival at the sampling location of water deriving from the precipitation event. This dynamic is similar to a piston flow model. These springs have strong influence on the discharge dynamic of Devil's Creek and are assumed to drain the weathered upper layer of the Permian clastic rocks (Figure 13.1). More detailed field surveys could focus on the abundance and thickness of this weathered layer.

(iii) As the proportion of event water in Zeleni Vir stream is very low (Figures 13.7 and 13.8) and its hydrologic response on the event very direct (Figures 13.6 and 13.8), it is assumed that the dynamic is only caused by surface runoff flowing into the stream in between the spring and the sampling location (Figure 13.1). Consequently no precipitation event water flowing through the aquifer was detected within the 48 hours of monitoring, unusual for a karst-dominated system. This inert reaction (>48 h) leads to the conclusion that the springs recharging the Zeleni Vir stream (Zeleni Vir´s discharge did not contribute to Zeleni Vir stream during the monitoring because it was captured in the pressure pipe line) are discharging from a fissured, but not karstified aquifer (Type 5-rock: Figure 13.5, Table 13.1). The results of hydrochemical analysis indicate a carbonate aquifer, most likely dolomitic. $\delta^{18}O$ values for the sampled water were used to estimate the height of the catchment. As the altitude of the Devil's Creek orographic catchment can be assigned to confine the Permian formation and ^{18}O depletion varies between 0.15 and 0.5 ‰ per 100 m rise (Clark & Fritz, 1997), the catchment location for Zeleni Vir stream was estimated using $\delta^{18}O$ data and found to be at the altitude were the contact of Triassic formations (*Karst floor 2*) and Permian rocks occurs (Figure 13.1). Further investigations would be desirable to gain more accurate information about location and dimensions of this aquifer.

The investigations show again, like other studies conducted in the Dinaric karst (e.g. Herak, 1980; Biondić *et al.*, 1997; Kresic & Stevanovic, 2010), that structural geological settings are affecting the karst hydrology of western Croatia on diverse scales.

ACKNOWLEDGEMENT

We thank Albrecht Leis (Joanneum Research, Graz) and Christian Müllegger (University of Vienna) for isotope analytics and advices. Jurica Stivićić (HE Zeleni Vir) provided the essential local support for the long-term sampling campaign at Zeleni Vir spring. Special thanks to Ronald W. Harvey (U.S. Geological Survey, Boulder, Colorado) for his constructive and detailed suggestions.

REFERENCES

Benischke R., T. Harum (1984) Computergesteuerte Abflussmessungen in offenen Gerinnen nach der Tracerverdünnungsmethode (Integrationsverfahren). *Steirische Beiträge zur Hydrogeologie* 36, 127–137.

Berden G., Peeters R., Meijer G. (2001) Cavity ring-down spectroscopy: Experimental schemes and applications [Review] *International Reviews in Physical Chemistry* 19(4), 565–607.

Biondic' B., Dukarić F., Kuhta M., Biondić R. (1997) Hydrogeological Exploration of the Rjecina River Spring in the Dinaric Karst. *Geologia Croatica* 50(2), 279–288.

Biondić B., Biondić R., Kapelj S. (2006) Karst groundwater protection in the Kupa River catchment area and sustainable development. *Environmental Geology* 49, 828–839.

Brand W.A., Geilmann H., Crosson E.R., Rella C.W. (2009) Cavity ring-down spectroscopy versus high-temperature conversion isotope ratio mass spectrometry; a case study on δ2H and δ18O of pure water samples and alcohol/water mixtures. *Rapid Communications in Mass Spectroscopy* 23(12), 1879–1884.

Clark I. D., Fritz P. (1997) *Environmental Isotopes in Hydrogeology*. CRC Press/Lewis Publishers.

Gupta P., Noone D., Galewsky J., Sweeney C., Vaughn B.H. (2009) Demonstration of high-precision continuous measurements of water vapor isotopologues in laboratory and remote field deployments using wavelength-scanned cavity ring-down spectroscopy (WS-CRDS) technology. *Rapid Communications in Mass Spectrometry* 23(16), 2534–2542.

Herak M. (1980) Sustav navlaka između Vrbovskog i Delnica u Gorskom Kotaru (Hrvatska) (The nappe system between Vrbovsko and Delnice in Gorski Kotar (Croatia). *Acta Geologica (Prirodoslovna Istraživanja)* 10(2), 35–51.

Hubbard E.F., Kilpartick F.A., Martens L.A., Wilson J.F.Jr. (1982) Measurement of time of travel and dispersion in streams by dye tracing. Techniques of Water-Resources Investigations of the US Geological Survey, Book 3: Applications of hydraulics, pp 44.

HGI – Geological Survey of Croatia (2011) Hydrogeological Map of Croatia, 1:300 000, Zagreb 2011.

Kresic N., Z Stevanović (2010) *Groundwater Hydrology of Springs – Engineering, Theory, Management, and Sustainability*. Elsevier, ISBN: 978-1-85617-502-9

McCuire K.J., McDonnell J.J. (2006) A review and evaluation of catchment transit time modeling. *Journal of Hydrology* 330, 543–563.

Pavuza R. & Traindl H. (1983) Über Dolomitkarst in Österreich. Die Höhle 34(1), 15–25, 7 Abb., Wien.

Savić D., Dozed S. (1985) Osnovna Geološka Karta SFRJ Delnice 1:100 000, Geoloski Zavod, OOUR Geologiju i Paleontologiju, Zagreb, Geoloski Zavod – Ljubljana.

Stadler H., Skritek P., Sommer R., Mach R. L., Zerobin W., A.H. Farnleitner (2008) Microbiological monitoring and automated event sampling at Karst springs using LEO-satellites, *Water Science Technology* 58(4), 899–909.

Strobl T., Zunic F. (2006) Wasserbau, Aktuelle Grundlagen – Neue Entwicklungen, Springer-Verlag, pp 92–95.

Vreca P., Bronić I.K., Horvatičić N., Barešić J. (2006) Isotopic characteristics of precipitation in Slovenia and Croatia: Comparison of continental and marine stations. *Journal of Hydrology* 330, 457–469.

Chapter 14

Scale-dependent evaluation of an unconfined carbonate system – Practical application, consequences and significance

Ádám Tóth & Judit Mádl–Szőnyi
Department of Physical and Applied Geology, Eötvös Loránd University, Budapest, Hungary

ABSTRACT

Understanding of karst systems at basin scale has only been highlighted in a few theoretical studies. This paper intends to demonstrate the applicability of the gravity-driven regional groundwater flow (GDRGF) concept in karst systems. For this purpose, the hydrogeological conditions of a golf course, situated in the Balaton Highland, Hungary, were examined at different scales. The goal was to define the appropriate scale and to reveal the effects of different structures; and to give prognoses for the possible impact of a planned drinking water well on the golf course. Field-scale study has shown that the discharge at the golf course is tectonically-controlled and it is fresh karst water. Cluster analysis of spring data resulted in heterogeneity- and structure-related groups at aquifer scale. The real underground flow path of spring groups could be interpreted only in GDRGF context at basin scale. The appropriate scale to solve the conflicts could be derived.

14.1 INTRODUCTION

Karst-related studies nowadays have to react to the future climate prognoses and deal with their consequences on water utilisation possibilities in these changing circumstances. The review of the most important approaches of karst modelling for water resource predictions by Hartmann *et al.* (2014) has revealed the difficulties of representing spatial heterogeneities in karst and the uncertainties due to restricted information on observed discharges in the system. These can cause uncertainties also in the evaluation of the effect of future climate changes on karst systems. The main conclusions of the review were that 'we need better methods to apply karst models at large scales' and 'we could make progress in the large-scale application of karst models' (Hartmann *et al.*, 2014 p.18). These conclusions clearly show the demand for understanding the processes on larger spatial scale in carbonates.

Karst systems are mostly integral parts of sedimentary basins. Therefore, the applicability of the gravity-driven regional groundwater flow (GDRGF) concept to the unconfined and adjoining confined carbonate systems on basin scale (large scale) can give a new insight into their processes as it was proposed by Mádl-Szőnyi and Tóth (2015) (Figure 14.1). The application of this approach has proved that the effect of heterogeneity and anisotropy on the flow pattern could be derived from hydraulic responses of the system on this scale. In addition, this study demonstrated that the evaluation of natural springs, as discharge phenomena (Tóth, 1971), reveals the nature, hierarchy, chemical and temperature distribution of topography-driven regional groundwater flow in carbonate systems at basin scale. The connection between large, basin-scale carbonate and shallow, aquifer-scale karst (Goldscheider & Drew, 2007) conceptual models were also proposed by the implementation of aquifer-scale model into the basin scale one as a local flow system (Figure 14.1).

The objective of this study was to demonstrate the usefulness of the GDRGF concept with the determination of the appropriate scales in solving karst-related problems, such as conflicts of interest regarding water utilisation. The study also demonstrates how the concept can be involved in practical solutions with the implementation of subsequent scale studies.

14.2   BASIC CONCEPT AND THEORETICAL CONSIDERATIONS

Thick carbonate systems can be characterised by hydraulically connected unconfined and confined subregions, where water table elevation differences induce GDRGF.

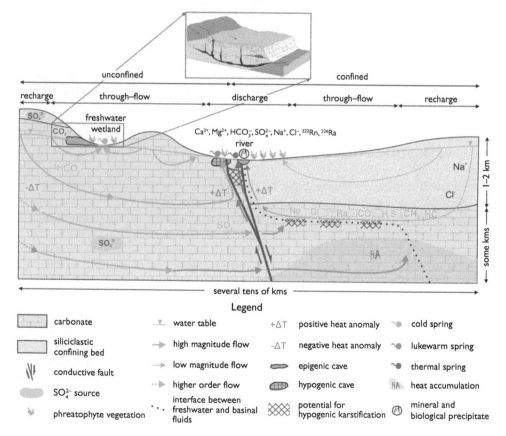

Figure 14.1 Conceptual Tóth-type GDRGF pattern of thick carbonate systems and its consequences on flow-related manifestations; the shallow karst aquifer (modified after Goldscheider and Drew 2007) is embedded into the regional flow pattern as a local system (Figure 14.10a of Mádl-Szőnyi and Tóth, 2015).

It has been revealed by theoretical considerations that hydraulic continuity is more effective in such systems than in siliciclastic sedimentary basins due to the more efficient hydraulic diffusivity (Mádl-Szőnyi & Tóth, 2015). The scale dependence of carbonates was discussed by Király (1975) with the distinction between laboratory, borehole and basin scale. This concept has been reformed by Hartmann *et al.* (2014) with the modification of the basin scale to catchment/aquifer scale and with the implementation of the temporal dimension of karstification into the concept. The difference in permeability increases from borehole to aquifer scale, depends on the evolutionary stage of karstification. However, the substitution of basin scale by aquifer scale is scientifically not acceptable, but borehole scale in Király (1975) can correspond to aquifer scale. At the basin scale, the field of interest is the karstified carbonate basin which is represented by a continuous porous environment with implemented fracture and fault zones (e.g. Wellman & Poetier, 2006).

It means that the previous concern regarding the representation of karstification can be resolved, because it is not necessary at basin scale. On this scale the hydraulic reactions of the system can be used to better understand the heterogeneity and anisotropy of the systems, including either hydraulic continuity or compartmentalisation.

Additionally, springs and other groundwater flow-related discharge features, like seepages, wetlands, epigenic and hypogenic caves, mineral and biological precipitates can be used to characterise subsurface flow conditions (Tóth, 2009), based on the concept of groundwater as a geologic agent (Tóth, 1999). Parameters and areal distribution of these features can inform us about their parent flow systems in thick carbonates.

14.3 PRACTICAL APPLICATION FOR A CASE STUDY

The local study area, a golf course, is situated in the unconfined part of the Transdanubian Range Unit, a few kilometres from Lake Balaton, Hungary (Figures 14.2a–b). The conflicts of interest in the area occur because a karst drinking water well is planned near (~1500 m) the golf course (ENVICOM2000, 2012). The question is how will the natural discharge on the golf course be influenced by the karst drinking water production well. The goal of this study is not to solve this local problem. This chapter intends only to demonstrate the importance of the understanding the appropriate scale in karst studies and to illustrate how the GDRGF concept can help to determine it.

14.4 GEOLOGY AND HYDROGEOLOGY OF THE STUDY AREA

The Transdanubian Range (TR) of Hungary, extending for 250 km in a north east – south west direction, is bounded by strike-slip and normal faults and consists of geologically complex hills and mountains (Haas 2001, 2012). The Transdanubian Range Unit (Figure 14.2a) is built up mainly of Triassic and Early Jurassic confined and unconfined carbonates, but beside carbonates Paleozoic metamorphic and Permian–Cretaceous sedimentary rocks can be found also in the region (Haas, 2001). The

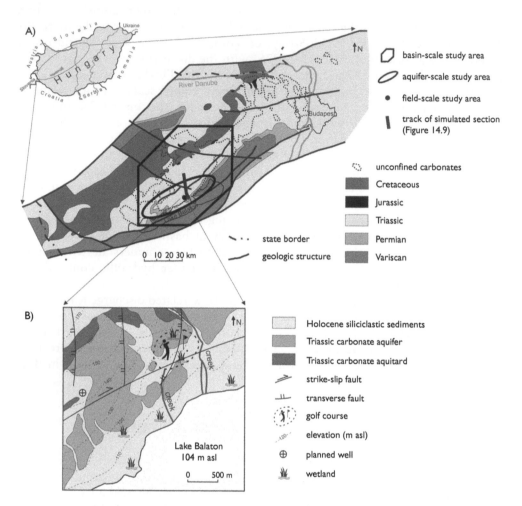

Figure 14.2 A) The Transdanubian Range Unit of Hungary, boundaries and distribution of Paleozoic and Mesozoic formations and major structural elements (modified after Haas (2001) after Fülöp *et al.* (1987)) in addition the delineation of unconfined and confined carbonates with the different study scales. B) Topography, hydrostratigraphy and the main structural elements of the field-scale study area. Location of the golf course and position of the planned water abstraction well are also indicated. Structures are adapted from Dudko (1991).

structural pattern is characterised by a large north east – south west trending synform accompanied by antiforms (Fodor, 2010; Haas, 2012).

The Transdanubian Range is the biggest karstified carbonate aquifer system in Hungary (Csepregi, 2007). Natural discharge takes place via cold, lukewarm and thermal springs as well as creeks and wetlands. Groundwater in the area is used mainly for public water supply, healing and heating purposes, and also as mineral water. The region is rich in brown coal and bauxite resources which can be found below the water table, therefore, for economic production, mine dewatering (between the 1950s to 1990s) was required which has caused significant water level decrease (Alföldi & Kapolyi, 2007).

The region of the golf course is located between 115–135 m elevation. Originally, it had gradually sloping relief but nowadays it is an artificially-formed hummocky small plateau with several tiny artificial lakes all but one of which are lined (Figures 14.2b and 14.3). The climate is continental with some Mediterranean influence (average annual precipitation is ~500–700 mm). The area of the golf course is characterised by natural diffuse discharge. East and south of the golf course small natural creeks can be found as discharge features. Historical records, memoirs and archaeological artefacts indicate that this site has been inhabited permanently from the 3rd century suggesting the area has had a constant and stable water supply for human beings (ruins of Roman Ages can be found here around a spring [Figure 14.3]) (http://www.orvenyes.hu).

Geologically the area is built up by unconfined Triassic carbonates and they are partially covered by Quaternary siliciclastic sediments (Figure 14.2b). At this site the Triassic carbonates can be distinguished according to their hydraulic behaviour. Undifferentiated dolomites and limestones are the main aquifer for karstwater in the region with hydraulic conductivity (K) of 10^{-5}–10^{-4} m/s. However, the cherty and tuffaceous carbonate formation has a lower K (~10^{-7}–10^{-6} m/s), therefore, it functions as aquitard (Tóth et al., 2014). The covering siliciclastic sediments have limited extension and they might have similar hydraulic properties to the Triassic aquifer depending on grain size distribution and degree of compaction. However, the region is highly influenced by tectonic events through geological evolution (Figure 14.2b).

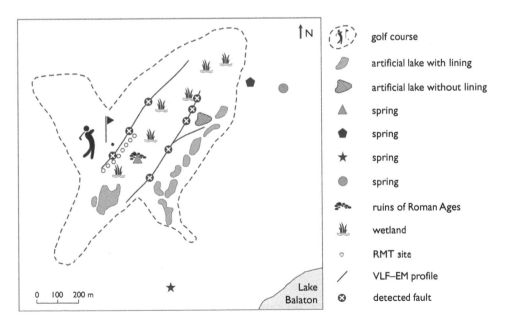

Figure 14.3 Discharge-related observations and location of electromagnetic measurements on the field-scale study area. (Different symbols of springs are introduced only for further visualisation purposes).

14.5 FIELD-SCALE STUDY

The goal of the field-scale study at the golf course was to characterise the natural discharge features. Water samples for chemical analysis were taken from the springs, geophysical measurements were carried out, as well as field observations. These data were complemented by archive data. The hydrostratigraphic and tectonic influence on discharge and the chemical character of springs were determined.

There are four springs that discharge in the vicinity of and within the golf course (Figure 14.3). Beside discrete natural discharge points, diffuse discharge also occurs forming a chain of waterlogged areas usually in local depressions of pot-holed terrain and around artificial lined lakes. These groundwater conditions together with the archive data prove the continuous existence and stability of water discharge in the area, a setting that is favourable for maintaining the fresh grass vegetation of the golf course.

Water samples of springs (indicated by circle, triangle, pentagon and asterisk) and lake without lining (indicated by grey dashed area) (Figure 14.3) were collected to assess the chemical character of the discharges. The low TDS values (760–835 mg/l) of all the water samples indicate the existence of fresh karst water characterised by calcium–magnesium and bicarbonate facies (Table 14.1, Figure 14.4). The highest TDS occurs in the spring (displayed by an asterisk) closest to the Lake Balaton.

These parameters represent relatively a shallow flow path and the magnesium concentration reflects interaction with dolomite. According to the archive dataset (Hungarian Spring Cadastre) the spring volume discharge rates in this area vary across a wide range with 0.16–0.33 l/s median values of discharge rate.

Electromagnetic geophysical measurements were carried out to determine and specify the hydrostratigraphic structure and see whether it has a modifying effect on the discharge pattern. Electromagnetic techniques are efficient tools for investigating shallow (~100 m) geological formations, and are commonly applied in hydrogeological research (e.g. Turberg et al., 1994; Nobes, 1996; Meju et al., 1999; Vereecken et al., 2005) because geological sequences (aquitards and aquifers) and structures (e.g. faults, fractures) can be investigated quickly, extensively and effectively (Tezkan et al., 2000; Gurk et al., 2001; Bosch & Müller, 2001). Radio-magnetotellurics (RMT) was applied to provide information about hydrostratigraphy and Very Low Frequency – Electromagnetics (VLF–EM) to detect tectonic elements and fractures, respectively (Figure 14.3).

Table 14.1 Measured field parameters of springs and surface waters.

	T[°C]	pH	EC (25°C) [μS/cm]	TDS [mg/l]	Discharge [l/s]
△	11.8	7.02	942	830	$\sim 1.6 \times 10^{-2}$
⬠	13.7	7.44	855	772	$\sim 8.3 \times 10^{-3}$
★	11.2	7.23	1027	835	$\sim 3.3 \times 10^{-1}$
●	12.3	7.09	903	811	$\sim 5 \times 10^{-2}$
◢	15.3	7.82	820	758	n/a

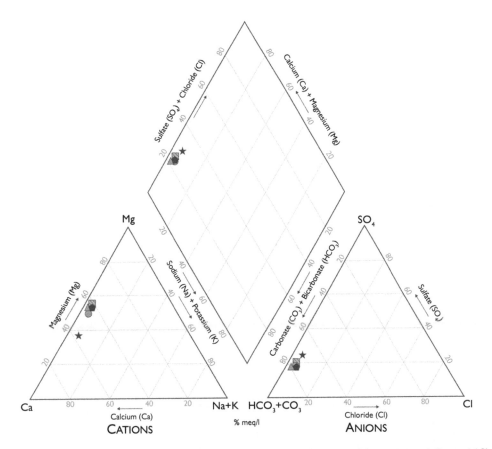

Figure 14.4 Piper diagram of water samples of the golf course (for location of data points see Figure 14.3).

According to the electrostratigraphic section (Figure 14.5) displaying RMT resistivity values in Ωm, carbonates with higher electrical resistivities (>300 Ωm) can be identified under a thin (~2–3 m) siliciclastic covering unit characterised by ~30 Ωm resistivity. There is a sharp resistivity contrast at ~100 m along the section. This anomaly can be interpreted as lithological change (carbonate and siliciclastic sediments) which might be attributed to a fault separating covered and uncovered carbonate parts. The existence of this tectonic structure could be confirmed by VLF–EM measurements (Figure 14.3). The results indicate a significantly structurally broken area which can be explained by the transverse fault system of the region (Figure 14.2b). This is an artificially influenced field and landscaping could have an influence on near-surface geological build-up as well.

Based on historical records and archive spring data the groundwater discharge of the golf course area appears to be stable and constant. Some springs were observed in the field and also wetlands with diffuse discharge can be found on the golf course (Figure 14.6). Hydrochemical facies indicate fresh karst water with similar low TDS values and similar chemical make up. Sharp lithological

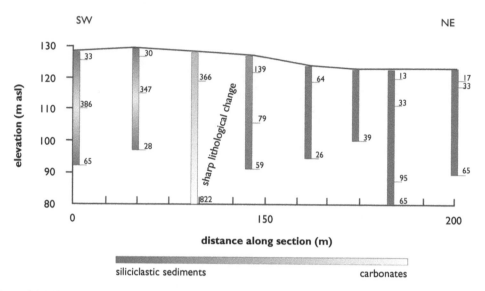

Figure 14.5 Electrostratigraphic section along RMT measurements indicating electrical resistivities in Ωm for each unit. See Figure 14.3 for the location of measured RMT data.

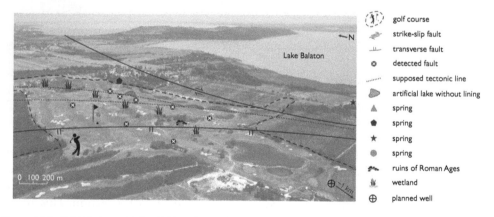

Figure 14.6 Conclusions of the field-scale study indicating tectonically-controlled discharge features on the golf course. Structures are adapted from Dudko (1991).

changes detected by electromagnetic geophysical measurements show north – south and north west – south east oriented fault systems which contribute to diffuse and point source discharges at the golf course, i.e. the discharge phenomena are tectonically-controlled.

The field-scale study showed the hydrogeological conditions around the golf course, however it did not inform in the determination of the source of the discharging groundwater which might be influenced by the planned water production. At this scale, the origin and supply of the flow systems for the wetlands and springs at the golf course cannot be shown, however, these questions can be answered by expanding the study area and lowering the scale.

14.6 AQUIFER-SCALE STUDY

In the next phase, the study area was expanded to a larger, aquifer-scale area (10s km × 10s km) of the Balaton Highland (Figure 14.2a). The delineation of the area was based on topographic considerations and the structural pattern, because the north west thrust fault is supposed to separate the Balaton Highland from the more elevated part of the Transdanubian Range (Csepregi, 2007). To understand the flow systems to the discharges at the golf course, the natural springs of the area were used to provide information about their parent flow systems. For this reason, archive data of springs (based on Hungarian Spring Cadastre) were examined and analysed by a descriptive statistical method to better understand the hydrogeological systems.

Cluster analysis (e.g. Kovács et al., 2012) was used to determine the possible groups of springs. The archive spring database of the Transdanubian Range was used for the basic data regarding springs, such as elevation, volume discharge, temperature and the most important water chemistry. The chemical parameters reflect different orders of flow systems that can be represented in the spring groups (Mádl-Szőnyi & Tóth, 2015).

The natural spring data were influenced by artificial mine dewatering (Alföldi & Kapolyi, 2007) therefore, only the data before 1960 were used to examine the natural flow conditions. The first parameter was the elevation of the spring orifice, the second one was the water temperature because there are cold and also lukewarm springs. The third parameter was the chloride content because it is a conservative natural tracer in water. Finally, the volume discharge of the springs was the fourth parameter, since springs fed by local flow systems display higher variability in discharge while higher order of flow systems are more stable (Bodor et al., 2014; Mádl-Szőnyi & Tóth, 2015).

The results of the cluster analysis show that the groups are more or less separated by elevation of spring orifice and water temperature (Figure 14.7). However, there are no significant changes in chloride concentration, discharge rates vary widely, therefore these two parameters cannot be applied for distinguishing spring groups.

The group of springs delineated by a triangle in Figure 14.7 are characterised by the lowest discharge elevation (135 m asl) and the highest temperature (12.5°C). The springs, indicated by a circle, discharge at higher elevation (~200 m asl) but are characterised by a consistent temperature of 11°C. The areal distribution of the other two groups does not show any definite pattern which can be related to groundwater flow systems. The springs on the field scale were very similar, with almost the same chemical composition, here they are part of different groups. Nevertheless, the position of springs are in good correlation with tectonic elements of the Balaton Highland area, i.e. on the aquifer scale the springs are separated into different groups supposedly due to the effect of geological heterogeneities and structures.

The origin of springs in this larger study area, the aquifer scale, have not been determined. Clustering the springs of the Balaton Highland region resulted in a geological heterogeneity-related grouping without any information about the parent groundwater flow systems. For this reason, the study area was further expanded to the Bakony–Balaton Highland region.

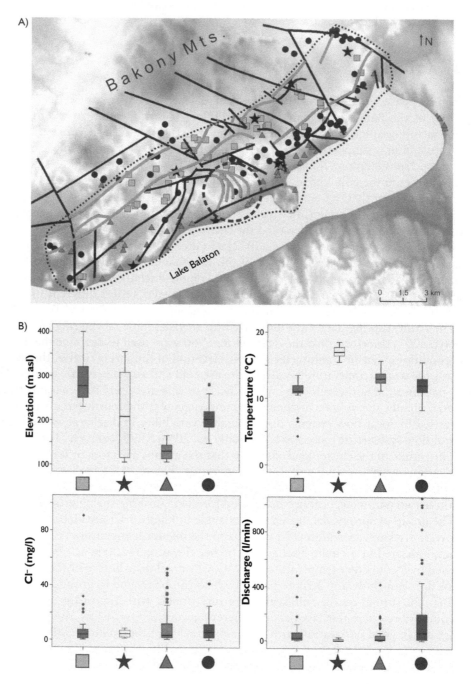

Figure 14.7 A) Areal distribution of spring groups on aquifer scale for the Balaton Highland (surrounded by dotted line). Dashed circle indicates the vicinity of field-scale study area. Tectonic elements of the Balaton Highland are adapted from Dudko (1991), transverse faults are indicated by grey colour. B) The box-and-whisker plots of the parameters.

14.7 BASIN-SCALE STUDY

The same statistical data analysis was done for the bigger study area (some 50 km × some 50 km) whih included the Bakony Mountains and Balaton Highland. In this case the parameters were more indicative of the hydrogeological settings (Figure 14.8). According to the elevation of the spring and the water temperature the springs form four distinctive groups: springs with relatively high water temperature can be found at low elevations around the Bakony–Balaton Highland area, while the coldest springs are situated at the elevated parts of the Bakony Mountains. The discharge rate of the springs still varies in a large range suggesting that there are springs with stable and high amounts of discharge and others with fluctuating discharge rates which may even dry up altogether. Variation of the chloride ion concentration is not significant because this region is unconfined and contribution from siliciclastic confining strata is small (Mádl-Szőnyi & Tóth, 2015).

These groups can be interpreted in the GDRGF context. The group indicated by triangles (Figure 14.8) represents the springs of the highest Bakony, above the Cretaceous marl in perched carbonate aquifers. Groups represented by circles and squares discharge at lower elevation (~150–300 m asl) with a slightly elevated temperature (~10–13°C). They relate to two different local flow systems. The group presented as asterisks has the highest temperature and it is characterised by the lowest elevation with a significant discharge rate (~8.33–16.66 l/s). These springs can be identified as fed by an intermediate flow system. In this basin-scale study the areal pattern of springs are more consistent with the topography-driven groundwater flow conditions. By lowering the scale, i.e. expanding study area, the areal pattern of the springs does not alter as demonstrated by Mádl-Szőnyi & Tóth (2015) for the whole Transnubian Range.

The relationship between spring groups of the field-scale study area and the simulated flow pattern were compared based with the scenario simulations of Mádl-Szőnyi & Tóth (2015, Figure 14.2a). The springs of the field-scale study indicate the same, supposedly local flow-related group (squares on Figure 14.7). Consequently, this scale is necessary to explain the source and flowpath of groundwater towards the golf course area, since GDRGF is the most important agent within this framework.

In the simulated section the northern boundary of the source area of the golf course a low permeability thrust fault is present (Figure 14.7). However, the simulated flow pattern displays that there is throughflow across this fault in spite of its low permeability, because the main hydraulic gradient is perpendicular to the fault. There is a deeper throughflow component as well, which can contribute with a deeper water component to the springs (the asterix group in Figures 14.3 and 14.4.).

14.8 DISCUSSION

The direct recharge area of the golf course can be found down-dip of the thrust fault between the Balaton Highland and the Bakony Mountains (~12 km in Figure 14.9). This direct recharge produces a large amount of water flowing towards the golf course. However, there is some throughflow across the thrust fault, from higher topographic regions (this additional recharge is at ~8 km) and a deep water flow contribution as

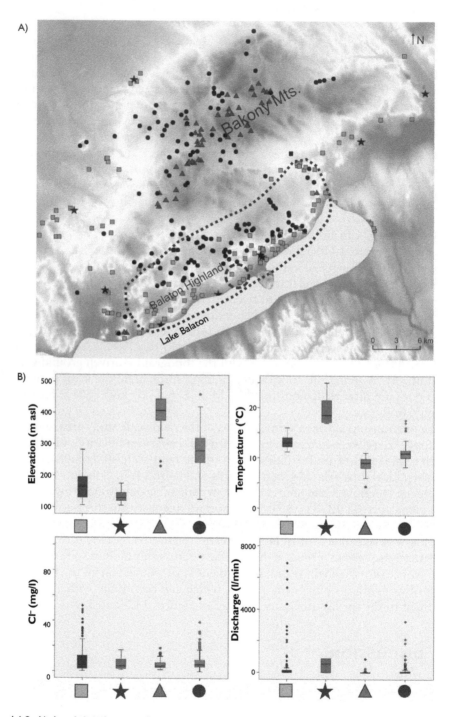

Figure 14.8 A) Areal distribution of spring groups on basin scale for the whole Bakony Mountains and Balaton Highland. Dashed circle indicates the vicinity of field-scale study area and dotted line surrounds the aquifer-scale study area. B) The box-and-whisker plots of the parameters.

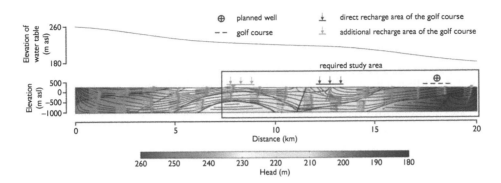

Figure 14.9 Flow distribution, heads (m), streamlines and Darcy's velocity vectors (magnitude-controlled) based on flow simulations along a section (see location on Figure 14.2a) of Mádl-Szőnyi & Tóth (2015; Figure 14.7c) and with the indication of the supposed recharge and discharge area of golf course and the projected location of the planned well.

well (Figure 14.9). The velocity vectors are displayed in magnitude, therefore, they do not represent the drop in discharge across the fault. The clustering of springs on the basin scale represents local flow systems (Figure 14.8) with Ca–Mg and HCO_3 water types. Only one spring displays a higher TDS and lower magnesium and higher sulphate and chloride concentrations, indicating some additional deeper water contribution (Table 14.1, Figure 14.4). The south east boundary of the discharge region of the golf course is the strike-slip fault (Figures 14.2b and 14.6). The discharge of the spring with elevated TDS (indicated by asterix) can be connected to this. There are two transverse faults which influence the delineation of the study area. The north east and south west boundary of the study area is a stream and the western transverse fault respectively (Figure 14.2b). Due to limited throughflow from the Bakony Mountains across the thrust faults and the efficient hydraulic continuity in the carbonates, the effect of the planned production well on the discharge at the golf course cannot be overlooked. The strike-slip fault in the south east and the thrust fault in the north west, as hydraulic boundaries, may contribute to this effect based on the supposed superposition of drawdown.

3D transient numerical simulation of the potential hydraulic effects of the new well require the field-scale study to be extended as far as the western transverse fault (Figure 14.2b) to understand the hydrostratigraphy and tectonics of the area between the well and the golf course.

14.9 CONCLUSION

The objective of this study was to demonstrate the usefulness of the GDRGF concept in solving karst-related problems in an unconfined carbonate basin. Theoretical research suggested that karst systems, as integral parts of sedimentary basins are characterised by gravity-driven flow of adjoining confined and unconfined carbonate systems. This study demonstrated the importance of springs and connected natural discharge features to reveal the nature, hierarchy, chemistry and temperature distribution of topography-driven regional groundwater flow. Hydraulic connection also occurs between large, basin-scale carbonate and shallow aquifer-scale karst features. On regional scale, hydraulic continuity or compartmentalisation of karst regions by structures or faults are more significant than the effects of local heterogeneities. In addition to these theoretical findings the applicability of GDRGF approach to solve practical questions in carbonate regions need to be assessed. One of the most important practical questions is the delineation of the appropriate study areas in such regions and to learn more about the effects of faults and hydraulic connectivity between different parts of the carbonate basin.

ACKNOWLEDGEMENTS

The authors appreciate the help of collaborating colleagues (Imre Müller, Szilvia Simon, Ferenc Zsemle, Tímea Havril) and students during field-scale study. Technical support of József Kovács, Gergely Hornyák, Petra Bodor and Soma Szathmári is also acknowledged. The access to the field-scale problem was provided by Tibor Sarlós and József Vers. Valuable comments of anonymous reviewers are appreciated. Language editing by Nicholas Robins is highly acknowledged.

REFERENCES

Alföldi L., Kapolyi L. (eds) (2007) Bányászati karsztvízszintsüllyesztés a Dunántúli-középhegységben. (Mining–dewatering in the Transdanubian Range). Geography Institute of Hungarian Academy of Sciences.

Bodor P., Tóth Á., Kovács J., Mádl-Szőnyi J. (2014) Multidimensional data analysis of natural springs in a carbonate region. *Extended Abstract, EAGE/TNO Workshop: Basin Hydrodynamic systems in Relations to their Contained Resources*. Utrecht, 6–8 May 2015.

Bosch F.P., Müller I. (2001) Continuous gradient VLF measurements: A new possibility for high resolution mapping of karst structures. *First Break* 19(6), 343–350.

Csepregi A. (2007) A karsztvíztermelés hatása a Dunántúli–középhegység vízháztartására. (The effect of water withdrawal on the water balance of the Transdanubian Range). In: Alföldi L., Kapolyi L. (eds) Bányászati karsztvízszintsüllyesztés a Dunántúli–középhegységben. (Mining–dewatering in the Transdanubian Range). Geography Institute of Hungarian Academy of Sciences, 77–112.

Dudko A. (1991) A Balaton-felvidék szerkezeti elemei (Tectonic elements of the Balaton Highland). Manuscript. Hungarian Geological, Mining and Geophysical Database, Budapest

ENVICOM2000 (2012) Balaton-felvidék vízbázis-kutatás előzetes vizsgálata (Preliminary investigation of drinking water resources in the Balaton Highland area). Report, 50.

Fodor L. (2010) Mezozoos–kainozoos feszültségmezők és törésrendszerek a Pannon–medence ÉNy-i részén – módszertan és szerkezeti elemzés. Akadémiai Doktori Értekezés. Magyar Tudományos Akadémia. (Stress–fields and structural settings in the NW Pannonian Basin – methods and structural analysis. DSc Thesis, Hungarian Academy of Sciences)

Fülöp J., Dank V., Ádám O., Balla Z., Barabás A., Bardócz B., Brezsnyánszky K., Császár G., Haas J., Hámor G., Jámbor Á., Kilényi É.Sz., Nagy E., Rumpler J., Szederkényi T., Völgyi L. (1987) Geologic map of Hungary without the Cainozoic, 1:500.000. Hung. Geol. Inst., Budapest.

Goldscheider N., Drew D. (2007) *Methods in Karst hydrogeology*. IAH International Contributions to Hydrogeology, Taylor & Francis, London.

Gurk M., Bosch F.P., Challandes N., Bigalke J. (2001) Integration of geophysical methods to study the fold geometry of the Tschera nappe, Eastern Switzerland. *Eclogae Geologicae Helvetiae* 94, 329–338.

Hartmann A., Goldscheider N., Wagener T., Lange J., Weiler M. (2014) Karst water resources in a changing world: Review of hydrological modeling approaches. *Reviews of Geophysics* 52(3), 218–242.

Haas J. (ed.) (2001) *Geology of Hungary*. Eötvös University Press, Budapest.

Haas J. (ed.) (2012) *Geology of Hungary*. Springer, Heidelberg.

Király L. (1975) Rapport sur l'état actuel des connaissances dans le domaine des caractéres physiques des roches karstique (Report on the current state of knowledge in the field of physical characteristics of karst rocks). In: Burger A., Dubertret L. (editors) *Hydrogeology of karstic terrains*. Int Union Geol Sci Ser B3, 53–67.

Kovács J., Tanos P., Korponai J., Kovács-Székely I., Gondár K., Gondár-Sőregi K., Hatvani I.G. (2012) Analysis of Water Quality Data for Scientists. In: Voudouris K., Voutsa D. (eds) *Water Quality and Water Pollution: Evaluation of Water Quality Data*. Rijeka: InTech Open Access Publisher, 65–94.

Mádl–Szőnyi J., Tóth Á. (2015) Basin–scale conceptual groundwater flow model for an unconfined and confined thick carbonate region. *Hydrogeology Journal* 23(7), 1359–1380.

Meju M.A., Fontes S.L., Oliveira M.F.B., Lima J.P.R., Ulugergerli E.U., Carrasquilla A.A. (1999) Regional aquifer mapping using combined VESTEM-AMT/EMAP methods in the semi-arid eastern margin of Parnaiba Basin, Brazil. *Geophysics* 64, 337–356.

Nobes D.C. (1996) Troubles waters. Environmental application of electric and electromagnetic methods. *Surveys in Geophysics* 17, 393–454.

Tezkan B., Hoerdt A., Gobashy M. (2000) Two dimensional magnetotelluric investigation of industrial and domestic waste sites in Germany. *Journal of Applied Geophysics* 44, 237–256.

Tóth Á., Mádl-Szőnyi J., Molson J. (2014) Comparison of simulated flow fields and clustering of springs for understanding deep carbonate groundwater systems. IAH International Congress, Marrakech, 15–19 September 2014.

Tóth J. (1971) Groundwater discharge: a common generator of diverse geologic and morphologic phenomena. *Bulletin of the International of Scientific Hydrology* 16(1–3), 7–24.

Tóth J. (1999) Groundwater as a geologic agent: an overview of the causes, processes, and manifestations. *Hydrogeology Journal* 7, 1–14.

Tóth J. (2009) Springs seen and interpreted in the context of groundwater flow-systems. GSA Annual Meeting 2009, Portland, 18–21 October 2009. *Geological Society of America Abstracts with Programs* 41(7), 173.

Turberg P., Müller I., Flury F. (1994) Hydrogeological investigation of porous environments by radiomagnetotelluric-resistivity (RMT-R, 12–240 kHz). *Journal of Applied Geophysics* 31, 133143.

Vereecken H., Kemna A., Munch H.M., Tillmann A., Verweerd A. (2005) Aquifer Characterization by Geophysical Methods. In: Anderson M.G. (ed) *Encyclopedia of Hydrological Sciences*. John Wiley and Sons Ltd.

Wellmann T.P., Poeter E.P. (2006) Evaluating uncertainty in predicting spatially variable representative elementary scales in fractured aquifers, with application to Turkey Creek Basin, Colorado. *Water Resources Research* 42(8), W08410. DOI: 10.1029/2005WR004431

Chapter 15

Characterization of karst system using modelling of rainfall-discharge relationship: Pireghar and Dimeh springs, Zagros Region, Iran

Zargham Mohammadi & Sepideh Mali
Department of Earth Sciences, Shiraz University, Shiraz, Iran

ABSTRACT

Hydrogeological models are an approximation of real systems. The structure of these models, includes several equations which are known as the system transfer functions that define the relationship between inputs and outputs. The transfer functions and the impulse response functions (i.e., Kernel function) of the karst springs, Pireghar and Dimeh, are identified by means of the ARX model. Daily rainfall data of the Chelgard and Farsan rain gauge stations and daily spring discharge of the Dimeh and Pireghar Springs during the period of 1999–2011 are selected as input and output, respectively in the modelling processes. The results show different contribution of the flow components, slow flow and quick flow, in the springs. The slow flow component in the Dimeh and Pireghar springs is 90% and 64% of the total spring discharge, respectively. Moreover, in order to evaluate spring discharge variations in response to a hypothetical drought conditions (assuming a decreasing trend in precipitation during the next twelve years), the spring discharge is predicted based on the calibrated model. The results reveal that, the Pireghar Spring will be more affected by drought events compared to the Dimeh Spring.

15.1 INTRODUCTION

Karst aquifers contain significant water reserves in many parts of the world, but development and management of these resources is difficult because of the complexity of the aquifers. Karst aquifers are mainly characterised by an irregular network of pores, fractures, fissures and solution conduits in various sizes and shapes. Such structures with inherent heterogeneous geometry cause complex hydraulic conditions and spatial and/or temporal variability of hydraulic parameters in these aquifers. After a rainfall event, water primarily recharges through large conduit, fissures and fractures and cause a large amount of water moves rapidly to the outlet of karst spring. But in smaller fractures and fissures, slow and predominantly laminar recharge occurs due to gradual drainage from them. These two processes form a quick flow and a slow flow component of a karst spring hydrograph (Denić-Jukić & Jukić, 2003).

Assuming a karst aquifer as linear, time-invariant and casual system, discharge of the karst spring at a time t may be represented by the superposition of three components as shown in Figure 15.1 (Denić-Jukić & Jukić 2003):

$$y(t) = y^D(t) + y^S(t) + y^Q(t) \tag{15.1}$$

Where $y^D(t)$, $y^S(t)$ and $y^Q(t)$ are named as a component of antecedent flow, slow flow and quick flow, respectively (Figure 15.1).

$y^D(t)$ is a component of the spring discharge resulting from antecedent rainfalls. It can be defined by using the superposition law and the property of independence of discharge events. It practically means that the component of discharge resulting from antecedent rainfall is equal to the spring discharge resulting of a period without rainfall. In this regard, the antecedent recession curve has the same form as the master recession curve. Consequently, the function $y^D(t)$ has the exponential form with the recession coefficient α :

$$y^D(t) = y_0 e^{-t/\alpha} \tag{15.2}$$

where y_0 is the karst spring discharge at the beginning of the period analysed.

$y^S(t)$ belongs to the water storage in the pore space and micro fissures which defined as the base flow. Generally groundwater flow in this part of the karst media known as the diffuse flow regime.

$y^Q(t)$ relates to quick recharge through sinkholes, conduit and large fissures and movement of groundwater in the karst system under the dominant turbulent flow regime.

The discharge component of $y^D(t)$ is the result of an initial storage in the karst system, while $y^S(t)$ and $y^Q(t)$ are the result of current rainfall (e.g., $x(t-\tau)$, $\tau \in [0,t]$). The relationship between the rainfall and the resulting karst spring discharge can be represented by the linear convolution integral between the rainfall and a transfer function, so Eq. 15.3 may be written as (Denić-Jukić & Jukić, 2003):

$$y(t) = y^D(t) + \int_0^t h^S(\tau)x(t-\tau)d\tau + \int_0^t h^Q(\tau)x(t-\tau)d\tau \tag{15.3}$$

where $h^S(\tau)$ and $h^Q(\tau)$ are the slow flow and quick flow transfer function, respectively.

One of well-known methods for characterising karst aquifers is modelling of the spring hydrograph based on the use of kernel functions (i.e., impulse response function). When the kernel function of a karst system is known, the output (i.e., spring discharge) can be determined for any specified input (i.e. rainfall events). If the internal structure of the system is known, the response function can sometimes

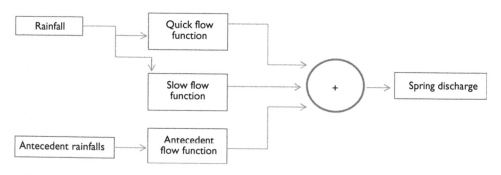

Figure 15.1 The superposition of three components to produce the spring discharge water.

be determined by analytical or numerical methods. However, the system cannot be described mathematically with a sufficient degree of accuracy, and in this case, the kernel function must be determined empirically from an observed set of output and input data (Neuman & de Marsily, 1976).

In order to simulate the karst spring hydrograph, three types of mathematical models can be applied: conceptual models, theoretical or physical models and empirical or black-box models. Very simplified physical interpretation of processes that control and transform rainfall events into spring discharge are the basis for the conceptual models. Generally, a series of linear or nonlinear reservoirs are assumed for conceptualisation of the natural processes in the real karst systems (Denić-Jukić & Jukić, 2003; Barrett & Charbeneau, 1997; Halihan & Wicks, 1998). However, the theoretical models are based on the reliable hydraulic laws which valid in different groundwater flow regimes, turbulent and laminar flow type, in the karst media (Denić-Jukić & Jukić, 2003; Teutch & Sauter, 1997; Eisenlohr et al., 1997). The theoretical models need extensive hydrological, geological and hydrogeological investigations to provide detailed information about inputs and outputs of the karst system. Since, our knowledge about the karst system is limited, often physical models are non-applicable; so a black or gray box method is used to model the karst system. Information about internal structure of the karst aquifers is not necessary in applying the black box models. The black box mode functions based on the relationship between input and output of the system. One of the most important black box model is the rainfall-runoff model based on the convolution integral between rainfall rates and system transfer functions. This model can be applied using a linear or nonlinear form of the convolution integral (Denić-Jukić & Jukić, 2003; Labat et al., 1999), but the linear form is more widely used (Denić-Jukić & Jukić, 2003). The solution of the linear form of the convolution integral is the transfer function that represents the unit response function of the karst aquifer (Denić-Jukić & Jukić, 2003). To apply the linear convolution integral in a karst system, two hypotheses are considered (Denić-Jukić & Jukić, 2003): (i) the system must be considered as time-invariant (i.e., the system transfer function is not modified with time), and the linearity of the system and consequently the property of proportionality and the superposition law are valid (i.e. the system input and output have the same scale ratio).

The ARX (Auto Regressive model with External input) model, which is the hybrid models (i.e., gray box model) with fixed internal structure is a well-known model for determination of the kernel function. In the ARX model, output values at any time are dependent on simultaneous input value and prior input and output records.

In this research, the ARX model has been used to identify the kernel functions of two karst springs, Dimeh and Pireghar, at the Zagros Region, Iran. The main objectives of this research are (1) characterisation of the karst system in the catchment area of the Pireghar and Dimeh springs base on the relationship between rainfall and spring discharge signals, and (2) evaluation of variation of spring discharge under a hypothetical drought crisis in the Pireghar and Dimeh Springs. Daily rainfall of the Chelgard and Farsan rain gauge stations and daily springs discharges of the Pireghar and Dimeh Springs over a period of 1999–2011 were used as inputs and outputs, respectively.

15.2   GEOLOGICAL AND HYDROGEOLOGICAL SETTING

The study area is located in the Zagros Mountains Range, which comprises a series of parallel north west-south east trending anticlines and synclines formed by compressional tectonics during the Miocene age. The stratigraphic and structural setting of the Zagros Mountain range is described in detail by Stocklin & Setudehnia (1977) and Alavi (2004). The Alpine Orogeny led to many thrust faults in the area. The major geological formations in the study area (Figure 15.2), in increasing order of age, include Pabdeh-Gurpi Formation (Palaeocene/Oligocene), Asmari-Jahrom and Shahbazan Formation (Oligocene to early Miocene), Razak Formation (Miocene), Gachsaran Formation (Miocene-Eocene), Mishan Formation (Middle to Late Miocene), Aghajari Formation (late Miocene to Pliocene), Bakhtiari Formation (late Pliocene-Pleistocene) and the recent alluvium. The Asmari Formation has the potential to form a karstic aquifer system. The Asmari-Jahrom and Shahbazan Formations includes limestone layers, thin layers of clay and gray limestone along with the fossils orbitolina and shale, marl and sandstone of Cretaceous age. The Upper Cretaceous and Tertiary geological formations comprise limestone layers which are suitable for the karst development.

The Asmari-Jahrom and Shahbazan formations outcrops mainly around Dime and Pireghar springs and the Saldoran and Zarab Mountains in this area are probably part of the catchment area of the springs (Figure 15.2). Large dissolution cavities have been observed in the massive limestone rocks of the Asmari-Jahrom and Shahbazan formations in the area, which could be considered as indications for development of karst. Different karstic phenomena such as springs, caves (e.g. Pireghar and Sarab

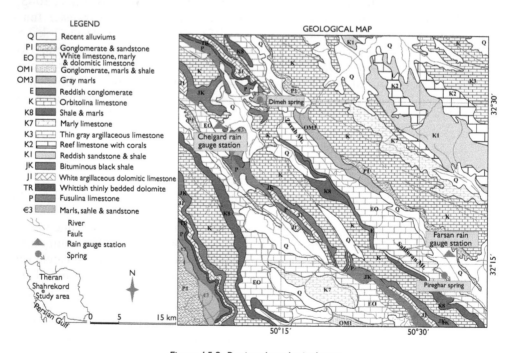

Figure 15.2 Regional geological map.

caves), karren, dissolution cavities, poljes, dolines, and dry valleys are observed locally. Several joint sets were extensively developed due to the Zagros Thrust Fault Zone, which provides favourable condition for surface recharge in the carbonate rocks consequent to the development of the karstic features.

The Dimeh and Pireghar springs are located in Zagros Region, Iran. A catchment area of 140 and 90 km² were determined for the Dimeh and Pireghar, respectively (Mali, 2014). Mean annual rainfall in the catchment area of the Dome and Pireghar springs is 542.5 and 1329.1 mm based on data from Farsan and Chelgerd rain gauge stations during the period of 1999–2011, respectively. Spring discharge in the Dimeh and Pireghar springs range from 1.5 to 4.3 and from 0.6 to 5.3 m³/s, respectively. The ratio of maximum to minimum spring discharge were computed 2.9 and 9.58 for the Dimeh and Pireghar springs. Mali & Mohammadi (2013) have shown that the recession coefficient value (α) of the Dimeh and Pireghar Springs ranges from 10^{-3} to 10^{-4} and 10^{-2} to 10^{-3}.

15.3 METHOD

The ARX model is one of the most popular models for modelling time-invariant linear systems. This model is one of the hybrid models (i.e., Gray box model) which characterise by specifying structure. Parameters of the ARX model are determined via calibration and verification process. According to the ARX model, any spring discharge record, $y_{(k)}$, at time k is dependent on values of simultaneous rainfall, $x_{(k)}$, and prior rainfall and spring discharge records as follows:

$$y_{(k)} = f\left[y_{(k-1)}, y_{(k-2)}, \cdots y_{(k-m)}, x_{(k)}, x_{(k-1)}, \cdots x_{(k-n)}\right] \tag{15.4}$$

Since f is a linear function, Eq. 15.4 could be rewritten as:

$$y_{(k)} = a_1 y_{(k-1)} + a_2 y_{(k-2)} + \ldots + a_m y_{(k-m)} + b_0 x_{(k)} + b_1 x_{(k-1)} + \ldots + b_n x_{(k-n)} + e_{(k)} \tag{15.5}$$

Where e (i.e., Error) belongs to probable noise in the measurements. By using of the Z-transform, Eq. 15.5 converts to:

$$Y(z) = (a_1 z^{-1} + a_2 z^{-2} + \ldots + a_m z^{-m})\, y(n) + (b_0 + b_1 z^{-1} + \ldots + b_n z^{-n})x(n) + E(z) \tag{15.6}$$

Rearrangement of Eq. 15.6 gives:

$$Y(z) = \frac{B(z^{-1})}{A(z^{-1})} X(z) \times Z^{-b} + \frac{1}{A(z^{-1})} E(z) \tag{15.7}$$

where

$$B(z^{-1}) = b_0 + b_1 z^{-1} + \ldots + b_n z^{-n} \tag{15.8}$$

$$A(z^{-1}) = a_1 z^{-1} + a_2 z^{-2} + \ldots + a_m z^{-m} \tag{15.9}$$

The polynomial $B(z^{-1})$ is the impact on the spring discharge from antecedent rainfall events and spring discharges. When coefficient n is large, it indicates that the spring discharge is dependent on the antecedent discharges. Polynomial $A(z^{-1})$ is the effectiveness of the current and prior rainfalls to impact the spring discharge. With a large coefficient m, prior rainfall impacts the spring discharge. $E(z)$ and Z^{-b} are the noise in the measurements and lag time between rainfall and discharge, respectively.

The modelling process follows two steps: first, determination of the parameters of the model including coefficients of n, m and b, and second, identification of the system order. The coefficients of the polynomials $B(Z^{-1})$ and $A(Z^{-1})$ are determined using the least-square algorithm.

If the response of a system to inputs is shown by h_t, the relationship between inputs and spring outputs in the time domain would be (Oppenheim & Willsky, 1997):

$$y_t = h_t \times x_t = \sum_{k=-\infty}^{\infty} h_k x_{t-k} \qquad (15.10)$$

h_t is known as the transfer function or impulse response (kernel function). The kernel function describes a logical relationship between rainfall and spring discharge in a karst system.

Two ARX models have been written in MATLAB for the Pireghar and Dimeh springs. The models were calibrated and verified based on rainfall and the spring discharge during 1999–2002 and 2002–2011, respectively. In the calibration process model parameters (e.g., coefficients of n, m and b) have been obtained according to the least-squares algorithm. The transfer functions of the quick flow, slow flow and antecedent flow are identified for the karst aquifer of Pireghar and Dimeh springs according to coefficients of n, m, b and polynomials $B(Z^{-1})$ and $A(Z^{-1})$. The kernel function of the Pireghar and Dimeh springs are obtained by superposition of three components of flow. The kernel functions have been used to predict the spring discharge under a hypothetical drought crises.

15.4   RESULTS AND DISCUSSION

Correlation coefficients between the observed and calculated spring discharge of the Dimeh and Pireghar springs are 0.93 and 0.96, respectively (Figure 15.3) which reflects an acceptable result. The difference between the observed and simulated values of the spring discharge in the Pireghar Spring is higher compared to the Dimeh Spring (Figure 15.3). Since the ARX model is a linear model, it seems that relation

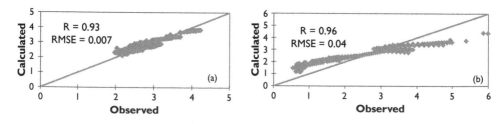

Figure 15.3 Observed and calculated discharge for (a) Dimeh Spring and (b) Pireghar Spring.

between the rainfall and the spring discharge in the Pireghar Spring is probably non-linear and the ARX model could not exactly simulate the transfer function of the karst system of the Pireghar Spring.

Figure 15.4 shows temporal variations of the observed discharges of the Dimeh Spring and the three components of spring discharge, including antecedent, slow and quick flow, calculated by the selected ARX model in the verification step. The antecedent recession curve decreases with an exponential trend ($y^D(t) = y_0 e^{-0.001t}$). It means that with a lack of recharge from rainfall, the spring will dry out during the next 3000–3500 days. The slow flow contributes to 90% of the total spring discharge and the contribution from the quick flow in the total spring discharge is low.

The recession trend of the antecedent flow in the Pireghar Spring shows the spring will dry out during the next 250–300 days, with no recharge from rainfall (Figure 15.5). The contribution of the slow flow is about 63% of total spring discharge in the Pireghar Spring (Figure 15.5). Contribution of the quick flow in the total spring discharge in the Pireghar Spring is greater than that in the Dimeh Spring.

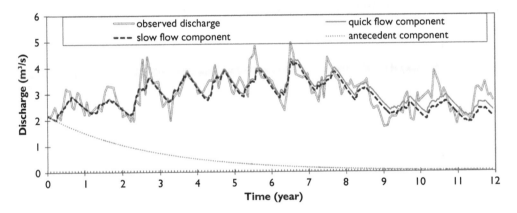

Figure 15.4 Temporal variations of observed discharges of Dimeh Spring and flow components as antecedent, slow and quick flow.

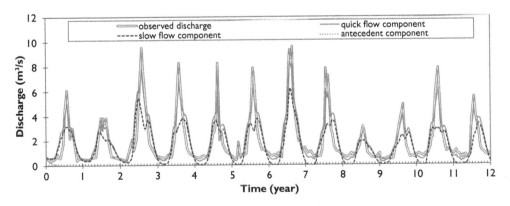

Figure 15.5 Temporal variations of observed discharges of Pireghar Spring and flow components as antecedent, slow and quick flow.

The response of the springs to rainfall events is quite different. The impulse response function (i.e. Kernel function) in the Pireghar Spring is abrupt with a sharp peak and shows less retention time in comparison to the Dimeh Spring (Figure 15.6). According to the shape of kernel functions, it seems that groundwater flow in Pireghar Spring and the Dimeh Spring is mainly controlled by the conduit and diffuse flow regimes, respectively (Figure 15.6). The rising and falling limb of the Kernel function in the Pireghar and Dime springs have different slopes due to different hydrogeological characteristics such as degree of karst development and/or size and shape of the catchment area of the springs.

Variation of the springs discharge in response to a hypothetical drought conditions (e.g., assuming a decreasing trend for rainfall during the next twelve years) is evaluated using predicted values by the calibrated models for the Pireghar and Dime springs (Figures 15.7 and 15.8). Discharge of the Dimeh and Pireghar springs has decreased to one third and one fifth of the mean annual discharge after about 8 years, respectively. Comparison of the spring discharge under assumed drought conditions with the current spring discharge reveals that discharge of the Pireghar and Dimeh springs decreases bb 59% and 36%, respectively. It seems that drought crisis is more

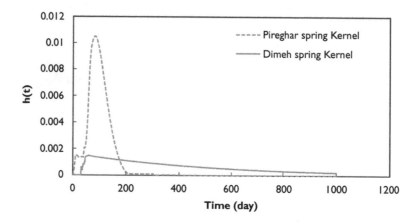

Figure 15.6 Impulse response function (Kernel function) in Pireghar and Dimeh springs.

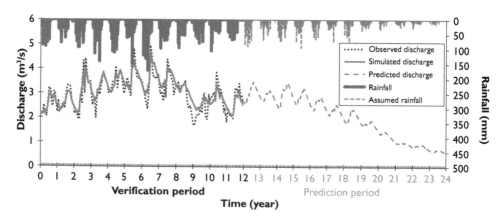

Figure 15.7 Calculated discharge of Dimeh Spring in response to real and assumed rainfall events.

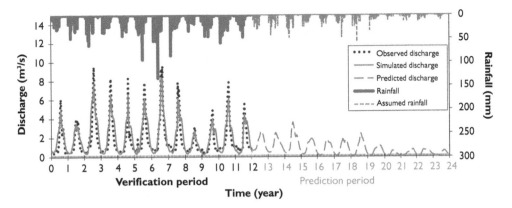

Figure 15.8 Calculated discharges of Pireghar Spring in response to real and the assumed rainfall events.

obvious in the Pireghar Spring with the dominant conduit flow regime than the Dime Spring with its diffuse flow regime. A karst spring with a conduit flow system is more vulnerable to drought.

15.5 CONCLUSIONS

The contribution of the slow flow in the discharge of the Pireghar and Dimeh springs is about 64% and 90% of the total spring discharge, respectively. The impulse response function in Pireghar Spring is abrupt with a sharp peak and includes a smaller retention time in comparison to the Dimeh Spring. It seems that the karst is probably developed to a lesser degree in the catchment area of the Dimeh Spring compared to that of the Pireghar Spring. Prediction of spring discharge reveal that the Pireghar Spring will be more vulnerable to drought crises than Dimeh Spring. In conclusion, the diffuse flow system in the catchment area of the Dimeh spring helps with storage and gradual releasing of stored groundwater during drought periods.

Comparison of the developed kernel functions for the springs confirms the conduit and diffuse flow regime in the catchment area of the Pireghar and Dime Springs, respectively. Lag time between rainfall events as input to the model and spring discharge as model output in the Pireghar Spring is lower than the Dimeh Spring. Even though the ARX model that was applied is a linear model, according to the results, it seems that the relationship between rainfall and spring discharge in both springs are not completely linear. In order to obtain a more exact simulation and prediction, a nonlinear model such as NARX is suggested.

ACKNOWLEDGMENTS

This research was supported financially by the Research Council of Shiraz University. The authors would like to thank the Chaharmahal and Bakhtiari Regional Water Authority for providing data.

REFERENCES

Alavi M. (2004) Regional stratigraphy of the Zagros fold-thrust belt of Iran and its pro-Foreland evolution. *American Journal of Science* 304, 1–20.

Barrett M.E., Charbeneau R.J. (1997) A parsimonious model for simulating flow in karst aquifer. *Journal of Hydrology* 196, 47–65.

Denić-Jukić V., Jukić D. (2003) Composite transfer functions for karst aquifers. *Journal of Hydrology* 274, 80–94.

Eisenlohr L., Bouzelboudjen M., Kiraly L., Rossier Y. (1997) Numerical versus statistical modelling of natural response of a karst hydrogeological system. *Journal of Hydrology* 202, 244–262.

Fazeli M.A., 2007. Construction of grout curtain in karstic environment case study: Salman Farsi Dam, Iran. *Environmental Geology* 51, 791–796.

Halihan T., Wicks C.M. (1998) Modelling of storm responses in conduit flow aquifers with reservoirs. *Journal of Hydrology* 208, 82–91.

Labat D., Ababou R., Mangin A. (1999) Linear and nonlinear input/output models for karstic spring flow and flood prediction at different time scales. *Stochastic Environmental Research and Risk Assessment* 13(5), 337–364.

Mali S. (2014) Prediction of karst aquifer behaviours by rainfall-spring discharge analysis: Case of Pireghar and Dime Springs, Shahr Kord M.S Thesis, Shiraz University, Shiraz.

Mali S., Mohammadi Z. (2013) Investigation of karstification by hydrograph and time-series analysis. *The 17th Symposium of Geological Society of Iran and the 1st International Congress on Zagros Orogen* Shahid Beheshti University, Tehran, Iran, Oct. 29–31.

Neuman S.P., de Marsily G. (1976) Identification of linear systems response by parametric programming, *Water Resources Research* 12(2), 253–262.

Oppenheim A.V., Willsky A.S. (1997) *Signal & System*. 2nd ed, Prentice Hall.

Singh V.P. (1988) *Rainfall-runoff Modelling, Hydrologic Systems, vol. 1.* Prentice-Hall, Englewood Cliffs, NJ.

Stocklin J., Setudehnia A. (1977) Stratigraphic Lexicon of Iran. *Geological Survey of Iran* Tehran, Iran, 376 pp.

Teutch G., Sauter M. (1997) Distributed parameter modelling approaches in karst – hydrological investigations. *Proceedings of the Sixth Conference on Limestone Hydrology and Fissured Media,* La Chaux de Fonds, Switzerland.

Part 3

The water flow in karst: From vadose to discharge zone

Chapter 16

Recharge processes of karst massifs in southern Italy

Francesco Fiorillo & Mauro Pagnozzi
Dipartimento di Scienze e Tecnologie, University of Sannio,
Benevento, Italy

ABSTRACT

The recharge processes have been evaluated for several main karst massifs of southern Italy, including the Terminio, Cervialto and Matese massifs, characterised by wide endorheic areas. The annual mean recharge has been estimated with GIS-basedtools, from regression of annual mean values of different ground-elevated rain gauges and thermometers. The recharge has been separated from endorheic areas and the other areas of spring catchment, and the ratio between the output spring and input rainfall has been estimated (the recharge coefficient). Endorheic areas are exploited for hydro-electrical purpose, especially in the Matese massif, and several hydraulic works have modified the natural recharge processes and the regime of karst springs.

16.1 INTRODUCTION

Along the Apennine chain, many karst massifs outcrop, which are generally hydraulically isolated by thick flysch sequences or other non-karstic rocks along their boundaries. These massifs constitute the main water resources of central and southern Italy, and feed several powerful karst springs.

Locally, due to the Mediterranean climate, the recharge generally occurs during autumn and winter; as a consequence, karst aquifers discharge water throughout the hydrological year, primarily in response to this climatic regime, but also to the geological-geomorphological setting and karst conditions (Fiorillo, 2009).

The recharge processes of Terminio, Cervialto and Matese massifs, (Southern Italy), are characterised by wide endorheic areas. Endorheic areas are closed depressions, generally characterised by a seasonal or permanent sinking lake located at the bottom; in these areas the runoff is completely adsorbed (White, 2002). The endorheic areas constitute important recharge zones of aquifers, generally hydraulically connected to one or more springs; practically, all the rainfall falling on these areas (net of evapotranspiration) infiltrates into the aquifer. Outside endorheic areas, part of the rainfall can leave the spring catchment by runoff processes, especially during intense storms, and are here referred to as 'open areas'. The open areas constitute the difference in area between the spring catchments and the endorheic areas.

Even if karst rocks allow a high rate of infiltration, runoff occurs especially along steep slopes and during high rainfall intensity events. Volumes are difficult to estimate, because they vary mainly as a function of several parameters such as the soil moisture conditions and the intensity and distribution of the rainfall; besides, snow accumulation and consequent snowmelt can have an important role in the runoff processes. However, as a karst aquifer is able to concentrate towards a single outlet (karst spring) the rainfall infiltration, the total output from these aquifers, can be measured, allowing the recharge processes to be assessed.

The most common hydrologic parameter used to estimate recharge is the ratio between the volume of spring discharge and the rainfall volume which falls in the spring catchment (Drogue, 1971; Bonacci & Magdalenic, 1993; Bonacci, 2001), computed generally at an annual interval. This 'rough' estimation can be improved, considering the evapotranspiration processes and distinguishing the endorheic areas and open areas, characterised by different recharge conditions (Fiorillo & Pagnozzi, 2013).

Most of the recharge occurs during autumn and winter in Mediterranean climate areas, due to the increase in the precipitation and decrease in temperature. Fiorillo *et al.* (2015) have estimated the amount of recharge on karst massifs, based on a daily time step model, calibrated against a long-term annual interval model. The amount of evapotranspiration, recharge, and runoff were estimated for Terminio and Cervialto massifs, and a similar approach has been used for the Matese massif (Fiorillo & Pagnozzi, 2015). However, due to hydroelectric exploitation of the main endorheic areas of the Matese massif since the beginning of last century, recharge estimation needs more specific assessment.

16.2 MAIN GEOLOGICAL AND HYDROGEOLOGICAL FEATURES

These massifs are primarily characterised by steep slopes and high elevations up to 2050 m a.s.l. for Matese massif and 1809 m a.s.l. for Picentini mountains (Figure 16.1). Flat zones are limited to the bottom of endorheic areas which induce higher and more concentrated water infiltration.

These karst massifs are made-up of a series of limestone and limestone-dolomite (Late Triassic-Miocene), which are between 2500 and 3000 m thick. Along the northern and eastern sectors, these massifs are tectonically overlapped on the terrigenous and impermeable deposits, constituting argillaceous complexes (Paleocene) and flysch sequences (Miocene). Along the southern and western sectors, these massifs are limited by normal faults and are covered by recent quaternary deposits. More detailed geological desriptions of the outcropping areas can be found in Parotto & Praturlon (2004) and related literature, and recent Geological Map of Italy, 1:50 000 scale (ISPRA, 2015).

Pyroclastic deposits of Somma-Vesuvius activity cover the Picentini mountains, with thickness up to several meters along the gentle slopes of the Terminio zone, and of few tens of meters along the steep slopes and the entire area of Mount Cervialto. These deposits, almost absent on the Matese massif, play an important role in the infiltration of water into the karst substratum.

16.2.1 Main karst springs

The karst massifs feed many karst springs with discharges up to thousands liters/second, and constitute the main water resource for Southern Italy.

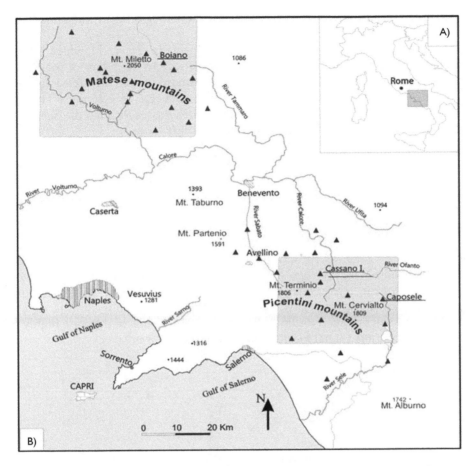

Figure 16.1 A) Study region in southern Italy – B) Map of the Western Campania region. Rectangular shaded area is detailed in figures 16.2 and 16.3. Triangles are thermometer/rain gauges used.

The Serino group is located in the valley of the Sabato River, along the north western boundary of the Picentini massif and it is formed by the Acquaro-Pelosi springs (377–380 m a.s.l.) and the Urciuoli spring (330 m a.s.l.). These springs are fed by Terminio massif (Civita, 1969, Fiorillo *et al.*, 2007) with an overall mean annual discharge of 2.25 m³/s. Local monitored piezometers have an upwelling flux in the alluvial deposits and flysch sequences, which comes from the below karst substratum of the Sabato valley, and supplies the springs. Roman aqueducts (first century AD.) were supplied by these springs and the Urciuoli spring was re-tapped between 1885 and 1888 by the Serino aqueduct, which is a gravity channel followed by a system of pressured conduits that is used to supply water to the Naples area. Additionally, the Aquaro and Pelosi springs were also re-tapped in 1934 by the Serino aqueduct.

The Cassano group is located in the Calore river basin along the northern boundary of the Picentini Mountains, and is formed by the Bagno della Regina, Peschiera, Pollentina and Prete springs (473–476 m a.s.l.). These springs are primarily fed by the

Terminio massif (Civita, 1969), with an overall mean annual discharge of 2.65 m³/s. In 1965 these springs were tapped to supply the Puglia region with water, and a gravity tunnel was joined to the Pugliese aqueduct.

The Caposele group is formed by the Sanità spring (417 ma.s.l.), which is located at the head of the Sele river basin along the north eastern boundary of the Picentini Mountains (Figure 16.2). This spring, which is primarily fed by the Cervialto mountain (Celico & Civita, 1976), has a mean annual discharge of 3.96 m³/s. The spring was tapped in 1920 by the Pugliese aqueduct, which passes through the Sele-Ofanto divide via a tunnel and supplies the Puglia region with water.

The powerful springs of Matese massif are located along the southern and northern side of the massif (Figure 16.3). The Torano spring (201 m a.s.l.), located along the eastern cliff of a karst canyon, is fed by the karst system of the central sector of Matese massif (Civita, 1969). It has a mean annual discharge of 2.04 m³/s. The Maretto spring (180 m a.s.l.) is located at the foot of limestone-dolomite slope, where there is a contact with less permeable debris, which covers the limestone rocks, and has a mean annual discharge of 0.99 m³/s. The Torano and Maretto springs were tapped during the 1960s by the Campano aqueduct, that supplies the Naples area.

The Grassano-Telese springs, are located along the southern side of the Mount Pugliano relief, near Telese village; this relief is separated from the Matese massif (Figure 16.3), but the springs have to be connected hydraulically with the wide area of the Matese karst system (Fiorillo & Pagnozzi, 2015). Some of them are highly mineralised, and the total annual mean spring discharge is 4.3 m³/s.

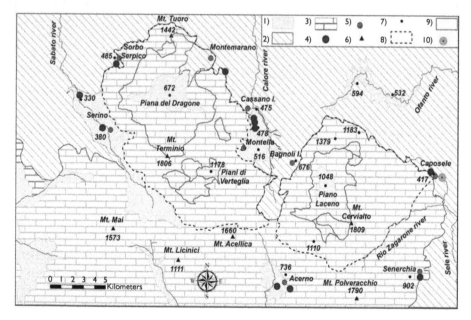

Figure 16.2 Hydrogeological sketch of north-eastern sector of Picentini Mountains (Fiorillo *et al.* 2015). 1) Slope breccias and debris, pyroclastic, alluvial and lacustrine deposits (Quaternary); 2) argillaceous complex and flysch sequences (Paleogene–Miocene); 3) calcareous-dolomite series (Jurassic–Miocene); 4) main karst spring; 5) village; 6) mountain peak; 7) elevation (m a.s.l.); 8) Cervialto and Terminio groundwater catchment; 9) endorheic area; 10) Caposele river gauge.

Along the Northern side of the Matese massif, near the village of Boiano (Figure 16.3) there are three main groups of springs (Civita, 1969): the Maiella group, west of Boiano, the Pietrecadute group located in the village, and Rio-freddo group, east of Boiano. These springs are fed by the karst system of the north-central sector of Matese massif and are located along debris deposits that cover the tectonic contact between limestone and flysch sequences. Altogether these springs have an annual discharge of about 2.80 m³/s. Most were tapped by the Biferno Aqueduct during the 1960s and supply the Molise region. A branch of this aqueduct is joined with Campano Aqueduct by a tunnel in the eastern side of Matese massif.

Other springs are located inside endorheic areas or in the high elevated zones, and their discharge from the massif is lower than from the basal springs.

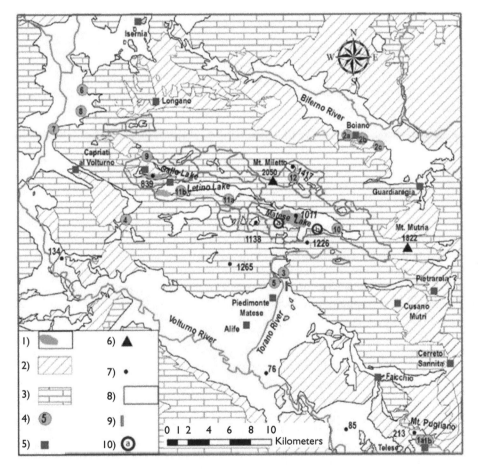

Figure 16.3 Hydrogeological sketch of Matese massif (modified from Fiorillo and Pagnozzi, 2015). 1) Slope breccias and debris, pyroclastic, alluvial and lacustrine deposits (Quaternary); 2) argillaceous complex and flysch sequences (Paleogene–Miocene); 3) calcareous-dolomite series (Jurassic–Miocene); 4) main karst springs, *1* – Grassano-Telese, *2* – Boiano, *3* – Torano, *4* – Ielo, *5* – Maretto, *6* – San Lazzaro, *7* – Torcio, *8* – Dei Natali, *9* – Capo Le Mandre, *10* – Matese lakes springs, *11* – Letino lake springs, *12* – Capo d'acqua; 5) village; 6) mountain peak; 7) elevation (m a.s.l.); 8) endorheic area; 9) Torano river gauge; 10) sinkhole: a-Brecce, b-Scennerato.

The long-term spring discharge measurements and the relation to climate variable have been analysed by Fiorillo & Guadagno (2012), and the hydraulic aquifer behaviour during droughts has been described by Fiorillo (2009) and Fiorillo *et al.* (2012).

16.2.2   Endorheic areas

These massifs are characterised by wide endorheic areas, which have an important role in the recharge processes. The origin of these endorheic zones is connected to tectonic activity during upper Pliocene-Pleistocene, which has caused a general uplift by normal faults, and the formation of grabens. Most of the grabbens are now typically poljes.

The Terminio massif is characterised by several endorheic areas (Figure 16.2), where the largest is the Piana del Dragone (55.1 km^2), the biggest polje of the Picentini mountains. Several sinkholes drain this endorheic area, and hydraulic works were carried out to limit flooding during the wet winter period, connecting a drainage system to the Bocca del Dragone sinkhole. Tracer tests testified the connection between this sinkhole and Cassano springs (Celico *et al.*, 1982), indicating that this area belongs to the recharge area of these springs.

The Cervialto massif is also characterised by several endorheic areas (Figure 16.2), where the largest is the polje of Piano Laceno (20.5 km^2). Here a small sinking lake exists, which increases in volume during the winter and spring period. As the Caposele spring can be considered the only large spring draining the Cervialto Massif (other springs has a considerable lower discharge), the Piano Laceno endorheic area is part of the recharge area of this spring. Due to a local cave system (Grotta del Caliendo) in the West side of the Piano Laceno polje, only a limited volume of water escapes from this endorheic area and is drained to the Calore valley.

The main endorheic area of the Matese massif is Lago Matese (Figure 16.3), which is the most important karst feature and constitute a wide polje. It occupies an area of 45 km^2 between 1000 and 2050 m a.s. l, and it is an important recharge area of the massif. A permanent lake exists, drained by several sinkholes covered by lacustrine sediments, and by two other sinkholes (Brecce and Scennerato) located along the southern side of the lake. Since the 1920s the lake has been exploited for hydroelectricity, by two distinct falls (479 and 353 m); before the hydraulic works of 1920s, the Brecce and Scennerato sinkholes drained the lake, and the range of the lake level was between 1007–1009 m a.s.l. The hydraulic works of the 1920s have isolated these sinkholes by earth dams, and the maximum water level of the lake reaches 1012 m a.s.l.; the maximum volume stored in the lake currently is 15 Mm^3. The hydropower system consists of two plants powered by two distinct falls. The amount of water discharged from the Lago Matese for hydropower purpose varies during the year, with a long period of no discharge during the summer and autumn.

On the Matese massif, two other main lakes are present: Lago Letino and Lago Gallo (Figure 16.3), which are both dammed as reservoirs. Before dam construction, the Sava river and Lete river drained the current Gallo and Letino lake catchments, respectively. Both rivers disappeared in swallow holes and emerged downstream of the current dams. Letino and Gallo lakes are joined by a tunnel, and feed a hydropower plant. Then, the water of Gallo lake feeds a second hydropower plant located in the Volturno plain (Capriati plant hydropower), with a fall of 654 m. Lago Gallo is also used as pump/storage reservoir.

16.2.3   Spring catchments

The definition of catchment boundaries of each of the karst springs is difficult in a karst environment, and induce erroneous estimation of recharge volumes. A useful approach is to associate the entire massif with one lumped system, and to consider the overall output from the spring outlets, without focussing the analysis on a single spring and its capture zone. This is the case of the Matese massif, where a complex karst system of 551.5 km^2 feeds numerous basal karst springs; 22% of this area is occupied by the endorheic areas.

In the case of the Cervialto massif, the definition of the spring catchment appears to be simpler, as it is drained only by the Caposele spring, and is bounded by impervious terrains; only along the southern part, can the spring catchment not be accurately defined. For the karstified limestone outcrops and the morphological features of the calcareous area with an elevation higher than that of the spring, the estimated recharge area is 110 km^2 and the endorheic areas occupy 25% of the spring catchment.

For the Terminio catchment several basal karst springs exist along its boundary, and the specific delimitation of each spring catchment is not easy; in this case, an overall estimation of the recharge and discharge outlet is considered. The endorheic areas occupy 39% of the spring catchment, which has area of 163 km^2.

16.3   GROUNDWATER RECHARGE MODEL

Fiorillo et al. (2015) proposed an model to estimate the groundwater recharge at the long-term scale, especially for large areas with strong morphological irregularities, and not completely covered by hydrological monitoring. Based on long-term annual mean data, the afflux (that is the total amount of meteoric precipitation), runoff, and recharge are computed in GIS environment at yearly intervals, to estimate the recharge and the runoff coefficient, for both the open areas and the endorheic areas. After estimating the annual mean recharge, daily recharge has been estimated. Starting from the beginning of the hydrological year, when field capacity has been reached, the daily rainfall is split in (daily) recharge and (daily) runoff.

The annual model provides the long term estimation of the recharge. It is based on GIS analyses, where the afflux, evapotranspiration and effective rainfall on the catchment are estimated. The spatial distribution of the rainfall allows estimation of the total amount of the afflux, F, in a specific area, A, by:

$$(F)_A = \frac{\sum_1^n P}{n} \times A \tag{16.1}$$

where n is the number of cells in the area A, and P is the annual mean rainfall assigned to each cell in as a function of its ground elevation, and deducted from the regression of available rain gauges, located at different elevations.

If the actual evapotranspiration, AET, is subtracted from the rainfall, the effective afflux, F_{eff}, in a specific area, A, is:

$$\left(F_{eff}\right)_A = \frac{\sum_{1}^{n} P_{eff}}{n} \times A \tag{16.2}$$

In endorheic areas, A_E, as the runoff cannot leave the catchment, the recharge amount, R, can be considered equal to the effective afflux, F_{eff}:

$$\left(R\right)_{A_E} = \left(F_{eff}\right)_{A_E} \tag{16.3}$$

If the water abstraction occurs inside the endorheic areas, the (net) recharge can be considered to be reduced:

$$\left(R\right)_{A_E} = \left(F_{eff} - Q_P\right)_{A_E} \tag{16.4}$$

where Q_P is the annual water amount abstracted from the endorheic areas. The value of Q_P has to be known in the case of abstraction from wells for human and agricultural activity; the amount of water subtracted for hydroelectrical purpose was estimated.

In the open areas the recharge amount, $(R)_{Ao}$, can be estimated assuming that all the groundwater flow feeds the spring discharges, Q_s, and no-flow boundaries occur towards the argillaceous, terrigenous and flysch sequences (impervious terrains). With this assumption, the total discharge, Q_s, from springs is:

$$Q_s = (R)_{A_E} + (R)_{AO} \tag{16.5}$$

Therefore, the recharge in the open areas is:

$$\left(R\right)_{Ao} = Q_s - \left(F_{eff} - Q_P\right)_{A_E} \tag{16.6}$$

and the total recharge on the catchment area, Ac, is:

$$\left(R\right)_{Ac} = (R)_{Ao} + (R)_{A_E} = Q_S - Q_P \tag{16.7}$$

Recharge can be expressed in term of a fraction of the effective afflux, providing the effective recharge coefficient, C_R; if water pumping does not occur, $Q_P = 0$, the following equations can be deducted:

$$\left(C_R\right)_{A_E} = 1; \quad \left(C_R\right)_{Ao} = \frac{(R)_{Ao}}{(F_{eff})_{Ao}}; \quad \left(C_R\right)_{Ac} = \frac{(R)_{Ac}}{(F_{eff})_{A_c}} \tag{16.8}$$

In general, these coefficients express the infiltration capacity of karst slopes and depend on the slope angle distribution, vegetation, soil type, and on the degree of karstification.

To estimate the fraction in the spring discharge provided by each endorheic area, the index, C_S, was computed:

$$(C_S)_{A_E} = \frac{\left(F_{eff} - Q_P\right)_{A_E}}{Q_S} \qquad (16.9)$$

and, as a consequence, the effective contribution to spring discharge open areas:

$$(C_S)_{A_O} = 1 - (C_S)_{A_E} \qquad (16.10)$$

The daily model is based on the balance of the water content of the soil the mantle provided by (Fiorillo & Wilson, 2004), transforming the daily rainfall to water content, up to field capacity, and then to recharge and runoff.

In particular, the daily rainfall, P_i, is divided into several parts:

$$P_i = AET + \Delta\Theta + R + RO \qquad (16.11)$$

where AET is the actual evapotranspiration, $\Delta\Theta$ constitutes the increase of water content of the soil, R is the recharge, and RO is the runoff.

During dry periods ($P_i = 0$), both R and RO are nil and Equation 11 reduces to:

$$-\Delta\Theta = AET \qquad (16.12)$$

This describes the soil moisture decrease as consequence of evapotranspiration processes at a daily scale, up to a minimum value of the soil moisture, Θ_{min}. The AET depends on the soil moisture, Θ:

$$AET = PET \text{ for } \Theta > \Theta_{min}; \ AET = 0 \text{ for } \Theta = \Theta_{min} \qquad (16.13)$$

where PET is computed adapting the *Thornthwaite* model (Thornthwaite, 1948) at daily scale (Fiorillo & Wilson, 2004).

During wet periods, the soil moisture can increase up to a maximum value, Θ_{max}, above which the water cannot be retained as retention water and has to percolate throughout the soils as runoff. The value Q_{max} can be approximated to the field capacity of the soil. Thus, when field capacity has been reached, recharge, R, and the runoff, RO, can occur. These two amounts, constitute the excess rainfall, P_{exc}:

$$P_{exc} = R + RO \qquad (16.14)$$

which are the part of the rainfall which is 'free' in the system (not transpired and not retained as soil moisture). In this model the amount $\Delta\theta + R$ constitutes the infiltration.

In the endorheic areas, A_E, as the runoff, RO, cannot leave the catchment, all the excess rainfall, P_{exc}, can be considered as recharge:

$$(P_{exc})_{A_E} = (R)_{A_E} \qquad (16.15)$$

In the open areas, A_O, the model assumes that runoff occurs if rainfall excess, P_{exc}, exceeds a specific threshold value, T_r, at a daily scale:

$$\text{for } P_{exc} \leq T_r, \ (P_{exc})_{A_O} = (R)_{A_O} \tag{16.16}$$

$$\text{for } P_{exc} > T_r, \ (P_{exc})_{A_O} = (R)_{A_O} + (RO)_{A_O}, \text{ and } (R)_{A_O} = T_r \tag{16.17}$$

The threshold value, T_r, corresponds to the infiltration capacity at a daily scale when the soil has reached field capacity. T_r is deducted in the simulation, for a specific hydrological year, the ratio between the annual recharge amount in open areas, $S(R)_{A_O}$, and annual excess rainfall, ΣP_{exc} (both computed summing daily values), is equal to $(C_R)_{A_O}$ computed at long-term annual scale (Equation 8). In particular, the threshold can be estimated by considering several hydrological years characterised by annual rainfall near the mean.

Some variables need to be fixed on the basis of in situ and laboratory tests, to allow simulation of the evapotranspiration and recharge processes on daily scale to be undertaken. The soil moisture capacity is needed, which depends on the thickness of the soil involved in the evapotranspiration processes and on the range of the soil moisture during the hydrological year $(\Theta_{max} - \Theta_{min})$.

In conclusion, the equations regarding annual and daily scale recharge, described in Fiorillo *et al.* (2015), are computed analysing the morphological features of the catchment area, which were divided in two main areas: the endorheic and open areas.

16.3.1 Data processing

The recharge processes have been estimated using hydrological data monitored for the period 1970–1999, except in the high-elevation station rain gauges (Figure 16.1).

The annual rainfall and the annual mean temperature depend strongly on the ground elevation; the regressions were found by Fiorillo (2011) for the Cervialto and Terminio massifs, and by Fiorillo & Pagnozzi (2015) for the Matese massif (Figures 16.4 and 16.5). These regressions, were used to estimate the actual evapotranspiration in a GIS environment (Turc, 1954).

Given Equations 16.4, 16.6 and 16.7, the recharge amount, has been estimated for each massif; the runoff, RO, which occurs only in the open areas, has been estimated by:

$$(RO)_{A_O} = (RO)_{Ac} = (F_{eff})_{A_O} - (R)_{Ao} \tag{16.18}$$

Tables 16.1 and 16.2 show the results for the long term annual scale obtained for each massif. The effective recharge coefficient computed for the open areas, $(C_R)_{A_O}$, appears to have comparable values for the Cervialto (0.66) and Terminio (0.67) massifs. The values of $(C_R)_{A_O} = 0.66$ has been used to estimate the recharge of the Matese massif, where the total outflow from the massif is partially known (Fiorillo & Pagnozzi, 2015).

For the entire spring catchment, the effective recharge coefficient, $(C_R)_{Ac}$, is a function of the endorheic areas, and it is maximum for the Terminio massif, where the Piana del Dragone has an important role in increase the recharge.

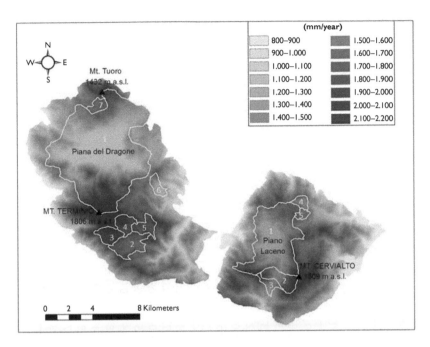

Figure 16.4 Effective rainfall distribution for Terminio and Cervialto catchments (Fiorillo *et al.*, 2015). Endorheic areas are distinguished by yellow line, and relative items are described in Table 16.1.

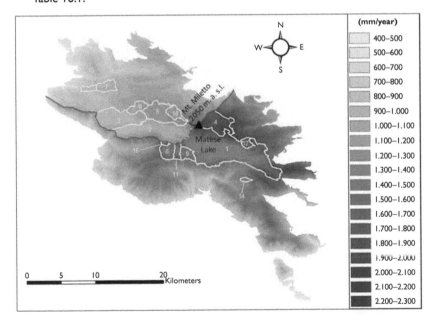

Figure 16.5 Effective rainfall distribution for the Matese massif (Fiorillo and Pagnozzi, 2015). Endorheic areas are distinguished by yellow line, and relative items are described in Table 16.2. The red line splits the upwind (southern) and downwind (northern) zones.

Table 16.1 Main hydrological parameters of Cervialto and Terminio catchments (modified from Fiorillo et al., 2015). F, afflux; F_{eff}, effective afflux; RO, runoff; Q_P, groundwater abstracted; C_R, effective recharge coefficient; C'_R, total recharge coefficient; C_S, effective contribute to spring discharge.

	Item	Area	Minimum elevation m (a.s.l.)	km²	F m³×10⁶	F_{eff} m³×10⁶	R m³×10⁶	Q_P m³×10⁶	RO m³×10⁶	C_R	C'_R	C_S
CERVIALTO	1	Piano Laceno	1047	20.5	43.8	32.9	32.9	0.0	0.0	1	0.75	0.256
	2	Piano Acernese	1168	3.3	7.4	5.8	5.8	0.0	0.0	—	0.78	0.045
	3	Piano dei Vaccari	1164	1.4	3.0	2.3	2.3	0.0	0.0	—	0.77	0.018
	4	Valle Rotonda	1156	1.1	2.3	1.8	1.8	0.0	0.0	—	0.78	0.014
	5	Raia dell'Acera	1246	0.7	1.5	1.2	1.2	0.0	0.0	—	0.80	0.009
	—	Closed areas.A_E	—	27.0	58.0	44	44.0	0.0	0.0	—	0.76	0.343
	—	Open areas.A_O	—	83	172.4	128.3	84.5	0.0	43.8	0.66	0.49	0.657
	—	Springs catch..A_C	—	110	230.4	172.3	128.5	0.0	43.8	0.75	0.56	1.00
TERMINIO	1	Piana del Dragone	668	55.1	103.9	71.6	65.3	6.3	0.0	0.91	0.63	0.383
	2	AcquadellePietre	1061	4.3	8.9	6.7	6.7	0.0	0.0	—	0.75	0.039
	3	Campolaspierto	1279	2.3	5.1	4	4.0	0.0	0.0	—	0.78	0.023
	4	Pianid'Ischia	1210	2.1	4.6	3.5	3.5	0.0	0.0	—	0.76	0.021
	5	Piano di Verteglia	1177	2.1	4.4	3.4	3.4	0.0	0.0	—	0.77	0.020
	6	Piana di Cetola	752	1.4	2.6	1.8	1.8	0.0	0.0	—	0.69	0.011
	7	PianaSant'Agata	1047	1.3	2.6	2.1	2.1	0.0	0.0	—	0.81	0.012
	—	Closed areas.A_E	—	68.6	132.1	93.1	86.8	6.3	0.0	0.93	0.66	0.509
	—	Open areas.A_O	—	94.3	180.2	125.0	83.7	0.0	41.4	0.67	0.46	0.491
	—	Springs catch.A_C	—	162.9	312.3	218.1	176.8	6.3	41.4	0.81	0.57	1.00

Table 16.2 Main hydrological parameters of Matese massif. F, afflux; F_{eff}, effective afflux; R, recharge; Q_P, pumped water; RO, runoff; C_R, effective recharge coefficient C'_R, total recharge coefficient; C_s, effective contribute to spring discharge.

Item	Area	Minimum elevation m (a.s.l.)	km²	F m³×10⁶	F_{eff} m³×10⁶	R m³×10⁶	Q_P m³×10⁶	RO m³×10⁶	C_R	C'_R	C_S
1	Matese lake	1007	45.6	91.2	69.0	0	23.7	0	0.66	0.50	0.104
2	Letino lake	870	22.4	33.8	22.8	0	7.9	0	0.66	0.45	0.034
3	Gallo lake	822	14.2	20.3	13.0	0	4.4	0	0.66	0.42	0.020
4	Campitello Matese	1417	9.6	21.9	17.9	0	6.2	0	—	0.82	0.027
5	Stampata	1316	9.5	15.3	11.0	0	0	0	—	0.72	0.025
6	Campo Rotondo	1150	3.7	7.5	5.7	0	0	0	—	0.76	0.013
7	Cannamate	976	3.4	5.0	3.2	0	0	0	—	0.65	0.007
8	Campo Figliolo	1100	3.1	4.8	3.3	0	0	0	—	0.69	0.008
9	Campo Braca	1135	2.9	5.8	4.3	0	0	0	—	0.75	0.010
10	V.Campitello	1301	2.9	4.7	3.5	0	0	0	—	0.73	0.008
11	Vallecupa	1171	1.9	3.9	3.0	0	0	0	—	0.77	0.007
12	Pianellone	1290	1.6	3.5	2.7	0	0	0	—	0.79	0.006
13	Torricella	851	1.5	2.1	1.3	0	0	0	—	0.63	0.003
14	Tagliaferro	1096	0.6	1.2	0.9	0	0	0	—	0.74	0.002
15	Selva Piana	1286	0.9	1.1	0.8	0	0	0	—	0.78	0.002
—	Closed areas A_E	—	123.8	221.8	162.5	0	42.2	0	0.78	0.57	0.276
—	Open areas. A_O	—	427.7	701.2	476.4	162.0	0	162.0	0.66	0.45	0.724
—	Springs catch. A_C	—	551.5	923.0	638.9	162.0	42.2	162.0	0.69	0.48	1.000

The recharge coefficient, C'_R, assumes values always <1, as it is based on the total amount of rainfall; considering the entire spring catchment, this value is 0.56 and 0.57 for Cervialto and Terminio, respectively, and provides a direct estimation of the annual mean fraction of the rainfall which contributes to recharge. This value is 0.52 for the Matese massif (Table 16.2).

The application of a daily scale model requires knowledge of some of the hydrologic parameters of the soil cover, which control the evapotranspiration processes and retain an amount of water before recharge can occur. In particular, the soil undergoing evapotranspiration depends also on the depth of plant roots and soil thickness. For the Cervialto massif, the soil thickness undergoing evapotranspiration processes is assumed $H=50$ cm, and the soil storage capacity is taken as $m=120$ mm (Fiorillo et al., 2015). However, because the soil mantle thickness cannot be considered constant, these values are an approximation.

The transformation of daily rainfall to daily recharge has been carried out in Equation 16.11, using a rain gauge characterised by annual mean rainfall similar to the mean of the rainfall height on the karst massif.

Figure 16.6 shows a graphic example of the simulation for the year 2011–2012, where the total cumulative rainfall reaches a value of 1769 mm (Figure 16.6A). In Figure 16.6B the daily rainfall is split according to Equation 11 for the daily scale recharge model and during the initial period (September–October), the daily rainfall is transformed into soil moisture increase, net of evapotranspiration. When field capacity has been reached ($\Theta_{max}=51\%$; Figure 16.6b), daily rainfall is able to provide

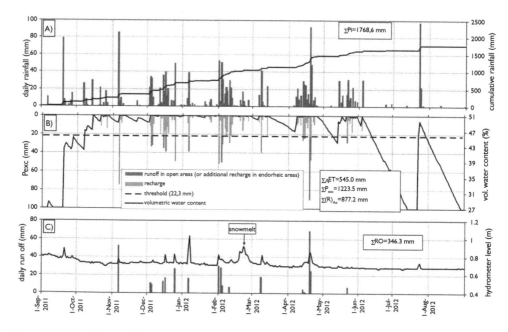

Figure 16.6 Hydrological characteristics of the 2011–12 hydrological year, Laceno rain gauge (1170 m a.s.l.) (modified from Fiorillo et al., 2015). A) daily and cumulative rainfall; B) volumetric water content, Θ, computed from daily hydrological balance ($\Theta_{min}=27\%$ and $\Theta_{max}=51\%$) and daily recharge (histogram); C) daily runoff (rainfall excess that exceeds 22,3 mm/day) and river discharge measured in Caposele village.

the daily rainfall excess amount, P_{exc} as well and the amount of recharge and runoff an be estimated from Equations 16.15, 16.16 and 16.17. The threshold of 22.3 mm (Figure 16.6B) has been deducted from the ratio of total recharge amount in open areas, $\Sigma(R)_{Ao}$, and total excess rainfall, ΣP_{exc}, equal to the effective recharge coefficient, $(C_R)_{Ao}=0.66$ (Table 16.1), for the hydrological year 1973–74 characterised by the long term average annual rainfall. In open areas, rainfall excess that exceed 22.3 mm is transformed into runoff.

16.4 DISCUSSION AND FINAL REMARKS

Several critical aspects in evaluating the afflux in the mountain areas and the estimation of the evapotranspiration were discussed by Fiorillo et al. (2015).

For the Matese massif, the runoff in the open areas is estimated at 34% of the effective afflux, F_{eff} (or 26% of the total afflux); this percentage is obtained using results for the Cervialto massif. The percentage (in term of total afflux) used is lower than values (30–35%) found by Selmo (1930) who carried out the hydrological balance of Matese lake for hydro-electrical purpose during the 1920s, and higher (7.4%) than Civita (1973) who assigned empirically the fraction of precipitation which infiltrates into the aquifer.

The daily scale model has allowed the threshold for the daily recharge to be assigned. This model considers a single point to estimate the recharge/runoff (1D model), located approximately in the median zone of the Cervialto catchment. Thus, the daily scale model provides an overestimation and underestimation of the recharge in the lower and higher elevation zones, respectively. However, as it is calibrated on the long-term annual scale recharge, it is a useful estimation of the 'mean' daily recharge in the catchment.

The total amount of recharge is a function of the rainfall distribution throughout the hydrological year, and varies during wet or dry years. During wet years, the runoff coefficient tends to increase and the recharge coefficient tends to decrease; opposite hydrological behaviour occurs during dry years. The simulation shown in Figure 16.6 has been replicated for several hydrological years, including dry and wet years, and for different thresholds of daily recharge. Figures 16.7A and 16.7B show the dependence of the recharge coefficient, $(C_R)_{Ao}$, from the maximum daily recharge (threshold), in different hydrological condition: dry, normal and wet years. During wet years the amount of runoff is higher than the mean, and $(C_R)_{Ao}$ reaches minimum values; as a consequence, the contribution from endorheic areas in recharging aquifers increases during the wet years.

The daily estimation of recharge for the massifs has to be connected to the mean of the entire catchment. Because of the wide range in ground elevation and the afflux on the massif, the recharge estimation of a specific sector of the massif needs to be calibrated as a function of the ground elevation and its exposure. However, similar results found by the application of the daily scale recharge model for Matese and Cervialto massifs allows this methodology to be transposed to other karst areas of the Apennine, characterised by similar climate and morphology. The model could also be used for other karst area world-wide, but needs a different hydrological setting to consider the local climate and geomorphologic context, soil cover and boundary flux.

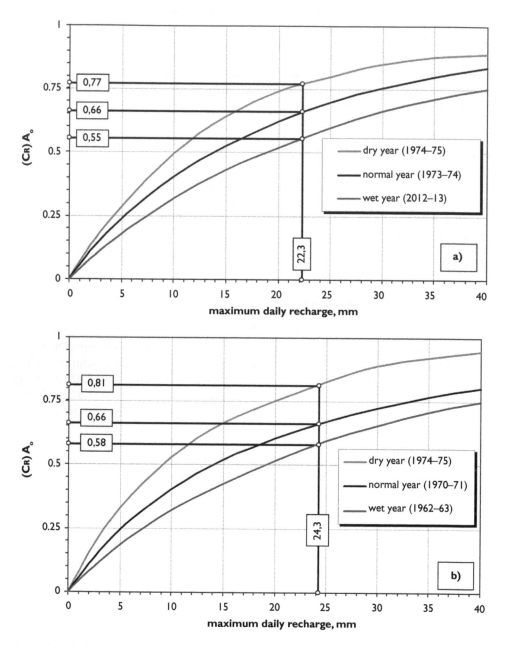

Figure 16.7 Recharge coefficient for open areas, $(C_R)_{Ao}$, function of threshold of daily recharge, for different hydrological years; a) Cervialto (modified from Fiorillo *et al.*, 2015); b) Matese (modified from Fiorillo and Pagnozzi, 2015).

REFERENCES

Bonacci O., Magdalenic A. (1993) The catchment area of the Sv. Ivan Karst spring in Istria (Croatia). *Ground Water* 31(5), 767–773.

Bonacci O. (2001) Monthly and annual effective infiltration coefficient in Dinaric karst: example of the Gradole karst spring catchment. *Hydrological Sciences Journal* 46(2), 287–300.

Celico P, Civita M. (1976) Sulla tettonica del massiccio del Cervialto (Campania) e le implicazioni idrogeologiche ad essa connesse. *Bollettino Società Naturalisti di Napoli* 85, 555–580.

Celico P, Magnano F, Monaco L. (1982) Prove di colorazione del massiccio carbonatico del monte Terminio-monte Tuoro. *Notiziario Club Alpino Italiano, Napoli* 46(1), 73–79.

Civita M. (1969) Idrogeologia del massiccio del Terminio-Tuoro (Campania). *Memorie e Note dell'Istituto di Geologia Applicata dell'Università di Napoli* 11, 5–102.

Civita M. (1973) L'infiltrazione potenziale media annua nel massiccio carbonatico del Matese (Italia meridionale). *Atti 2° International Conference Acque Sotterranee*, Palermo, 129–142 pp.

Drogue, C. (1971) Coefficient d'infiltration ou infiltration eficace, sur le roches calcaires – *Actes du Colloque d'Hydrologie en Pays Calcaire*, Besancon, 15, 121–130.

Fiorillo F., Wilson R.C. (2004) Rainfall induced debris flows in pyroclastic deposits, Campania (southern Italy). *Engineering Geology* 75(3–4), 263–289.

Fiorillo F., Esposito L., Guadagno F.M. (2007) Analyses and forecast of the water resource in an ultra-centenarian spring discharge series from Serino (Southern Italy). *Journal of Hydrology* 36, 125–138.

Fiorillo F. (2009) Spring hydrographs as indicators of droughts in a karst environment. *Journal of Hydrology* 373, 290–301.

Fiorillo F. (2011) The Role of the Evapotranspiration in the Aquifer Recharge Processes of Mediterranean Areas – In Evapotranspiration – From Measurements to Agricultural and Environmental Applications, InTech, October 2011, Rijeka (Croatia), 373–388 pp.

Fiorillo F., Guadagno F.M. (2012) Long karst spring discharge time series and droughts occurrence in Southern Italy. *Environmental Earth Sciences* 65(8), 2273–2283.

Fiorillo F., Revellino P., Ventafridda G. (2012) Karst aquifer drainage during dry periods. *Journal of Cave and Karst Studies* 74(2), 148–156.

Fiorillo F, Pagnozzi M. (2013) The role of endorheic areas on recharge processes of karst massifs. *Internationl Sympsium on Hierarchical Flow Systems in Karst Regions, September 2013, Budapest*, Abstract volume, p. 65. ISBN 978-963-284-369-8

Fiorillo F., Pagnozzi M., Ventafridda G. (2015) A model to simulate recharge processes of karst massifs. *Hydrological Processes* 29, 2301–2314.

Fiorillo F., Pagnozzi M (2015) Recharge processes of Matese karst massif (southern Italy). *Environmental Earth Sciences*, DOI 10.1007/s12665-015-4678-y

ISPRA (2015) *Geological Map of Italy, 1:50.000 Scale*. Istituto Superiore per la Protezione e la Ricerca Ambientale (ISPRA) – Ministry of Environment: Rome. http://www.apat.gov.it/Media/carg/

Parotto M, Praturlon A. (2004) The southern Apennine arc. In *Special Volume of the Italian Geological Society*, Crescenti U, D'Offizi S, Merlino S, Sacchi L (eds). Proceedings of the 32nd International Geological Congress, Di Virgilio, Chieti, Florence, 34–58.

Selmo L. (1930) Note idrologiche sul lago del Matese. *L'Energia Elettrica*, 7, 190–199.

Thornthwaite, C.W. (1948) An approach towards a rational classification of climate. *Geographical Review* 38, 55–94.

Turc L. (1954) Le bilan d'eau des sols. Relations entre les precipitations, l'evaporation et l'ecoulement. *Annales Agronomiques*, 5, 4–24.

White WB. (2002) Karst hydrology: recent developments and open questions. *Engineering Geology* 65, 85–105.

Chapter 17

Water balance analysis of a vadose stream to discern hillslope hydrology in bare karst area (South West China)

Guanghui Jiang[1,2], Fang Guo[1,2], Kwong Fai Andrew Lo[3], Xiaojiao Guo[1], Xiaoping Gong[1] & Changjie Chen[1]

[1]Key Laboratory of Karst Dynamics, Institute of Karst Geology, Chinese Academy of Geological Sciences, Guilin, P.R. China
[2]The International Research Centre on Karst under the Auspices of UNESCO, Guilin, P.R. China
[3]College of Science, Chinese Culture University, Taipei, Taiwan

ABSTRACT

Epikarst in South West China is important to the hydrological regime and consequent environmental problems in the area. A slope with bare epikarst in the Fengcong area near Guilin has been studied to understand the hydrological function of the surface layer. A physical model is set up based on data from multiple methods of field observation, tracer tests and monitoring. The water budget is used for comparing different hydrological links such as canopy interception, soil moisture deficit, runoff generation and leakage/infiltration in the slope. Although in the top layer of the slope, epikarst zone and soil infiltration dominates, but surface runoff still happens during extreme storms and changes into mostly vadose flow within a very short distance. Results from this study show that the drainage area of a vadose stream varies depending on climate. In the monsoon period it is about 491 m², which is larger than that in typhoon period. However the average rainfall interception depth, which is the threshold for flow generation in both periods, is about 16 mm. The threshold derives mainly from the canopy interception, soil moisture deficit, and epikarst storage.

17.1 INTRODUCTION

Groundwater is plentiful in the karst areas in South West China. By contrast, there is a shortage of surface waters. The karst environment is usually dry and suffers from drought every year (Yuan, 1991; Guo et al., 2013). Residents in the karst mountain areas sometimes have to pump groundwater from deep wells or karst windows to combat water shortage. However, the cost is usually high because of the great thickness (100–1000 m) of the vadose zone. It is usually difficult to find groundwater within the complex karst hydrogeological setting. In practice, the drilling failure ratio reaches 1 out of 3 wells.

Due to the lack of rivers and the high cost of groundwater extraction, small water tanks have been traditionally used to harvest rainwater and runoff from hill slopes. Since 1999, in keeping with the Chinese Western Development Policy, thousands of new water tanks were built in the Fengcong areas. Although these small tanks are popular, they are vulnerable to water quality degradation and dry out in extreme droughts. To improve the supply, tanks could be replenished more frequently with new water from the hill slopes.

Hill slopes in the karst are often selected as rainwater collection areas and supply water for the tanks. To optimise management and to guarantee the quantity and quality of water in the tanks, the hydrological processes on the slopes need to be studied. Hill slopes in the Fengcong areas are usually convex, steep, with little soil and low vegetation. Epikarst plays a key role in runoff generation and infiltration. The epikarst not only increases the water storage capacity in the surface layer (Williams, 1983), but also provides paths for rapid recharge. The morphology of the epikarst zone exerts a significant impact on the hydrological process of each slope. This study provides a closer insight into the epikarst and vadose zone by monitoring vadose streams in a cave, and analyses the water balance by selecting a distinctive slope with a developed cave. The results enhance understanding of the environmental problems in karst areas.

17.2 THE STUDY AREA

Yaji Karst Experimental Site is about 8 km away from Guilin City, Guangxi Zhuang Autonomous Region in South West China. It is an experimental research site for karst study that started in 1986 (Yuan *et al.*, 1990). The site is located in the border region of Fenglin and Fengcong, both of which are typical landforms in Guilin and famous throughout the world (Sweeting, 1995).

The climate in Guilin is characterised as subtropical Asia Monsoon. It is hot and wet in summer, cool and moist in winter, with an annual temperature of 19.2°C and an annual rainfall of 1935 m. The precipitation in Guilin outs in two distinct periods. One rainy season occurs between winter and spring, when the sun moves towards the north and the moist winds from the ocean are strongest. It meets over Guilin with the cool and dry wind from the north continent, and being well-matched in strength, leads to a long rain period. Another rainfall period occurs during the summer and autumn. It is often dry with several short rainstorms caused by typhoon.

A cave named Xiaoyan is developed in the vadose zone at the Yaji Experimental Site. It is located in the Fengcong areas with the only entrance in the middle of a west-facing slope. The Xiaoyan cave extends from west to east, with a total length of 100 m and a variable width of one to 20 m. It occurs in massive limestone of Devonian age and is controlled by fractures. The shape of the cave is wide in the middle, but narrow in the tail (Figure 17.1).

There is vadose water drainage from the ceiling at three places in the cave, located at the entrance and in the middle of the cave. The first one XY1, located at the entrance, is 4 m above the floor. The second one XY3, is 2 m high and flows along a stalagmite from the top to the floor. The third one XY5, slumps in a hole at the ceiling 20 m high, and then flows quickly in a film, forming a 10 m long flowstone. At the end of the flowstone the stream becomes free falling. The thickness of the ceiling rock is 40 m.

The slope where the cave developed is 175 m a.s.l., being 152 m a.s.l. at the foot and 327 m a.s.l. at the topographic divides. The cave entrance altitude is 197 m a.s.l.,

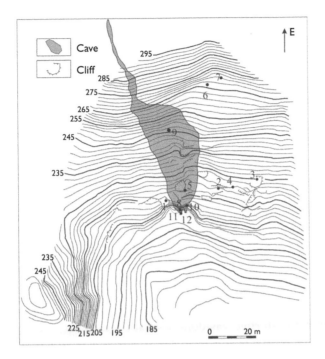

Figure 17.1 The slope with cave (shaded area) and survey points 1 soil moisture monitoring; 2 lysimeter; 3 surface flow I; 4 surface flow II; 5 surface flow III; 6 tracer injection I; 7 tracer
injection II; 8 tracer injection III; 9 cave stream XY5; 10 cave stream XY3; 11 cave stream
XY1; 12 cave entrance.

while the groundwater level is 150 m a.s.l.. The vegetation on the rocky slope is shrub
dominated by *Loropetalum chinense* and *Bauhinia championii (Benth.) Benth*. Its
root grow in fractures and small holes. Soil fill in the open fractures is clayey. On the
slope, there are many cliffs and small ledges between the cliffs. The cliffs are shaped
by weathering and are controlled by vertical fractures. The cliffs range from 1 to 5 m
high, with some conduits and flowstone developed, indicating that they play a part in
the flow discharge in the epikarst zone.

17.3   METHODS AND METHODOLOGY

The terrain of the slope and the cave were mapped in order to show the surface karst
features and the projection of the cave on the slope. A 3 m long trench was dug in
a small platform in front of a cliff to take samples for assessing the soil physical
properties in the profile, including soil depth, structure, color and density. A mini-
permeameter was used to test the permeability of the soil. These tests were repeated
at three randomly selected points. There was no perennial flow at the surface. Short-
duration runoff in the slope was determined by field observation during storm events.
Hydrological monitoring of the surface flow was carried out several times.

Soil water content monitoring sensors (Frequency Domain Reflectometry (FDR)) H21 were installed when the trench was re-covered. Four sensors of the H21 type were buried in the soil at depths of 10, 50, 90 and 140 cm. The soil moisture at 10 cm represented the top layer mainly affected by climate and vegetation. The layer at 50 cm depth has a soil colour change from brown to red, indicating the root transition zone. The interface of soil and rock was located at 140 cm depth.

The water content in the soil profile was also measured by soil sampling. Samples were obtained by a manual spiral drill every 10 cm. Usually eight samples were taken in each profile. Then they were collected in a bag and sealed in the field before they were transported to the laboratory. The samples were weighted and dried to obtain water content. A total of eleven profile samplings were collected before and after rainfall events between April and August 2012. Sample locations were selected randomly on the slope.

Hydrological monitoring was conducted in the surface of the slope and in the cave. For the surface flow, it was monitored in three different locations several times over during May and June, 2012. The discharge was measured by a rectangular trough with a right angle weir at one end, and a bucket with a measuring cylinder. The manual monitoring interval was one hour, while automatic measurement by pressure sensor was at a five minutes interval.

Vadose flow and drip water in the cave were monitored by manual or automatic methods. Manual monitoring was done several times at an interval of several hours or days, and lasting one to three days. A bucket together with measuring cylinder was used for vadose water discharge measurement. However, at the XY5 vadose stream, the biggest one in the cave, specially designed equipment was used. A rectangular platform made of a steel plate with an area of 60 m^2 was installed under the stalagmite for water collection. All scattered drip water and flow water from the flowstone was concentrated on the platform. The flow was piped into a special cylinder bucket with a hole for discharge, where a sensor was installed to record the water level at hourly intervals. The experimental relation between the level and the discharge into the bucket could be calculated. However, when flooding occurred, the discharge was too large to measure by the bucket and a right angle weir was used. The discharge through the weir was calculated by an experimental equation provided by the Chinese National Standard (MWR, 1993).

Tracer tests were designed to determine the flowpath on the surface of the slope and the vadose zone, and to look for the corresponding drainage area of vadose stream. Three tracer tests were carried out on the slope. For the first test, the injection points were in a surface flow channel in the upper slope, which had water in flood, and along this path tracer was seaping into the vadose zone through fractures. A total of 200 gm of uranine was injected in a small dug pit in the path in May 2012. All surface flow on the slope and water in the cave was sampled once a day. The flow duration and sampling work lasted for about 7 days. The second test was carried out in November 2013, 18 months after the first one. Again, 200 gm of uranine was put into a pit, which was 10 m away from the first location. The tracer was detected by a field flourometer installed at XY5, at the same time samples were collected in the surface flow and vadose streams for measurement by fluorescence spectrophotometer in the laboratory. In the third tracer test, 200 gm of uranine was injected into a fracture in October 2013. Then 1 litre of water was used to wash it. The only detection point was in the XY1 vadose stream under the injection point. One sample was taken after each subsequent rainfall event.

Rainfall was measured at a raingauge station in the Yaji Experimental Site. Precision of the measurements was about 0.2 mm. Intercepted rainwater from the canopy was obtained by the rain gauges. Three rain gauges were installed under the canopy. The rainfall amount was compared with the gauge station in Yaji. The difference was the amount of rainwater intercepted by the canopy.

A simple micro-lysimeter was used to measure evaporation in the soil. The lysimeter was made of PVC cylinders that had a diameter of 10 cm and lengths of 5, 15, and 50 cm. A slightly disturbed soil pillar was put in the cylinders, and the nylon net was fixed at one of the ends. After weighing at 8 a.m., the cylinder was put back in the pit where the soil pillar was taken. The cylinders were weighed again at 7 p.m. in the same day. The weight difference represents evaporation in the day and night. The test was performed continuously for seven days during July and August, 2012.

17.4 STUDY RESULTS

17.4.1 Terrain on the slope and its flowpath

The slope surface is irregular, full of rocks, and the terrain is slightly convex in the upper part but concave in the lower part. The gradient gradually increases from foot to peak. The parallel cliffs in a south-north orientation are the main feature of the landscape, while there is a small platform in front of the cliffs. The formation of the cliffs and platforms is controlled by fractures also in along south-north direction, which is related to a fault in the same direction crossing in front of the cave. The cave, developed below the slope surface, has an area of 1133 m². The thickness of the rock in the cave ranges from several meters at the entrance to more than 100 m at the end of the cave.

In the upper slope, the biggest cliff stands about 5 m high, and this is the source of runoff or flow generation on the slope. There is an intermittent stream from the cliff during rainstorms, forming many flowstones on the cliff. The stream goes down the slope but spreads into several segments. However, tracking these tributaries is difficult for the slope is covered with thorns. However, the flowpath can be delineated with tracers. The tributaries split into three parts. One of them sinks down and becomes vadose streams. Another continues to runoff in the surface. And the other part of water enters into the epikarst zone.

The first tracer test was done on May 29, 2012. The tracer was injected in the flow channel close to its source in the cliff. Seven days after rainfall, the green color of dissolved uranine was found in all of the surface streams in the down slope, vadose streams and drip water in the cave. It indicates that the convex shape of the slope surface enhances the diffusion of the runoff.

The tracer concentration in XY5 was still higher than the background value when the second tracer test was conducted in January 2014, indicating that the tracer was stored in the epikarst zone. The tracer releases slowly because of buffering from a large number of fissures and the soil fill in the fractures in the epikarst zone.

In the second tracer test, the injection point was close to the first one. However, in this test the tracer was only detected in XY5. Several samples taken from the surface flow and drip water in the cave contained no tracer. The breakthrough curve of the tracer in XY5 only lasted for four months. It recovered after being washed by several storms in the rainy season, suggesting that the tracer cannot enter the epikarst at this time. But, there is a negative peak in the concentration due to the dilution of baseflow in the epikarst by surface flow leakage. It may be concluded that the tracer injected in the first test was still stored in the vadose zone. However, it is still unclear how the tracer enters into the epikarst in the first tracer test and where the open joint and shaft is connected with the cave in the slope surface. The topography in the slope is not the only factor which controls the flow. The subsurface space formed by karst processes can also strongly affect the flow regime.

17.4.2   Hydrology of the vadose streams

The discharge of XY5 is permanent but highly variable. The highest discharge measured reached 500 ml/s, while the smallest one was about 0.4 ml/s, i.e. a difference of about 1,000 times. The large change in the vadose stream is similar to the karst springs nearby and is also common for karst areas in South West China. The vadose water in its high discharge state has a short residence time, usually only one to two days. Then the vadose stream maintains a small discharge with small changes in flow. The great discharge difference implies that there is a new supplemental source to the vadose flow when flooded. In other times, it is fed by seepage water only.

In the second tracer test, the change of tracer concentration in XY5 can only be found in the high discharge stage of the vadose stream. It proves that the tracer is transported to vadose flow by surface runoff, as it is injected at the surface flow path. It may be concluded that the added discharge to XY5 in flood conditions originates from leakage of surface runoff. The recharge area of XY1 is in the upper part of its discharge point, which is proved by the third tracer test. The source area of XY3 probably lies on top of it, but this needs to be confirmed.

The XY1 has a very limited drainage area, and the thickness of its corresponding vadose zone is only about 10 m, where the widely open fractures cutting limestone in the upper epikarst are about 2 m in depth. The next layer below the open fractures is thin patchy soil and massive rock. In the third tracer test, uranine was injected in the open fracture, and then washed by one litre of distilled water. The tracer flows along the wall of the fracture and infiltrates into the soil. Ten days after injection, the tracer was found in XY1 with a high concentration following a rainfall event. The tracer can always be detected thereafter once the vadose stream appears. The third tracer test shows that the vadose stream of XY1 is recharged from the upper epikarst zone and has a short transport distance and residence time for water and solute, and it is only buffered by the soil. The intermittent XY1 has the most sensitive response to rainfall among all the vadose streams. Its discharge varies from 0–130 ml/s. However the duration only lasts several days.

The drainage area of XY3 is not very clear. It is close to a short-lived waterfall at the cave entrance and thus may be originated from surface flow. However, sometimes when the vadose stream XY3 appears, the waterfall does not occur. The stream forms a stalagmite. It has a medium delayed time, and disappears for several months during the year. The discharge of the stream has only been measured roughly because it is spread into many segments by the stalagmite. The measured flow discharge ranges from 0–1 ml/s.

The three vadose streams provide an opportunity to understand the behaviour of vadose flow under the slope. Although they are all related with surface flow and epikarst, the hydraulic relationships and flow regime have a slight difference. Probably, in the perched vadose zone system, the XY3 is in the downstream of XY5, while the XY1 is isolated. It may imply that flow generated on the slope is concentrated at different levels and moves along independent pathways. The vertical vadose flow is transformed from lateral flow and surface flow and breaks the integrity of flow on the slope.

17.4.3   Hydrology in the hill slope of the surface

The surface of the hill slope is rocky and partially covered with soil and vegetation. The soil and rock groups are spatially complicatedly, but generally the soil is thin. Rainwater falling on the slope surface is reduced by the thin canopy and soil interception. The rock surface retention is small as well, such that it can be ignored. Rainwater interception by shrub canopy averaged 5 mm throughout the measurements by three rain gauges placed under the canopy. Rainwater retained by soil changes in time and space depend on soil depth and moisture deficit. The uneven distribution of the soil and bare rock on the slope make analysis difficult.

Eleven soil profiles were randomly selected to obtain soil profile moisture and depth time series in the slope. The soil profiles in the platform in front of the cliffs have a depth and width of one to several meters. The soil block is partly surrounded by rock, so it can obtain recharge from the surface and rocky boundary and discharge from the free edge. As a result the soil is easily saturated and dried, and there is often perched water at the soil bottom. The water content in the soil varies from 9 to 22%. The soil with a thickness of 1,000 mm may have the largest moisture deficit of 169 mm, considering the moisture changing from 9% to 22%, and the density of soil of 1.3 g/cm^3. But for the cliff-platform model in the presence of rock and soil, the runoff yield under excess infiltration (Hortonian flow, Ford & Williams, 2007) in the rock surface may recharge the nearby soil in the platform. This may lead to soil saturation at the bottom of soil. This is confirmed by soil content monitoring by FDR. The process is important for slope hydrology, because it accelerates water saturation in the soil and leads to interflow generation. It can also buffer surface flow.

The interception capacity of the slope surface is estimated by the response of XY5 to rainfall. When there is new water supplied, its physical and chemical index of discharge may change. The threshold value for recharge can be calculated by regression statistics of a large number of rainfall events (Jiang et al., 2008). The maximum threshold record is 120.6 mm rainfall within 30 hours, which happened in November 2012 with seven sunny days and three dry months occurring in the autumn before the rainfall event. The soil evaporation rate can be 5 mm/day in high moisture conditions and 1 mm/day in low moisture conditions. It dries quickly in sunny days, resulting in high soil moisture deficit.

17.5   DISCUSSION

17.5.1   Physical model for hydrological process in the slope

The hydrological processes in the slope and the related karst features are examined in detail. In order to understand water cycle model, a hypothesis is proposed. In one case, flow in the slope prefers to vertical movement as all rainfall may infiltrate or leak into the vadose zone. In another case, all rainfall will transform into lateral flow in the epikarst once interception is satisfied. The hydrological regime of a karst hillslope may lie somewhere between the two end numbers. The relative importance between infiltration/leakage and lateral flow is controlled by geological conditions, rainfall and the initial water condition.

On the slope the infiltration and leakage will occur along fissures, fractures, joints and shafts in the rock. However, there is little evidence to judge where the infiltration/leakage happens. Sometimes the lateral flow also occurs in the epikarst zone. Under this condition, water balance analysis in the slope offers an opportunity to understand the hydrological function of the karst land surface.

In the slope, the top layer includes three parts, i.e. surface, epikarst and vadose zone (Figure 17.2). The surface is composed of bare rock, soil and vegetation. Rain falls on the different interfaces. After canopy interception in path 2, rainwater arrives at rock and soil in path 1 and 3. Because the karstified rock has high permeability and low storage capacity, most of the rainwater goes to Hortonian runoff. Therefore, some of the runoff flows into the surrounding soil in path 4 and another enters the epikarst by preferential paths, such as shafts, joints and expanded fractures in path 5. Soil water will penetrate into the epikarst by fissures in path 6. Vegetation grows in the

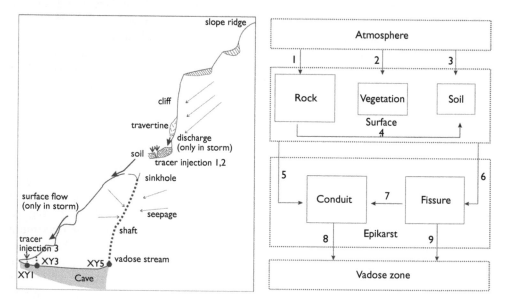

Figure 17.2 Schematic diagrams of the water cycle in the bare karst slope: left: slope profile; right: diagram of flow.

soil and epikarst, where the roots absorb water. Epikarst water enters into the vadose zone by flows and seepage in the paths 8 and 9, while some of them water returns to the surface through epikarst springs, and some returns to the atmosphere by evaporation from the epikarst. When the flow lines from paths 1, 5 and 8 in the karst slope are overwhelmed, drought or flood will occur. The relative weakness of flow paths 3, 6 and 9 makes vegetation grow slowly and the ecosystem becomes fragile.

17.5.2   The water balance model

Rainfall in the slope transforms into soil water, epikarst water, surface runoff, seepage and shaft flow or vadose flow. Usually the balance of the components is difficult to measure and calculate. However, it is important to understand the flow transform in the slope, because it informs water resources management, and helps with droughts and floods. The water balance analysis becomes much easier in the slope where a cave is developed, as it is convenient to monitor seepage and vadose flow directly in the cave. But there are still a lot of challenges, such as the boundary and drainage area determination of a small vadose stream.

In path 1 (Figure 17.2), the Hortonian flow in bare rock can be expressed as:

$$R = L(P - E),$$

considering the rock surface to be impermeable.

In this equation, R means Hortonian flow, and L represents the proportion of bare rock in slope surface. P is rainfall, and E is evaporation which also includes moisture absorption by rock and vegetation canopy.

R changes further into two components (Figure 17.3). One is Q, which means quick flow through shaft or conduit to the epikarst by path 5. Another is S, referring to flows into soil or fissures by path 4. So the equation is written as:

$$R = Q + S$$

The amount of Q entering the epikarst will be reduced by infiltration in voids such as the conduit wall and by algae and litter. The porosity of the rock surface enhanced wherever lichens are present (Cao *et al.*, 1998). Q should needs to be corrected as,

$$Q = Q_0 - h \cdot q$$

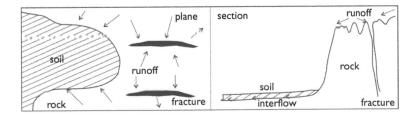

Figure 17.3 Model of flow paths in the slope land surface, the plane and in section.

Where Q_0, h, and q are the initial value of Q, the distance, and water absorption by conduit per unit length, respectively.

Interflow in the slope IF by path 6 can be expressed as:

$$IF = S - D$$

In the equation D is the soil water deficit. It can be calculated by the equation:

$$D = \theta_f - \theta_1$$

where θ_f is soil field capacity, and θ_1 is initial water content in soil.

Rainfall falling on the soil surface will infiltrate completely because of the high permeability of the soil. There is no Hortonian flow in the soil, but the rainfall recharge will enhance soil water content. The relation is expressed as:

$$(P - E)(1 - L) = \theta_1 - \theta_0$$

θ_0 means the water content before rainfall.

Finally the water balance in slope can be expressed by the following equation:

$$P - E = Q_0 + IF + \theta_f - \theta_0 + h \cdot q$$

The threshold value for runoff on the slope will satisfy the water deficiency in the soil and epikarst, so it can be defined as:

$$T = \theta_f - \theta_0 + h \cdot q$$

Where T is the threshold value for runoff generation.

Here, the fissure storage is ignored because of the slow change in storm periods. The quick flow in the conduits and the interflow usually mix and transfer between each other. They can, therefore, be combined, and represented by M:

$$M = Q + IF$$

Then the balance equation can be expressed as:

$$P - E = M + T$$

The equation can also be changed into:

$$M = P - E - T$$

For a vadose stream with a drainage area of A, the water balance equation changes into:

$$M = A(P - E - T)$$

where M is in volume units. It can be obtained by measuring the discharge of a vadose stream.

17.5.3 Determination of the drainage area for a vadose stream

The vadose stream XY5 is taken as an example for the water balance analysis. The discharge of the vadose stream and the corresponding rainfall are plotted in a graph. The result shows a good linear relationship between them, so they satisfy the water balance equation. The fitted linear equation can be used to calculate the drainage area A and the value of $E+T$. A is equal to the gradient of the fitted straight line, the $E+T$ is the ordinate at the origin. Two fitted lines are obtained from the graph. They have the same origin ordinate and but different gradients, indicating that the drainage area of the vadose stream is changeable, while the $E+T$ keeps stable. Further analysis indicates that the area A in the rainy seasons is about 491 m², while it is 200 m² in the dry seasons (Figure 17.4). This is probably because there is no runoff generation in some of the dry soil covered areas for a long time.

17.5.4 Water balance evaluation in vadose stream

The equation in Figure 17.4 shows the water balance of a vadose stream during a flood event, when the runoff generation in the slope surface feeds into the vadose stream. This lead to an abrupt increase in the discharge of the vadose stream. The corresponding recharge area for the added runoff and the threshold value of runoff generation are determined by regression statistics. However, in many rainfall events, the runoff will not generate, as a result the discharge in the vadose stream does not change. Under this condition, the threshold value for runoff generation is usually

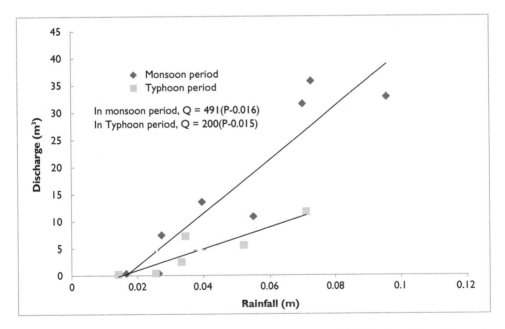

Figure 17.4 Relationship of rainfall and vadose stream XY5 discharge. The fitted results can be used to determine drainage area and runoff decay.

bigger than that during storms. The statistics also shows that the discharge of XY5 does not change until rainfall reaches 50 mm in average.

The vegetation will take up water from the soil and the epikarst. A test of evapo-transpiration was carried out in the karst environment in Guilin. The result shows that a tree will take up 600 mm water in a normal year (Huang *et al.*, 2013). The vegetation in the slope usually suffers from drought, and it is not because of rain-fall shortage but because of insufficient storage in the slope. Runoff feeding into the vadose stream has in a large proportion compared to the total discharge (about 70% of flow) in the vadose stream in one year. This provides a good reason why the karst environment is so easy to dry out and to flood.

Among the components of the water balance for a vadose stream, the initial water content and soil mass are related to the threshold values. In bare karst the soil stored in fractures has a critical role for preferential recharge. After soil erosion occurs the flow may generate more easily and quickly and in large amounts.

17.6 CONCLUSIONS

A hillslope with a cave developed below provides a good opportunity to examine the flow pathways in the vadose zone. Both surface and subsurface flow can be observed and monitored, which provide a basis for water budget analysis especially in a narrow layer close to the slope surface. The hydrological process, such as canopy interception, soil evaporation and soil moisture can be monitored and evaluated by various meth-ods. Through establishing a water balance equation, a linear relationship between rainfall and discharge of a vadose stream can be determined. Two important parame-ters, the drainage area and the threshold value for flow generation are obtained from the correlation analysis. The results show that a vadose stream contains preferential recharge which is related to the drainage area of the slope surface.

The hydrological functions of a typical karst rocky slope comprise a large pro-portion of quick flow, a large ratio of subsurface flow and a small ratio of surface runoff. The soil water and seepage in the rock is not large and changes quickly. The hydrological characters of the hill slope make the karst surface environment dry and fragile. Water resources in the slope are not small but they are difficult to use due to rapid drainage to the vadose zone.

The high proportion of vadose flow in the hydrological balance of the slope is consistent with the geomorphologic character of mountain areas. The water balance model for the slope in the bare karst areas provides a significant understanding of the environmental problems in karst areas.

ACKNOWLEDGEMENTS

Financial support was provided by the Chinese National Natural Science Foundation (41102161, 41172231, and 41472239), Key Project of Guangxi Natural Science Foundation (2013GXNSFDA019024) and the Chinese Geological Survey project (1212010916063). We would like to thank the reviewers who read the first draft of this paper for their constructive comments.

REFERENCES

Ford D., Williams P. (2007) *Karst hydrogeology and geomorphology*. Wiley, Chichester.

Guo F., Jiang G.H., Yuan D.X., Polk J.S. (2013) Evolution of major environmental geological problems in karst areas of Southwestern China. *Environmental Earth Sciences* 69(7), 2427–2435.

Jiang G.H., Guo F., Wu J.C. (2008) The threshold value of epikarst runoff in forest mountain area. *Environmental Geology* 55, 87–93.

Huang Y.Q., Li X.K., Zhang, Z.F. (2013) Seasonal changes in Cyclobalanopsis glauca transpiration and canopy stomatal conductance and their dependence on subterranean water and climatic factors in rocky karst terrain. *Journal of Hydrology* 402(1–2), 135–143.

MWR (Ministry of Water Resources of People's Republic of China). 1993. *Code for liquid flow measurement in open channel.*

Sweeting M.M. (1995) *Karst in China: its Geomorphology and Environment*. Springer-Verlag: Berlin.

Williams P.W. (1983) The role of the subcutaneous zone in karst hydrology. *Journal of Hydrology* 61, 45–67.

Yuan D.X. (Editors) (1991) *Karst in China*. Geological Publishing House: Beijing.

Yuan D.X., Drogue C., Dai A. D. (1990) Hydrology of the karst aquifer at the experimental site of Guilin in southern China. *Journal of Hydrology* 115, 285–296.

Cao J. H, Wang F. X. (1998) Reform of carbonate rock subsurface by crustose lichens and its environmental significance. *Acta Geologica Sinica* 72(1), 94–99.

Chapter 18

Hydraulic behaviour of a subthermal karst spring – Blederija spring, Eastern Serbia

Vladimir Živanović, Veselin Dragišić, Igor Jemcov &
Nebojša Atanacković
University of Belgrade, Faculty of Mining and Geology,
Department of Hydrogeology, Belgrade, Serbia

ABSTRACT

Karst aquifers are characterised as a highly heterogeneous media which affects the behaviour of the groundwater outflow regime. The Blederija subthermal karst spring is located on the south-eastern part of the Miroč mountains. Complex quantitative analysis has been carried out in order to obtain new insights into the hydraulic mechanism of the discharge regime. This is characterised by frequent mixing of a 'normal' and higher temperature waters. Binary karst hydrogeological system and the location of ponor zones causes a specific variation of the catchment area, which is reflected in the discharge regime properties of the karst aquifer. Two different approaches of types of hydrograph separation were applied in order to identify the interconnected flow system e.g. base flow and fast (direct) flow components. The first one was based on the discharge rate and second on the water temperature. Both analyses gave very similar results, and indicate a complex hydraulic mechanism, caused by a pressure pulse through the karst hydrogeological system.

18.1 INTRODUCTION

Karst spring hydrographs are often used for characterisation of a karst aquifer. Discharge analysis and correlation with the changes of climate parameters can be used to understand the functioning of the karst system (Jemcov & Petric, 2010). This is particularly important for gravitational springs where discharge variates significantly and the influence of climate is pronounced. Deep siphonal springs have a smaller discharge range and lower temperature variations. But, there are some karst subthermal springs with significant changes in capacity and temperature which result from mixing with colder karst groundwaters. In such circumstances, time series regime analysis can help to understand the discharge mechanism as well as the recharge conditions. This methodology has been applied at the Blederija subthermal karst aquifer.

Blederija is a karst spring which drains the Miroč karst massif in eastern Serbia. The karst aquifer is formed in fractured and highly-karstified massive Upper Jurassic limestones (Stevanovic *et al.*, 1996). A significant portion of the spring catchment area is made up of low permeable Cretaceous clastic rocks, enabling the formation of a network of surface streams, which sink as they pass through the karstic area (Figure 18.1). At times of high-water flows, most of the surface streams sink via the ponor Cvetanovac. At the discharge zone, a cold spring and a subthermal spring (called

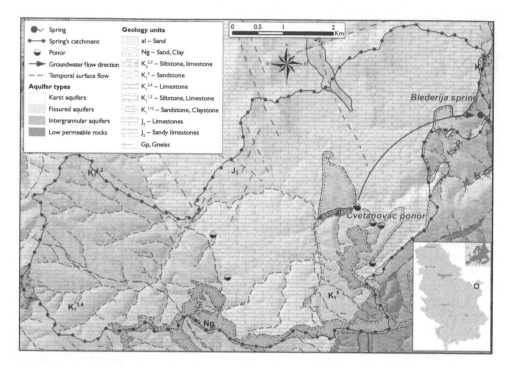

Figure 18.1 Hydrogeology map of the Blederija Spring catchment area.

mixed spring in this paper due to the results of regime monitoring) are located 10 m apart. The average discharge of the Blederija springs is 280 l/s (Prohaska *et al.*, 2001).

18.2   ANALYSED DATA

Monitoring was carried out for three years in order to understand the recharge and discharge mechanism of Blederija spring (Figure 18.2). Temperature was measured at three locations (Blederija mixed spring, Blederija cold spring and Blederija stream 10 m below the confluence of the mixed and cold spring), surface water level and flow of the Blederija stream were measured, weather measurements: precipitation, air temperature and wind-speed were also recorded. The measurements period was March 2011 to April 2014.

The water temperature was measured every 5 days with digital and mercury thermometers. In order to assess the capacity of Blederija stream, a water level gauge was installed and read daily. Flow measurements were also conducted several times in order to obtain a useable rating curve. Flow measurements represent a total flow of both the cold and mixed springs, and only occasional measurements of subthermal springs were performed during low-flow period.

In the vicinity of the spring, a meteorological station recording precipitation, air temperature and wind speed and direction measurement was also installed. Precipitation data were also collected from the Miroč rainfall station located near the

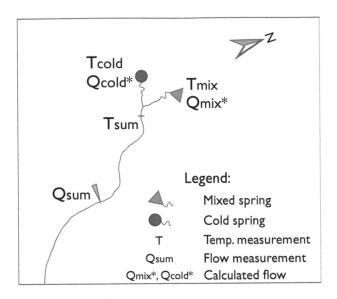

Figure 18.2 Sketch map of measuring points.

upper parts of the spring's catchment. Data were also collected from the Negotin and Crni Vrh meteorological stations where long-term digital climate event monitoring is undertaken.

Beside the regime monitoring, field explorations were carried out, including tracer test. Temperature and flow data were used for hydrograph separation. Time series analysis was used to understand hydraulic behaviour of the Blederija karst system. Together the data fed into the conceptual model of the aquifer.

18.3   APPLIED METHODS AND RESULTS

The hydrograph separation method was applied in order to identify the base flow and the fast flow components. There are numerous methods available to separate baseflow from measured stream /spring flow hydrographs, but they are all based on two concepts (Kyoung *et al.*, 2010). One is based on a simple automated smoothing and separation rule (Sloto & Crouse, 1996), and the second on a digital filtering method that provides consistent results (Nathan & McMahon, 1990). A recursive digital filter was used – Eckhardt filter separates baseflow and fast flow components (Eckhardt, 2005):

$$b_t = \frac{(1 - BFI_{max})\alpha + b_{t-1}(1 - a)BFI_{max}Q_t}{1 - \alpha BFI_{max}} \tag{18.1}$$

Where b_t is the filtered baseflow at the t step; b_{t-1} is the filtered base flow index at the *t-1* step; BFI_{max} is the maximum value of long-term ratio of baseflow to total flow; α is the filter parameter; and Q_t is the total flow at t step.

All data were processed using the BFI+3 module of HydroOffice 2015 (Gregor, 2010). To avoid uncertainties of estimation of the main parameter for the BFI_{max} recession curve analysis, occasional flow measurements of the subthermal karstic spring were used. Based on the 5 day data level, BFI_{max} 0.45 (for the subthermal spring) and parameter α of 0.8 were estimated (Figure 18.3).

In the next step, a different method of hydrograph separation was performed. This method is based on the water temperature, because direct measuring of the spring discharge was not possible due to the narrowness of the spring discharge area. Spring discharges were calculated according to heat and mass continuum equations (Kobayashi, 1985):

$$Q_{sum} = Q_{cold} + Q_{mix} \ (or \ m_{sum} = m_{cold} + m_{mix}) \tag{18.2}$$

$$m_{sum} \cdot c \cdot T_{sum} = m_{cold} \cdot c \cdot T_{cold} + m_{mix} \cdot c \cdot T_{mix} \tag{18.3}$$

where Q_{sum} is the Blederija stream flow, Q_{cold} is the Blederija cold spring discharge, Q_{mix} is the Blederija mixed spring discharge, m is water mass, c is specific heat and T is water temperature.

Combining the upper two equations, hydrograph separation was obtained (Figure 18.4) based on the following relation:

$$Q_{mix} = Q_{sum} \cdot (T_{sum} - T_{cold})/(T_{mix} - T_{cold}) \tag{18.4}$$

Comparative analysis of two different methods of hydrograph separation showed similar behaviour (Table 18.1) and a high correlation coefficient (0.98) of

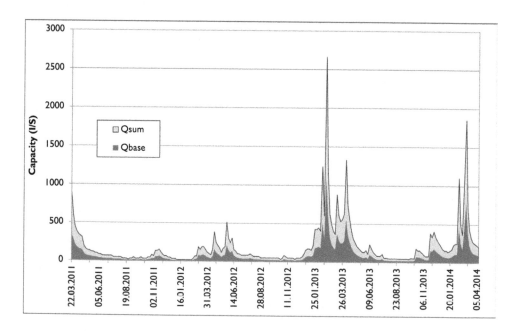

Figure 18.3 Hydrograph separation, base-flow and fast flow components.

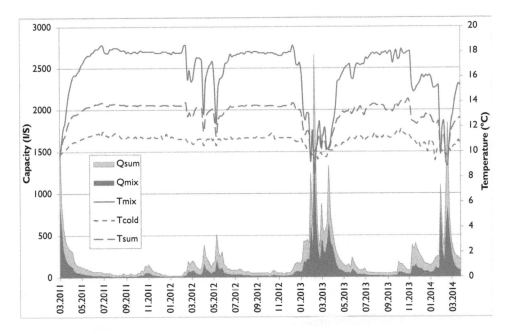

Figure 18.4 Hydrograph separation based on thermal equilibrium.

Table 18.1 Descriptive statistics of summary discharge rate of Blederija spring and separated flow components (l/s).

	Mean	Minimum	Maximum	Variance	Std. Dev.
Q_{sum}	203.0	17.00	2661	88700	297.8
Q_{base}	91.3	7.65	1089	16540	128.6
Q_{mix}	100.1	5.91	1544	30217	173.8

hydrograph components Q_{base} and Q_{mix}. This confirms the applicability of the method based on water temperature. Some discrepancy of the data between separated components, Q_{base}, and Q_{mix}, reflects the different methods applied and their parameter calibration.

The temperature measurements have shown some unexpected results. The temperature of the Blederija mixed spring in dry periods reached as high as 18.5°C. In the periods of snow melting, the mixed spring capacity significantly increases and temperature decreases by almost 10°C. In previous studies (Dragisić *et al.*, 1992; Stevanović *et al.*, 1996), this phenomenon was explained by simple mixture with the groundwater from Blederija cold spring. However, in the highest water flow periods, the temperature of the mixed spring falls below the temperature of Blederija cold spring ($T_{mix} = 8.1°C$, $T_{cold} = 9.7°C$). This suggests that the main cause of the lowering of the water temperature of the Blederija mixed spring might not be the mixture with the water from the cold spring. The temperature decrease could be due to mixture with groundwater coming from the eastern part of the catchment, where large portion of surface water recharges the karst aquifer via the Cvetanovac ponor. Dye test conducted in December 2013 has confirmed this. The tracer (2 kg of uranine) was injected in the ponor Cvetanovac with a river flow rate of 4 l/s (Figure 18.5). The tracer was detected 2.5 days later, but only at the Blederija mixed spring, and not at the cold spring. During that time, the karst groundwater carrying the tracer travelled 3060 m along a hydraulic gradient of 0.044 with an average flow velocity of 1.4 cm/s.

In order to properly analyse the recharge-discharge relationship, the input component of the karst system (precipitation) was converted to effective precipitation. In order to separate the various processes in the air, vegetation and the soil from the processes within the karst aquifer system, the interception on vegetation cover, and of snow and snowmelt were assessed to define the effective precipitation, or the quantity of precipitation that reached the ground (Jemcov & Petric, 2009).

One of the methods to study karst aquifer behaviour is time series analysis of the recharge and discharge data as functions of the karst hydrogeological system (Jemcov & Petric, 2010). The analysis has been carried out as complementary method to the other field methods of karst groundwater exploration such as the dye test.

The univariate analysis characterises the individual structure of time. The auto-correlograms of all of three functions (Figure 18.6.) exceeds the confidence limits for approximately 13–15 time lag (~65–75 days), which implies that the system is well structured, and storage is significant.

The slope of autocorrelogram Q_{cold} initially drops quickly for less than 15 days, and afterwards remains on the same trend. This bimodal behaviour indicates the

Figure 18.5 Dye test – verified connection of the Cvetanovac ponor and Blederija mixed spring.

duality of the outflow regime. The same occurs in autocorrelograms Q_{sum} and Q_{mix} (but less pronounced), and indicates a zone of mixing of water with 'normal' and subthermal components.

The bivariate analysis considers the transformation of the input to the output signal. The cross-correlation function (CCF) of the Blederija source for all considered components (Figure 18.7.) shows non-symmetrical behaviour and a high level of influence of effective precipitation on outflow components, particularly within the initial time lag (5 day). This suggests that all of the analysed components recharge from the binary karst system. A function for all three outflow components becomes insignificant after 55 days, and thereafter it exceeds the level of significance because

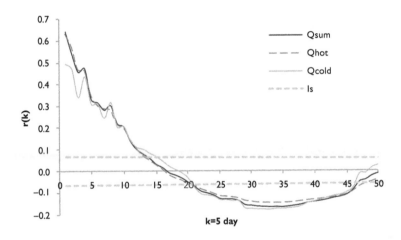

Figure 18.6 Autocorrelation function of discharge components of the Blederija source. Legend: Q_{sum} – Blederija stream flow, Q_{cold} – Blederija cold spring Q_{mix} – Blederija mixed spring.

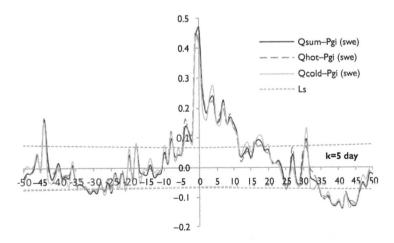

Figure 18.7 Cross-correlation function of effective precipitation and flow components of the Blederija karst source. Legend: Pgi (swe) – effective precipitation – transformed from the measured values corrected for wind influence and interception by vegetation, and finally melted snow.

of the influence of the Q_{mix} components, and the hydraulic behaviour of the outflow components. Constant variation of cross-correlation functions implies non-homogenous karst hydrogeological system, with different responses in various parts of the systems with frequent interchanges of 'normal' and subthermal waters. A strong influence of the effective precipitation on the Q_{mix} component, similar to the Q_{cold} component, indicates a process of continuous interference of 'normal' and subthermal waters. Moreover, a common attenuation effect for the Q_{mix} component is not quite so obvious, and suggests possibilities of different development of the various levels of the karst channel system (upper and lower).

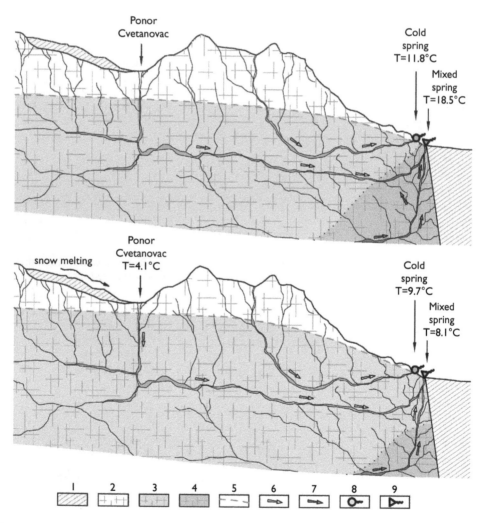

Figure 18.8 Conceptual model of Blederija spring discharge during the low (up) and high (down) water regime. (1) low permeable rocks, (2) karstified limestone (vadose zone), (3) karstified limestone (saturated zone with cold water), (4) karstified limestone (saturated zone with thermal water), (5) groundwater level, (6) cold groundwater flow direction, (7) thermal groundwater flow direction, (8) cold spring, (9) mixed spring.

18.4 CONCEPTUAL MODEL AND CONCLUSION

Applied analysis and field exploration were complementary, and both were needed to obtain the conceptual model. There is no doubt that water from the higher parts of groundwater system has an effect on the decrease in temperature of the lower subthermal groundwater system. There are two independent upper groundwater karst systems, coming from the north and the east, and only the water from the east system effects the lowering of the water temperature of the subthermal spring. Therefore, the application of the field techniques such as dye test along with other methods, e.g. isotope analysis, are still essential to the exploration of the karst hydrogeological system. The complex hydraulic relation between the cold and the subthermal groundwater in different groundwater stages is presented schematically in the conceptual model (Figure 18.8).

During the low water flow regime, the Cvetanovac River dries up and diffuse infiltration occurs. During low water period, groundwater flows slowly, especially in the deeper parts of the karst system. As a result of small spring capacities the water temperatures become higher. The temperature of the mixed spring goes up to 18.5°C. In the periods of snow melt, punctual infiltration becomes pronounced, and intensive recharge with very cold water ($T = 4.1$°C) occurs resulting in significant increase of both spring flows, but also a significant decrease in water temperatures. Dye test proves this phenomenon of temperature inversion for Blederija springs.

Time series analysis confirmed these results, especially the bimodal behaviour, with strong diffuse infiltration in the binary karst hydrogeological system. This indicates different responses in various parts of the systems with frequent interchanges of 'normal' with subthermal waters.

ACKNOWLEDGMENTS

This research was supported by the Ministry of Education, Science and Technological Development (as a part of the Projects No. 43004 and No. 176022).

REFERENCES

Dragišić V., Stevanović Z., Filipović B. (1992) The occurrences of deep siphonal circulation of karst aquifer of Miroč mountain (NE Serbia). *Theoretical and Applied Karstology* 5, 115–120.

Eckhardt K. (2005) How to construct recursive digital filters for baseflow separation. *Hydrological Processes* 19, 507–515.

Gregor M. (2010) BFI+ 3.0 modul. HydroOffice 2015. URL: http://hydrooffice.org; last visited 10/10/2015.

Jemcov I., Petric M. (2009) Measured precipitation vs. effective infiltration and their influence on the assessment of karst systems based on results of the time series analysis. *Journal of Hydrology* 379(3–4), 304–314.

Jemcov I., Petric M. (2010) Time Series Analysis, Modelling and Assessment of Optimal Exploitation of the Nemanja Karst Springs, Serbia. *Acta Carsologica* 39(2), 187–200.

Kobayashi D. (1985) Separation of the snowmelt hydrograph by stream temperatures. *Journal of Hydrology* 76, 155–162.

Kyoung J.L., Youn S.P., Jonggun K., Yong-Chul S., Nam W.K., Seong J.K., Ji-Hong J., Bernard A. E. (2010) Development of genetic algorithm-based optimization module in WHAT system for hydrograph analysis and model application. *Computers & Geosciences* 36(7), 936–944.

Nathan R.J., McMahon, T.A. (1990) Evaluation of automated techniques for baseflow and recession analysis. *Water Resources Research* 26(7), 1465–1473.

Prohaska S., Ristić V., Dragišić V. (2001) Groundwater budget and dynamical reserves estimation of the Miroč karst massif (in Serbian) Proračun bilansa i dinamičkih rezervi podzemnih voda karstnog masiva Miroč. *Vodoprivreda* 33(189–194), 35–40.

Sloto R.A., Crouse M.Y. (1996) HYSEP: a computer program for stream flow hydrograph separation and analysis. *U.S. Geological Survey, Water-Resources Investigation Report* 96-4040, Reston, VA, 46 pp.

Stevanović Z., Dragišić V., Dokmanović P., Mandić M. (1996) Hydrogeology of Miroč Karst Massif, Eastern Serbia, Yugoslavia. *Theoretical and Applied Karstology* 9, 89–95.

Chapter 19

Delineation of the Plitvice Lakes karst catchment area, Croatia

Hrvoje Meaški, Božidar Biondić & Ranko Biondić
University of Zagreb, Faculty of Geotechnical Engineering Varaždin,
Croatia

ABSTRACT

The observed Plitvice Lakes Catchment area is located in the central part of the Dinaric karst, in Croatia. It is the largest catchment area within the Plitvice Lakes National Park. Due to its specific hydrogeological conditions, it can be divided into several subcatchmens. These were determined through the synthesis of research results that includes hydrogeological characteristics of the catchment area, hydrochemical and isotopic analysis of water from springs, and hydrological analysis of the surface water system. The result is a division of the Plitvice Lakes catchment into three main subcatchment areas: Matica, Plitvica and Jezera (Lakes).

19.1 INTRODUCTION

The Plitvice Lakes catchment area is approximately 152 km². It is a part of the Korana groundwater body (Biondić *et al.*, 2013), which belongs to the Black Sea catchment area, and is located in the central part of the Dinaric karst in Croatia. The Plitvice Lakes catchment is the largest catchment within the Plitvice Lakes National Park whose total surface area is 296.85 km². The wider Plitvice Lakes catchment area has been designated a national park since 1949 due to its outstanding natural beauty. For the same reason the Plitvice Lakes National Park has been included on the UNESCO List of World Heritage Sites since 1979. This was the first area in Europe to be included in this list due to the attractive phenomena created by water. It is the largest, oldest and most visited Croatian national park.

The most prominent surface water occurrences within the Plitvice Lakes catchment are cascading lakes of various sizes. The biodynamic process of predominant tufa barrier growth created those lakes and nowadays there are 16 lakes, whose surfaces and forms are constantly changing over time (Figure 19.1). The largest lakes are Kozjak and Prošćansko. Beside the well-known lakes system, there are numerous permanent karst springs in the Plitvice Lakes catchment, of which the most important are the Crna Rijeka, Bijela Rijeka and Plitvica.

Nowadays the Plitvice Lakes National Park receives more than one million tourists per year; urbanisation is encroaching upon the administrative boundaries of the Park, causing the human impact to be pronounced in some parts. Therefore, for the protection of the Plitvice Lakes National Park it is now no longer advisable, nor sustainable, to restrict protection only to the narrow area around the lakes system but to expand the protection zone out to the entire Plitvice Lakes catchment. It is necessary

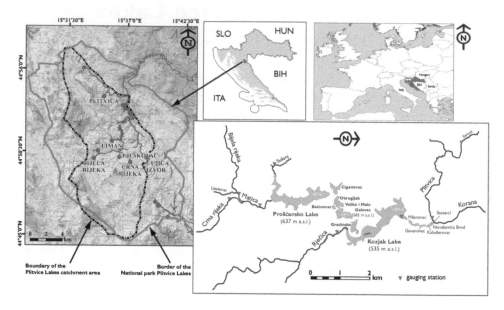

Figure 19.1 Location of the National Park Plitvice Lakes with position of the Plitvice Lakes catchment area (left). Sketch of the lakes system (right).

to observe Plitvice Lakes as a unique hydrogeological unit, composed of a series of interconnected small subcatchment areas. This delineation is not simple or unambiguous, but it can be divided into three subcatchments: Matica, Plitvica and Jezera. Never before have these areas been delineated in such way (Meaški, 2011).

19.2   METHODS AND RESULTS

Determination of the catchment area is the basis for all water balance calculations. It enables the estimation of groundwater and surface water volumes, as well as the identification of possible sources, the direction and flow of water, and the transportation of potential contamination in water.

Delineation of the subcatchment areas was carried out using three basic sets of data: hydrogeological, hydrometeorological and hydrochemical. Hydrogeological data represent the basic data that are needed to define the conceptual position of the catchments and subcatchments. Hydrological and meteorological analyses were used to quantify the presumed hydrogeological catchments, which were then confirmed by iteration. Hydrochemical data were used for verification; these are mainly discrete data that are influenced by conditions of spatial data in the catchment.

19.2.1   Hydrogeological research

The Plitvice Lakes catchment is a part of the Dinaric karst area in which the dynamic of groundwater is related to the process of karstification of carbonate rocks of Mesozoic age. Geotectonic and geodynamic evolution of the Dinarides is well-researched

(Dimitrijević, 1982; Herak, 1986, 1991; Pamić *et al.*, 1998; Korbar, 2009) and it can be noted that the evolution of the Dinarides has had a significant impact on these processes. Hydrogeological characteristics and water permeability assessments of layers (Figure 19.2) were determined according to existing structural data and the lithological composition of rocks.

The geological map for the area (Polšak *et al.*, 1967; Velić *et al.*, 1970), the Geological Map of Plitvička jezera (Polšak, 1969) and other relevant and available data relating to the geology of investigated area were used (Bahun, 1978; Biondić & Goatti, 1976; Biondić, 1982; Herak, 1962; etc.).

Data collected in the field, the position of sources and sinks, karst geomorphological features as well as the results of tracing tests from the wider area were taken into consideration during the hydrogeological analyses of the Plitvice Lakes catchment. Research of the direction of groundwater flow in the Plitvice lakes has been carried out on several occasions, either through systematic scientific research or as part of other research. By the end of 2012 in the Plitvice Lakes catchment, 17 tracer tests for groundwater flow were carried out (Biondić *et al.*, 2008; Meaški, 2011).

19.2.2 Water balance determination

Spatial and temporal distribution of rainfall, particularly effective rainfall, runoff and evapotranspiration are the next basic elements that are needed to calculate the water balance of the presumed catchment area.

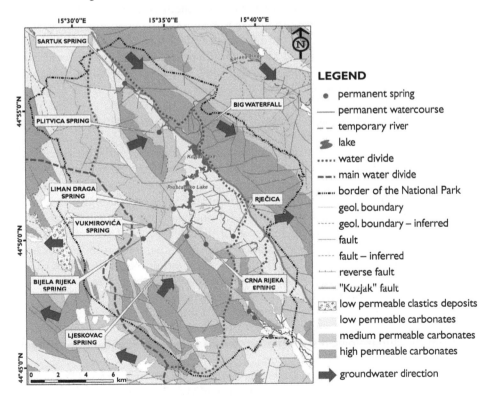

Figure 19.2 Hydrogeological sketch of the Plitvice Lakes catchment.

The spatial distribution of rainfall was obtained from the available rainfall data recorded in the reference period from 1961 to 1990 (DHMZ, 2009). Point data were interpolated using the Spatial Analyst Tools module within the ArcGIS software package. The mean annual precipitation in the Plitvice Lakes catchment varies from 1224 to 1475 mm (see P in Table 19.1), or approximately 1350 mm for the whole Plitvice Lakes catchment. Isohyets highlight a north west to south east zone coinciding with the dominant mountains in this area (Meaški, 2011).

Effective precipitation is a part of the total precipitation that can reach the saturated zone and recharge the aquifer. Ultimately, this is the volume of spring discharge. Usually it can be obtained through the application of various empirical formulae. In the Plitvice Lakes catchment the empirical formula of Žugaj (1995) was applied, which is especially suitable for the Dinaric karst catchments in Croatia. Values varied from approximately 790 mm in the southern part of the Plitvice Lakes catchment to approximately 880 mm in the middle and northern part of the Plitvice Lakes catchment (see Pe in Table 19.1 and Figure 19.2). The possible deviation of obtained result is $\pm 18\%$.

The runoff coefficient (sometimes called effective infiltration coefficient) was determined also from the precipitation data. This coefficient is defined as the ratio of effective precipitation to total precipitation and it includes all processes that can occur in some catchments during transformation of the total precipitation into effective precipitation. Therefore, it is used very often in scientific and theoretical hydrological analysis (Bonacci & Jelin, 1988; Žugaj, 1995; Bonacci, 1999, 2001). The calculated runoff coefficient for the whole Plitvice Lakes catchment is approximately 0.65 (see c in Table 19.1).

The analysis of the spatial distribution of air temperature includes the correlation between altitude and normal annual air temperature at certain measuring stations in the wider area for the reference period from 1961 to 1990. This is a linear regression equation, which represents the vertical temperature gradient, i.e. a drop in air temperature of 0.5°C per 100 m of altitude (Meaški, 2011). This is in line with a similar approach that was made in previous research related to this part of Croatia (Zaninović et al., 2004).

In order to calculate evapotranspiration the empirical formula of Turc (1954) and Coutagne (1954) were used. When using the Coutagne method it is essential that the correct equation is chosen, in accordance with Coutagne parameter λ. In the Plitvice Lakes catchment the Coutagne equation: precipitation $\geq 0.5/\lambda$, was selected. The resultant average values of evapotranspiration, based on both empirical formulae, are in the range of 450 to 500 mm (see Et in Table 19.1). These methods have been used for a very long time for estimation of evapotranspiration in various areas, including karst. As these are empirical formulae, they should be applied with caution when used in a certain area. In the Plitvice Lakes catchment area several methods were tested, and ultimately this two were selected as the most appropriate.

The average annual evaporation from the lake surfaces has also been calculated. The calculation is based on an empirical relationship (Meyer, 1915) using measurements from the climatological station located near Kozjak Lake. The value obtained is 422 mm $\pm$ 19 mm (Meaški, 2011).

For the hydrologic analysis available data on discharges measured at water gauging stations within the Plitvice Lakes catchment (DHMZ, 2009) were used (see $Q_{av.81-08}$

Table 19.1 Calculated hydrometeorological values for the Plitvice Lakes catchment (PLC) compared to measured flow rates (period from 1981 to 2008).

[1]Description	% of PLC	Area (km²)	P (mm)	Pe (mm)	Et (mm)	c (Pe/P)	Q_{calc} (m³/s)	[2]$Q_{av.81-08}$ (m³/s)	[3]g.s.
Bijela River with Ljeskovac	13	20	1224	774	449	0.63	0.50	0.448	2
Crna River	41	62	1245	791	454	0.64	1.55	1.36	1
Matica (direct)	1	2	1273	798	475	0.63	0.042	no data	3
Subcatchment MATICA	**55**	**84**	**1247**	**788**	**459**	**0.63**	**2.09**	**2.09**	**3**
Upper lakes (direct catchment)	10	10	1376	893	483	0.65	0.276	no data	4
Sušanj	3	5	1257	797	460	0.63	0.125	0.055	5
Rječica River	8	12	1455	969	486	0.67	0.383	0.426	3
Matica inflow							+2.09	2.09	
evaporation from the lakes							−0.025	no data	
water intake from the Kozjak Lake							−0.100	no data	
Subcatchment JEZERA	**18**	**28**	**1363**	**886**	**477**	**0.65**	**2.75**	**2.65**	**6**
Sartuk	7	10	1475	998	477	0.68	0.325	0.094	8
Plitvica spring	18	28	1374	919	455	0.67	0.807	no data	
Plitvica River (direct catchment)	2	3	1452	959	493	0.66	0.077	no data	
Subcatchment PLITVICA	**27**	**41**	**1434**	**959**	**475**	**0.67**	**1.21**	**0.655**	**7**
PLC TOTAL	**100**	**152**	**1348**	**878**	**470**	**0.65**			

[1]For location of specific subcatchments, see Figure 19.7.
[2]Calculation is based on flow measurements at gauging stations.
[3]For location of specific gauging station (g.s.), see Figure 19.4.

in Table 19.1). It should be noted that most of the gauging stations only started operating in the early 1980s, at a time when the majority of streams in the Croatian karst region had sustained long dry period (Žugaj, 1995). Therefore, to obtain a sufficiently long time series of annual flow, analyses of homogeneity of time series were carried out using Wilcoxson (1945) nonparametric test. The length of the hydrological series is validated according to the error size of the variation coefficient, using the equation by UNESCO (1982) and Kritsky-Menkel (1961).

Time series of the flow as well as identification of possible changes in the hydrological regime were analysed using the Rescaled Adjusted Partial Sums (RAPS) method (Garbrecht & Fernandez, 1994). Biondić (1999) and Bonacci & Andrić (2008) carried out similar analysis of hydrological changes in other karstified terrains using the RAPS method.

The cyclicality of the hydrological regime in the Plitvice Lakes catchment was analysed using RAPS (Figure 19.3). In 50 years of hydrological observations of the Plitvice Lakes catchment area, exchange of three wet and three dry periods were detected, i.e. every 16 years there is an extreme wet and dry period. However, for now it cannot yet be seen a continuous trend of decline or increase in the amount of water in the Plitvice Lakes catchment. It also should be noted that the individual annual extremes seen in Figure 19.3 do not need to fit into the perceived frequency of the hydrological regime and that the calculation are based on hydrological observation related to gauging station marked with number 9 (see Figure 19.4).

The final analysis of the water balance shows that the Matica flow represents a major contribution of water to lakes system, approximately 2.09 m³/s of water, more than all the other inflows put together. From this amount of water, about 74% is from Crna River and about 24% of water is from Bijela River (Figure 19.4).

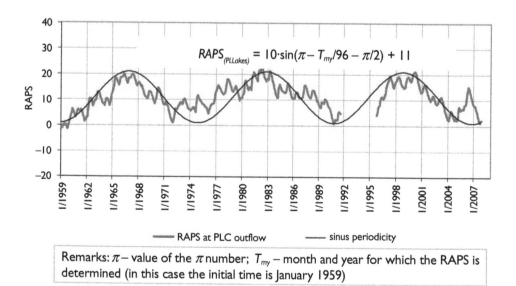

$$RAPS_{(Pl.Lakes)} = 10 \cdot \sin(\pi - T_{my}/96 - \pi/2) + 11$$

——— RAPS at PLC outflow   ——— sinus periodicity

Remarks: π – value of the π number; T_{my} – month and year for which the RAPS is determined (in this case the initial time is January 1959)

Figure 19.3 The RAPS analysis of mean monthly flow in the Plitvice Lakes catchment outflow, with marked periodicity.

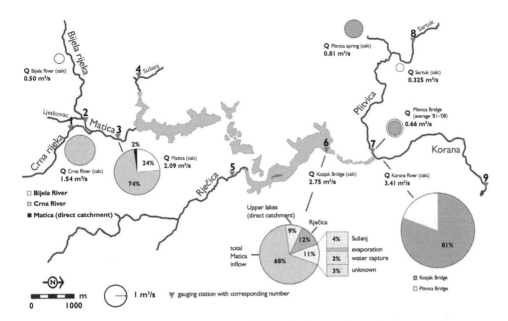

Figure 19.4 Visual comparison of the calculated flow in the Plitvice Lakes catchment in the period from 1981 to 2008.

Details and more results from testing the homogeneity of time series of annual flow at water gauging stations in the basin lakes, the length of individual datasets, as well as calculations of the error coefficient variation can be found in Meaški (2011).

19.2.3   Hydrochemical analyses

Measurement of basic physico-chemical parameters (water temperature, pH, electrolytic conductivity, and dissolved oxygen concentration), water sampling and laboratory analysis of chemical composition in the Plitvice Lakes catchment were carried out within the international research project during two hydrological years 2005–2007 (Biondić *et al.*, 2008). All observed points are shown in Figure 19.7, and some of the results in Table 19.2.

Alkalinity values are shown as concentration of HCO_3^- ions, because the concentration of all other ions that constitute the total water alkalinity in the Plitvice Lakes are negligible due to the acidity of the water (pH) (Stumm & Morgan, 1981; Appelo & Postma, 1994). The higher concentration of HCO_3^- at springs and streams is a consequence of the lack of precipitation of calcite. In downstream parts of rivers and lakes, HCO_3^- concentration gradually decrease due to the deposition of tufa sediments along the lake system.

The chemistry of groundwater is frequently dependent on the source rocks through which the fluid flows. The simple comparison of the HCO_3^- concentration and the Ca^{2+}/Mg^{2+} ratio in source waters shows that the area is built of predominantly carbonate sediments. However, due to some differences in the hydrochemistry, different catchment areas could be identified (Figure 19.5).

Table 19.2 The average concentrations of major ions at certain observation points (period from 2005 to 2007).

¹Description & mark	Ca^{2+}	Mg^{2+}	Na^+	K^+	HCO_3^-	SO_4^{2-}	Cl^-	NO_3^-	TDS
	shown as % of total dissolved solids								meq/l
Bijela Rijeka spring (BR-I)	28	22	0.4	0.1	48	0.4	0.8	0.6	10.55
Crna Rijeka spring (CR-I)	36	14	0.4	0.1	48	0.5	0.7	0.5	9.08
Matica River (M-R)	33	17	0.3	0.1	48	0.3	0.6	0.4	9.46
Rječica River (R-R)	26	24	0.2	0.1	49	0.2	0.4	0.3	11.16
Plitvica spring (PL-I)	32	18	0.2	0.1	48	0.6	0.4	0.8	9.78
Sartuk spring (SAR-I)	27	24	0.2	0.1	48	0.1	0.4	0.6	10.26
Ljeskovac spring (LES-I)	33	18	0.2	0.1	47	0.2	0.5	0.7	10.31
Rječica (RJ-I)	27	24	0.1	0.1	49	0.0	0.4	0.0	11.22
PLC average based on all data (Biondić *et al.*, 2008)	**29**	**21**	**0.3**	**0.1**	**49**	**0.4**	**0.5**	**0.4**	**9.73**

¹For location of specific water sampling points, see Figure 19.7.

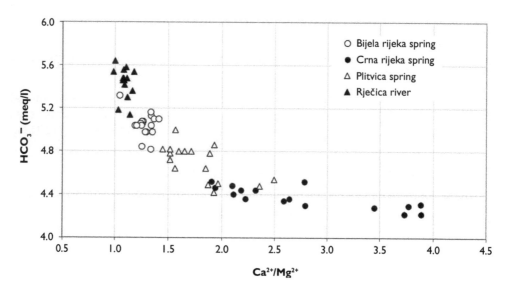

Figure 19.5 Comparison of the HCO_3^- concentration and the ratio of Ca^{2+} and Mg^{2+} in source waters (period from 2005 to 2007).

Sampling and analysis of stable isotopes of water was carried out, which aimed to complete the picture of the natural water system; the identification/verification of the mean recharge altitude of springs, the determination of the mean residence time of groundwater as well as understanding the processes of ionic changes in water and evaporation. The Local Meteoric Water Line (LMWL) was also obtained (Figure 19.6). The content of stable isotopes $\delta^{18}O$ and δD in precipitation was determined from a composite sample of rainwater for the period 2003–2007.

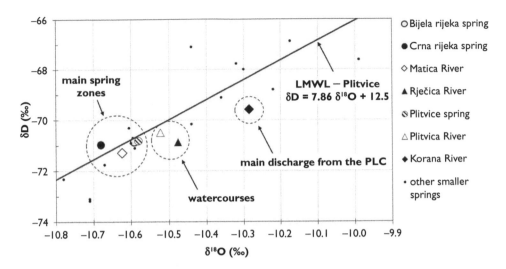

Figure 19.6 Analysis of $\delta^{18}O/\delta D$ values for all major and some minor sources and streams in the Plitvice Lakes catchment area (period from 2005 to 2007).

More details about results regarding hydrochemical analysis can be found in Biondić *et al.* (2008) and Meaški (2011).

19.3   DISCUSSION & CONCLUSIONS

The Plitvice Lakes catchment is a unique hydrogeological catchment because all waters are directed toward the lakes system, or the source area of the Korana River. However, due to its specific hydrogeological conditions it can be divided into three main subcatchment areas: Matica, Plitvica and Jezera (Lakes).

Determination of smaller subcatchments was carried out through the synthesis of research results that include the hydrogeological characteristics of the area, hydrochemical and isotopic analysis of groundwater, and hydrological analysis of the surface water system. It should be noted that the final subcatchment surface area and its divides needs further confirmation through new and more detailed hydrogeological investigations.

19.3.1   Matica subcatchment area

The Matica subcatchment includes all sources and watercourses that gravitate to the Matica River, which then inflow into Lake Prošćansko. The most important are the Crna River, Bijela River and Ljeskovac and accordingly the subcatchment can be divided into three hydrogeological parts, (Figure 19.7). A natural barrier for discharge from this karst catchment is the low permeable dolomite of Triassic age. The total outflow from the subcatchment is measured on the Matica River at the gauging station

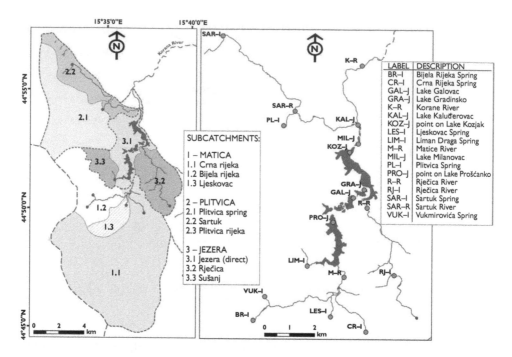

Figure 19.7 Main subcatchments in the Plitvice Lakes catchment (left). Name and position of major water sampling points used for hydrochemical and isotopic analysis (right).

'Matica' (marked with number 3 at Figure 19.4). Its hydrochemistry is a mixture from the main watercourses (Table 19.2).

Hydrogeological part of Crna River consists of Liassic and Malmian ages limestones and dolomites that vary in the permeability (Figure 19.2). The main discharge zone is the Crna Rijeka spring (Figure 19.8, Table 19.1). Emerging from the spring is an approximately 2.5 km long surface stream of the same name, which connects to the Bijela River and forms the Matica River. Along its course, the Crna River does not receive significant surface runoff. Total outflow is measured on the Crna River at the gauging station of the same name (marked with number 1 at Figure 19.4).

Hydrogeological part of Bijela River almost entirely consists of Upper Malmian age dolomites. It is a difficult catchment to analyse due to the lack of reliable data on the direction of groundwater flow. The main discharge zone is the Bijela Rijeka spring (Figure 19.8, Table 19.1), and the largest part of the catchment extends north west from this zone, along the divide between the Adriatic and Black Sea catchment areas (Biondić & Goatti, 1976; Dešković *et al.*, 1981, 1984; Kuhta, 2010). From the spring emerges an approximately 4.5 km long surface stream called the Bijela River. Along its course, it receives numerous small tributaries, e.g. water from the Vukmirovića spring (Figure 19.2), and just before the junction with the Crna River the Ljeskovac Stream also flows in.

The Ljeskovac Stream belongs to the hydrogeological part Ljeskovac (Figure 19.7). This is the only watercourse located between the Bijela and the Crna River. It begins as

Figure 19.8 Crna Rijeka spring (left) and Bijela Rijeka spring (right) (October 2005).

a series of minor karst springs, approximately 2 km before the junction with the Bijela River. Higher parts of the basin consist of Upper Malm age carbonate sediments, while the lower parts mainly consist of Doger age limestones. The total outflow from the Bijela River and Ljeskovac Stream is measured on the Bijela River at the gauging station of the same name (marked with number 2 at Figure 19.4).

19.3.2   Plitvica subcatchment area

The Plitvica subcatchment includes all waters that gravitate to the Plitvice River and the Big Waterfall (78 m height), the highest waterfall in Croatia. There it merges with waters that come from the lakes system. This area is called Sastavci, also known as the beginning of the Korana River (Figure 19.10). The total outflow from this sub-catchment is measured on the Plitvica River at the gauging station 'Plitvica most' (marked with number 7 at Figure 19.4). This subcatchment could be further divided into hydrogeological parts of Sartuk, Plitvica Spring and Plitvica River (Figure 19.7).

The hydrogeological part of Sartuk includes the area drained by the Sartuk Stream. This is the small watercourse, which is formed after the merging of several small springs in the area whih consists of poorly permeable dolomite of Triassic age (Figure 19.2). Division is based partly by the topographic setting, and partly by the parallel contact of less permeable carbonate rocks of Liassic age and highly permeable rock of Doger age. In general, there are plenty of major water loss indications in this part of the Plitvice Lakes catchment (Table 19.1).

Hydrogeological part of Plitvica Spring consists of Lower Cretaceous and Doggerian age limestones, which here represent the main aquifers. Hydrogeological conditions are affected by the significant longitudinal fault known as Kozjak fault (Figure 19.2), which separates Doggerian and Malmian ages carbonate sediments.

Hydrogeological part Plitvica River includes the surface catchment area of the watercourse itself (Figure 19.9). The separation of this part is due to the specific problems of water loss along the main river flow. These are most pronounced at the Big Waterfall, which during summer dry periods almost dries up (Biondić *et al.*, 2010; Meaški, 2011).

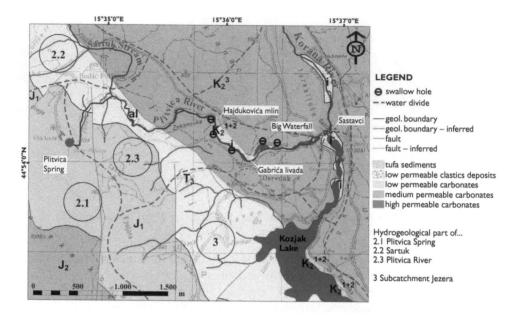

Figure 19.9 Hydrogeological sketch of the Plitvice River area.

19.3.3   Jezera subcatchment area

The Jezera subcatchment includes all the surface water inflow into the lakes system. The water from the Matica River flows in this subcatchment. The most significant direct inflows are the Rječica River and the Sušanj stream. This subcatchment can, therefore, be divided into three hydrogeological parts (Figure 19.7).

The hydrogeological part of Jezera covers the area of the lakes themselves and their direct inflow area. It consists of low permeable dolomites of Triassic age, which also form the base of the Upper Lakes and the largest part of the Kozjak Lake. The eastern part of this area is confined by highly permeable limestones of Upper Cretaceous age. According to tracer test experiments, those areas belong to the Klokot Spring catchment are located in the neighbouring state of Bosnia and Herzegovina (Biondić *et al.*, 2008).

Hydrogeological part of Sušanj (Figure 19.7) mainly consists of lower permeable dolomite of Malmian age. The surface water drains to the Sušanj stream, which is ephemeral upstream while downstream it has a permanent flow. This stream receives water from a permanent source in Limanska Draga shortly before entering Lake Prošćansko.

The hydrogeological part of Rječica (Figure 19.7) is located in an area consisting of weakly permeable dolomites of Triassic age. There are no greater karst springs in this area but there is a well-developed surface network of streams, which drain towards permanent surface flow called the Rječica River. This watercourse flows into the Kozjak Lake at approximately 535 m a.s.l. and the topmost part of the catchment is located at approximately 840 m a.s.l.

Figure 19.10 The Big Waterfall and Sastavci (source area of the Korana River).

ACKNOWLEDGEMENTS

A great deal of data was collected with the project 'Sustainable utilisation of water in the pilot area Plitvice Lakes'. The project was carried out within the Kompetenznetwerk Wasserressourcen Gmbh, co-financed by the Plitvice Lakes National Park and the Austrian government. The authors wish to thank all associates from the Joanneum Research institute (Graz, Austria) and the Research Centre 'Ivo Pevalek' (Plitvice Lakes National Park, Croatia), who have made significant contributions and enabled the successful completion of the project.

REFERENCES

Appelo C.A.J., Postma D. (1994.) *Geochemistry, groundwater and pollution*. Balkema, Rotterdam, 536 p.

Bahun S. (1978) Model razvoja hidrogeologije nekih polja u Dinarskom kršu. Zbornik radova 9. kongresa geologa Jugoslavije, Sarajevo, 2, 855–861.

Biondić B. (1982) Hidrogeologija Like i južnog dijela Hrvatskog primorja (in Croatian). PhD thesis. Faculty of Science, University of Zagreb, Zagreb, 214 p.

Biondić B., Biondić R., Meaški H. (2010) The conceptual hydrogeological model of the Plitvice Lakes. *Geologia Croatica*, 63(2), 195–206.

Biondić B., Goatti V. (1976) Regionalna hidrogeološka istraživanja Like i Hrvatskog primorja (in Croatian). Archive of the HGI-CGS, Zagreb.

Biondić B., Zojer H., Yehdegho B., Biondić R., Kapelj S., Meaški H., Zwicker G. (2008) Mountainous lakes: Sustainable utilization of water in the pilot area Plitvice Lakes, Final report. Varaždin (Croatia) – Graz (Austria) – Plitvice Lakes (Croatia). Archive of the Faculty of Geotechnical Engineering, University of Zagreb, Varaždin.

Biondić D. (1999): Status of surface waters quantity at Upper Kupa catchment area (in Croatian). In: Biondić, B., Kapelj, S., Biondić, D., Biondić, R., Novosel, A., Singer, D. (2002): Studija ugroženosti sliva Gornje Kupe. Archive of the HGI-CGS, Zagreb.

Biondić R., Biondić B., Rubinić J., Meaški H. (2013) Quality and quantity status and risk assessment of groundwater bodies in the karst areas of Croatia. In: Maloszewski P., Witczak S., Malina G (editors) *Groundwater quality sustainability*. IAH Selected papers on Hydrogeology 17, 163–171.

Bonacci O. (1999) Water circulation in karst and determination of catchment areas: example of the Zrmanja River. *Hydrological Sciences Journal* 44(3), 373–386.

Bonacci O. (2001) Monthly and annual effective infiltration coefficients in Dinaric karst: example of the Gradole karst spring catchment, *Hydrological Sciences Journal*, 46(2), 287–299.

Bonacci O., Andrić I. (2008) Sinking Karst Rivers Hydrology – Case of the Lika and Gacka (Croatia). *Acta Carsologica* 37(2–3), 185–196.

Bonacci O., Jelin J. (1988) Identification of a karst hydrolgic system in the Dinaric karst (Yugoslavia). *Hydrological Sciences Journal* 33(5), 483–497.

Coutagne A. (1954) Quelqes considération sur le pouvoir évaporant de l'atmosphère, le déficit d'écoulement effectif et le déficit d'écoulement maximum. 360–369.

Dešković I., Pedišić M., Marušić R., Milenković V. (1981) Značaj, svrha i neki rezultati hidrokemijskih, hidroloških i sanitarnih istraživanja površinskih i podzemnih voda Nacionalnog parka Plitvička Jezera (in Croatian). *Vodoprivreda* 13 (69–71) Beograd, 1–3, 7–19.

Dešković I., Marušić R., Pedišić M., Sipos L., Krga M. (1984) Neki najnoviji rezultati hidrokemijsko-hidroloških istraživanja voda na području Plitvičkih Jezera (in Croatian). *Vodoprivreda* 16 (88–89) Beograd, 2–3, 221–227.

Dimitrijević M. (1982) Dinarides – an outline of the tectonics. *Earth Evolution Science* 1, 4–23.

DHMZ (2009) Hydrological and meteorological data for the Plitvice Lakes area for the period 1997–2007. Archive of the Državni hidrometeorološki zavod (DHMZ), Zagreb, Croatia.

Garbrecht J., Fernandez GP. (1994) Visualization of trends and fluctuations in climatic records. *Water Resources Bulletin* 30 (2), 297–306.

Herak M. (1962) Tektonska osnova hidrogeoloških odnosa u izvornim područjima Kupe I Korane (s Plitvičkim jezerima). Referati V. savetovanja geologa FNR Jugoslavije, Beograd, 3, 17–25.

Herak M. (1986) A New Concept of Geotectonics of the Dinarides. Acta Geologica 16(1), 1–42.

Herak M. (1991) Dinaridi – mobilistički osvrt na genezu i strukturu. *Acta Geologica*, 21/2, HAZU, Zagreb.

Korbar T. (2009) Orogenic evolution of the External Dinarides in the NE Adriatic region: a model constrained by tectonostratigraphy of Upper Cretaceous to Paleogene carbonates. *Earth-Science Reviews*, 96, 296–312.

Kritsky S.N., Menkel M.F. (1961) On the regularities of long term river flow fluctuations and on the methods of estimation of lingering droughts. *Hydrological Sciences Journal* 6(2), 34–40.

Kuhta M. (2010) Vodoistražni radovi u cilju zaštite izvora Krbavica – II. faza (in Croatian). Archive of the HGI-CGS, Zagreb.

Meaški H. (2011) Model zaštite krških vodnih resursa na primjeru Nacionalnog parka 'Plitvička jezera' (in Croatian). PhD thesis. Faculty of Mining, Geology and Petroleum Engineering, University of Zagreb, Zagreb, 211 p.

Meyer A.F. (1915) Computing Run-Off from Rainfall and Other Physical Data. Trans. ASCE, 79, 1056–1224.

Pamić J., Gušić I., Jelaska V. (1998) Geodynamic evolution of the Central Dinarides. *Tectonophysics* 297, 251–268.

Polšak A. (1969) Geološka karta Plitvičkih jezera, M 1: 50 000. Kartografski laboratorij Geodetskog fakulteta. Archive of the Faculty of Science, University of Zagreb, Zagreb.

Polšak A., Šparica M., Crnko J., Juriša M. (1967) Osnovna geološka karta SFRJ M 1:100000, list Bihać, L 33–116. Archive of the HGI-CGS, Zagreb.

Stumm W., Morgan J.J. (1981) *Aquatic Chemistry*, 2nd ed. John Wiley and Sons, New York, 109 p.

Turc L. (1954) Calcul du bian de l'eau evaluation en fonction des precipitations et des temperatures. IASH Symposium, Rome, 111(38), 188–202.

UNESCO (1982) *Methods of computation of low streamflows*. Paris.

Velić I., Bahun S., Sokač B, Galović I. (1970) Osnovna geološka karta SFRJ, M 1:100000, list Otočac, L 33–115. Archive of the HGI-CGS, Zagreb.

Wilcoxon F. (1945) Individual comparisons by ranking methods. *Biometrics*, 1, 80–83.

Zaninović K., Srnec L., Perčec Tadić M. (2004) Digitalna godišnja temperaturna karta Hrvatske. Hrvatski meteorološki časopis, 39, 51–58.

Žugaj R. (1995) Regionalna hidrološka analiza u kršu Hrvatske. Hrvatsko hidrološko društvo, Monografije br. 1, Zagreb, 139 p.

Part 4

Engineering, sustainable use and protection of water in karst

Chapter 20

Creating environmental impact indicators in dynamic karst system – Dinaric karst case example

Zoran Stevanović
University of Belgrade – Faculty of Mining and Geology,
Centre for Karst Hydrogeology of the Department of Hydrogeology,
Belgrade, Serbia

ABSTRACT

Environmental indicators are a tool to measure results of human activities and their impacts on nature and biodiversity. Water dependent ecosystems are under increasing pressure from human activities and in karst they are exposed to greater potential hazard because of the high vulnerability of karst aquifers. The aquifers of the Dinaric karst, one of the largest karst systems in the world, are abundant in water reserves; however there are several important potential threats to environmental and water quality values which have been identified through environmental and socio-economic surveys conducted within the DIKTAS project. Five strategic Environment Quality Targets are envisaged in this project: two (Groundwater sustainably used; Water and soil quality controlled and improved) have a direct relationship with the karst aquifer, while the other three (Agricultural, industrial and tourist sectors sustainably developed; Energy used efficiently and in a sustainable manner; Measures to protected nature formulated and applied) have an indirect relation to the karstic aquifers. A list comprising 25 different indicators was derived to assess pressures on groundwater quantity and quality as well as resulting pressures on dependent ecosystems in selected transboundary aquifers. Estimating indicator values will support sustainable water use and protection of ecosystems in specific karst environment.

20.1 INTRODUCTION

A concept – sustainable development – resulted from the work of the World Commission on Environment and Development (Brundtland Commission, 1987), and soon became one of the most successful approaches to be applied in many years. Since then, many international events, international and bilateral conventions, and international and national laws have been adopted taking into consideration the importance of sustainable development and have further developed the basic principles. For instance, water, as a source of life for humans, flora and fauna is an essential component of the environment. A concerted effort by land managers, industry, catchment groups, the community, and environmental groups is required in order to protect water resources for future generations. The concept supports strong economic and social development, which cannot be improved with measures that destroy the environment, and which helped to shape the international community's attitude towards economic, social and environmental development.

Managing water quality requires a catchment-based approach to management planning, with staged actions required to reach long-term goals (Levy, 2002). Effective natural resource management also requires recognition of the influence catchments and land uses have on the quality of water resources. The causes of many water quality problems are broad-scale in origin but to be remedied they often need regional solutions with action taken at the local level.

20.1.1 Pressures on water quantity and quality, related responses and the role of indicators

At present, most indicator reports compile sets of physical, biological or chemical indicators. They generally reflect a systems analysis view of the relations between the environment and human activities. But many of the relationships between the human system and the environmental system are not sufficiently understood or are difficult to capture in a simple framework.

The sustainable development indicators of the United Nations Commission on Sustainable Development (UNCSD) follow a three-part framework: Driving forces an extended version: the Driving forces – Pressures – State – Impact – Responses (DPSIR) framework (Figure 20.1).

According to this systems analysis view, social and economic developments exert *Pressure* on the environment and, as a consequence, the *State* of the environment changes. This leads to *Impacts* on human health, ecosystems and materials that may elicit a societal *Response* that feeds back on the *Driving forces* or on the *State*, or *Impacts* directly through adaptation or curative action.

Smeets and Weterings (1999) consider that the DPSIR framework is useful in describing the relationships between the origins and consequences of environmental problems, but in order to understand their dynamics it is also useful to focus on the internal links between DPSIR elements (Stevanović, 2011). In order to meet this information need, environmental indicators should reflect all elements of the causal chain

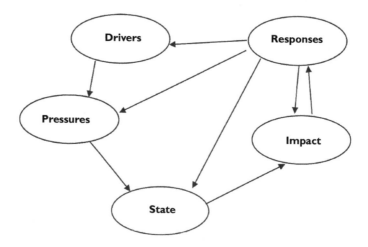

Figure 20.1 DPSIR scheme.

that links human activities to their ultimate environmental impacts and the societal responses to these impacts.

The indicators in general should be used to measure achieved results of human activities or to define the gap between *desired* and *executed*. For example, we may use indicators to assess the status of environment and present geo- and bio diversity components. An *Indicator* can also be defined as a parameter used to provide a measure of the quality of water or condition of an ecosystem.

According to the UNECE report (2007) there are indicators of several types based on the questions they answer:

- Descriptive indicators answer the question: 'What are the pressures on the environment and how is the quality of the environment developing?'
- Performance indicators answer the follow-up question: '...and is that relevant for policy goals?' Generally these indicators use the same variables as descriptive indicators but are connected with target values.
- Eco-efficiency indicators answer the question, 'Have we become more efficient in our economic processes?' Eco-efficiency indicators link driving forces with state or pressure indicators. Efficiency refers to pressures such as emissions or natural resources use.
- Policy-effectiveness indicators answer the question, 'What has been the effect of policy?' (in terms of policy effects, for instance, reduced pollution due to strict application of the principle *Polluters pay*).
- *Welfare indicators* are connected with the question: '... and are we on the whole better off?' and ask for a balance between economic, social and environmental progress.

Smeets and Weterings (1999) also state that most countries and international bodies currently develop performance indicators for monitoring their progress towards environmental targets. These performance indicators may refer to different kinds of reference conditions/values, such as:

- national policy targets;
- international policy targets, accepted by governments;
- tentative approximations of sustainability levels.

According to the Swedish Ministry of Sustainable Development (2004) the *Values* on which the interim targets are based include the following: Human health, Biodiversity and the natural environment, the Cultural environment and cultural heritage, Long-term ecosystem productivity, and Effective management of natural resources. Regarding the *Criteria*, they should be: clear and accessible, capable of being followed up in the short term, form part of an all-inclusive structure, and capable of serving as a basis for local environmental work and efforts to achieve the objectives.

Indicators should be used as a tool to compare *current status* versus *desired situation* (proposed or required) (Figure 20.2).

Targets should be set for each key management issue and may include a range of scales and time frames, as some issues take longer to address than others.

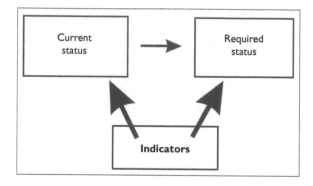

Figure 20.2 Scheme showing role of the indicators.

Levy (2002) indicates the following possible targets:

- Visions or goals for the region: *long-term targets* that cannot be reached immediately but are needed to protect the environmental values of the water body. They may be numerical, descriptive or relative statements (e.g. no net loss);
- Resource condition targets: specific, measurable and time-bound targets used to work towards the overall goal or vision for the water body. They may be *medium term targets* (10–20 years) required for regional planning (e.g. 30% reduction in phosphorus loads by 2006); or
- Management action targets: targets that are linked closely to management actions. They are usually measurable and time-bound but with shorter time spans (1–5 years) than the regional targets (e.g. 70 km of riparian vegetation is to be fenced within five years).

The quality targets can be explained as numerical levels or descriptive statements that must be met within a specified timeframe to protect and maintain environmental values. Water is of course one of the most important environmental values and water related indicators commonly cover: water availability and use, chemical quality and level of water protection.

Water quality targets can also be set for physical, chemical and biological components of aquatic ecosystems. Once defined, water quality targets become indicators of management performance and progress towards management goals or attainment of environmental values.

When it comes to groundwater as one of specific environmental values the recent study of Foster and Macdonald (2014) indicates water security as a term which includes three main indicators:

1. Groundwater storage availability: an indicator of 'buffer capacity' to support water-supply abstraction, which might be constrained by current groundwater resource status (aquifer water-level and salinisation trends) and connectivity to surface water (since in some aquifers small storage changes can have marked impacts on river baseflow, springflow and aquatic ecosystems).

2. Groundwater supply productivity: a measure of how easy it is to abstract groundwater from an aquifer, which relates to its depth and the aquifer transmissivity and any evidence of reducing productivity (primarily due to falling water levels).
3. Groundwater pollution protection: the effectiveness of pollution control and aquifer protection measures, and evidence of deteriorating quality trends.

A wide variety of environmental indicators are presently in use. These indicators reflect trends in the state of the environment, monitor the impact of human activities (Figure 20.3) and evaluate progress made in realising environmental policy targets. As such, environmental indicators have become essential to policy-makers.

Indicators are thus powerful tools for making important dimensions of the environment and of society visible and for enabling their management (Dahl, 2012). In addition,

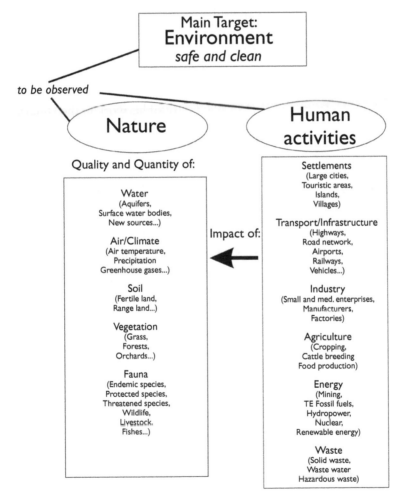

Figure 20.3 Scheme of main nature components and human activities to be observed towards the creation of a safe and clean environment.

environmental indicators are a key tool for environmental reporting and could be used as a powerful tool to raise public awareness on environmental issues. Appropriately chosen indicators, based on sufficient time-series data, can show key trends, help describe causes and effects of environmental conditions, and track and evaluate policy implementation (UNECE, 2003). Providing information on driving forces, impacts and policy responses is a common strategy to strengthen public support for policy measures.

20.1.2   Karst as specific media and brief on Dinaric karst environment and water

Among other groundwater-bearing media the karstic aquifer is specific in many terms, especially in heterogeneity, discharge variability, vulnerability to pollution and low attenuation capacity. The classical karst terminology recognises a karstic region as an area mainly consisting of compact and soluble carbonate or evaporitic rocks in which appear distinctive surficial and subterranean features, caused by solutional erosion. Karst is a complex and specific environment and accordingly requires a specific approach in protecting its natural values, features and species.

The Dinaric system (Dinarides) represents the geologically heterogeneous, south European orogenic belt of the Alpine mountain chain (Alpides) (Figure 20.4). The Dinaric region is a karst holotype. It is a fact that not only was the term 'karst' born in the area, but also Jovan Cvijić (1893) did most of his work in the Dinaric karst and created the foundation of a new scientific discipline – karstology (Ford, 2005).

The Dinaric karst is a mountainous region with a prevalence of highly karstified rocks and large karstic poljes and valleys created in tectonic depressions (Mijatović, 1983, 1984). Karstified zones have been drilled at depths up to 2000 m (Milanović, 2000, 2005) The Dinaric region contains all types of karst landforms and features including uvalas and poljes as the largest karstic forms. Along with its richness in various karstic features, the Dinaric region is by far the best endowed in Europe in water resources, but they are unequally distributed throughout the year (Bonacci, 1987; Stevanović et al., 2012). In the Dinaric region of ex-Yugoslavia and Albania there are more than 200 springs with a minimum discharge of over 100 l/s (Komatina, 1983; Eftimi, 2010).

Because karst has always been of central interest to hydrogeologists, investigations have expanded to a larger scale, thanks to numerous projects that have included the construction of large and medium dams. In the 1970s several such dams were built with the support of ex-Yugoslavian companies and experts (Croatia, Herzegovina, Montenegro) and for the first time successful results were achieved in the karst. Technical applications for control and regulation of karst aquifers through the construction of galleries, batteries of wells, and groundwater reservoirs (storage) represent an important contribution to international hydrogeological science (Milanović, 2000). All these activities changed the natural environment and had a significant impact with both positive and negative implications.

The DIKTAS Project (Protection and Sustainable Use of the Dinaric Karst Transboundary Aquifer System) was implemented in the period of 2011–2014 with GEF funding and support from UNDP and UNESCO IHP. The appearance of several new sovereign states from what was once Yugoslavia has established complex transboundary inter-linkages that impact on water use and water sharing for domestic supply, power generation, and agriculture (Kukurić, 2011). This is one of the reasons

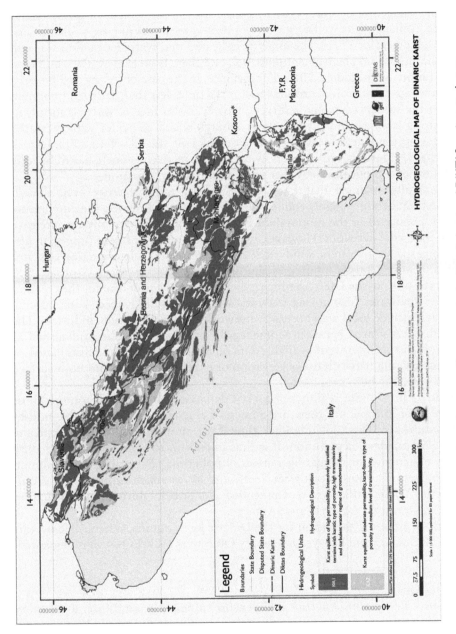

Figure 20.4 Distribution of main karstic aquifer systems in the Dinaric karst and DIKTAS project study area.

why the DIKTAS project was initiated and included three former Yugoslav countries (Croatia, Bosnia & Herzegovina, Montenegro) and Albania.

After delineation of several transboundary aquifers (TBA) of major concern, further analysis included the characterisation and development of conceptual models. Groundwater budgeting of TBAs created a base for the assessment of groundwater reserves and availability, as well as for proposals and measures aiming to ensure sustainable development of TBAs. The analysis indicated that water extraction was still far below the aquifer's replenishment potential, and there is no evidence of significant over-exploitation in the studied TBAs. For instance, in the case of Cetina and Neretva TBAs, the average extraction of groundwater is ten times less than the total minimum discharge of the springs (dynamic reserves). However, shortage of water is locally in evidence during summer and early autumn months which coincides with increased demands during the tourist season. That principles of the EU Water Framework Directive (WFD) regarding ecological flow for downstream consumers has to be fully respected, further complicates the water and environmental situation.

It is generally assumed that karst water quality is satisfactory even though Dinaric karst aquifers are highly vulnerable to pollution. However, the catchments are sparsely populated in the mountainous areas and there is virtually no intensive farming or industrial activities. However, when pollutants are present (mines, industrial and domestic waste waters, solid waste dumps, fertilisers), deterioration of water quality in unconfined karstic aquifers is almost assured.

One of the tasks of the DIKTAS project was to prepare a proposal for the creation of a new Groundwater Monitoring network which will fully respect karst specific behaviour and include local water users (waterworks, dams, irrigation, industry). The Cijevna/Cemi TBA shared between Montenegro and Albania has been identified as the most problematic in terms of available data on water resources, and installation of a modern monitoring network for observation of climate elements, and both surface and groundwater has been proposed.

Therefore, the Dinaric region has a very dynamic water budget, one of most intensive in all of Europe, but there are numerous challenges for sustainable utilisation of groundwater. These include high annual variation of natural flows and the vulnerability of aquifers to pollution. It is thus important to (1) improve the quality of water by eliminating or mitigating sources of pollution, (2) regulate the minimum spring discharges, (3) ensure ecological flows, and (4) establish proper water monitoring systems. These actions are recommended as a priority during implementation of the DIKTAS Strategic Action Plan (SAP).

To assist in attaining the vision for the Dinaric karst aquifer system, five overarching Water Resources and Ecosystem Quality Objectives (WR/EQO) were suggested. These are:

Groundwater Quantity
Water Resource/Ecosystem Objective A: To ensure sufficient groundwater availability in dry periods, especially for water supply and to support environmental flow.
Groundwater Quality
Water Resource/Ecosystem Objective B: To maintain and improve (where required) quality of karst groundwater in the Dinaric region.

Protection of Groundwater Dependent Ecosystems (GDE)
Water Resource/Ecosystem Objective C: To ensure protection of GDE, specific features and their ecosystem services for the future.
Equitable Use
Water Resource/Ecosystem Objective D: To support equitable use of groundwater resources.
Capacity Building
Water Resource/Ecosystem Objective E: To raise awareness and capacity building related to karst water and dependent ecosystem.

20.1.3   Environmental quality targets in Dinaric karst

In the process of proposing Environmental Impact Indicators the role of the stakeholders is very important. The stakeholders and community should be asked what they want to protect and should confirm any proposal. They also need to identify the water resources required now and for the future, and those requiring protection from the effects of pollution or degrading activities.

Water quality targets in every region are influenced by environmental, social and economic considerations, which in most cases will be unique to that region. Targets should also, where possible, consider current conditions, and long-term trends in water quality (Levy, 2002). In the Dinaric karst region there are several important facts influencing environmental and water quality values, which have been identified through environmental and socio-economic survey and described in Transboundary Diagnostic Analysis. Some of the major facts are:

- With the exception of Croatia, which is already an EU member, the other three project countries have lower income levels in comparison with the EU28 average.
- The percentage of the countries' agricultural area ranges from 24% to 47% while the GDP percentage of the agriculture sector in the participating countries varies from 8% to 18%.
- The population growth rate is low or negative. Across the region, there is a trend of migration from remote, rural areas towards urban areas and industrialised zones. Small settlements are extremely dispersed, and a number of settlements in rural areas are already abandoned.
- Hydropower plays a central role for energy production in all DIKTAS countries, and hydro power generation from the Dinaric karst system plays a significant role in the national economies.
- Historic industrial sites are one of the main sources of pollution in the region. The main method of waste disposal is in unlined landfills and there is still no system for the safe management of hazardous waste, especially in Albania and Montenegro.
- Waste and wastewater pollution have been identified as major threats to the protection of the Dinaric karst aquifer and eco-systems. Sewage systems discharge to improvised permeable septic pits, smaller adjacent surface streams or depressions.

- All DIKTAS countries are considered to have abundant groundwater resources at their disposal. However, during the summer period water shortages may occur, particularly in tourist areas along the Adriatic coast.
- The region is abundant with pristine nature areas, which are often vulnerable and under threat. Yet none of the countries recognised the vulnerability, complexity, and importance of integrated protection of the karst environment through national policies. The percentage of protected surface to the total area of the country varies from 0.5–12.4%.
- Finally, as already highlighted, monitoring systems in all four countries are not satisfactory.

A main objective of the DIKTAS project is to facilitate the equitable and sustainable utilisation of the transboundary water resources of the Dinaric Karst Aquifer System, and protect the unique groundwater dependent ecosystems that characterise the Dinaric Karst region of the Balkan Peninsula, the environmental quality targets should be focused mainly on natural reserves, but all other environmental issues which indirectly impact karstic aquifers, or depend on them, must also be taken into consideration, evaluated and improved.

In total, five strategic Environment Quality Targets (EQT) are suggested. Two have a direct, and three have an indirect relation to the karstic aquifers or are dependent upon them:

Direct EQT:
1. Groundwater sustainably used.
2. Water and soil quality controlled and improved.

Indirect EQT:
3. Agricultural, industrial and touristic sectors sustainably developed.
4. Energy used efficiently and in sustainable manner.
5. Measures to protect nature formulated and applied.

To improve the environmental situation in the Dinaric karst region and achieve the targets the following measures and activities are recommended:

Water must be used in a sustainable manner, by reducing consumption and losses in current water utilities. Some new sources should be opened and some others regulated to improve their minimum flows and safeguard water dependent eco-systems. The drinking water quality should be preserved from any kind of pollution, and sanitary conditions improved in general: the air must be clean enough not to represent a risk to human health or to animals, plants or cultural assets. Solid waste should be sustainably managed with greater emphasis on source reduction, recycling and recovery. Waste water treatment should be widely applied in industry along with the principle 'polluter pays'. The energy saving and new sources of renewable energy such as hydropower, solar, wind and sub-geothermal waters should become dominant. Food production should be based mostly on organic agriculture with minimal use of pesticides and herbicides. Animal husbandry as an agricultural component is essential to the long-term conservation and development of natural and cultural assets in the agricultural landscape. The value of pasture land and forests for biological production must be protected, at the same time as biological diversity is safeguarded. Biological diversity must be preserved for the benefit of present and future generations. Special attention should be paid to strict protection of threatened and endemic species.

However, all these protection measures should not limit further development of tourism and recreation but they have to be oriented and organised more ecologically and environmentally.

Establishing a proper system for monitoring water quantity and quality, continually evaluating the status of environmental impact indicators, improving local technical capacity and raising awareness of the local population on the importance of water savings, protection from pollution, sustainable development and tourism, remain major tasks beyond this phase of the DIKTAS project.

Not all of the EQT are easily and fully achievable because they largely depend on national GDPs and economic growth, and they should have a long-term effect. For this reason a phased approach is proposed, which, similar to the concept of the EU Water Framework Directive, includes an interim stage of 15 years with partial EQTs. Table 20.1 contains the list of these partial EQTs and related actions and indicators. It is important to note that more attention has been paid to direct EQTs, because of the main objective of DIKTAS.

20.1.4 Environmental impact indicators in the Dinaric karst region

By evaluating policy in South East Europe, Stritih *et al.* (2007) highlighted the main concerns which are applicable to Dinaric karst:

* How to secure a high level of protection of surface and groundwater, preventing pollution and promoting sustainable water use?
* How to secure funds for investments for water infrastructure and protection from pollution?
* What is appropriate institutional structure and division of responsibilities in water management?

UNECE (2003) stated that environmental monitoring systems are crucial for environmental policy: they are the 'eyes and ears' for policy makers, researchers, and the public seeking to understand and improve the environment. National environmental policies, institutions and funding mechanisms provide the context for monitoring systems. National policies can also specify goals for their development. Moreover, providing information to support national policies is a key objective of monitoring systems.

Water dependent ecosystems are essential components of the watersheds which are under increasing pressure from human activities. In karst, dependent ecosystems are exposed to greater potential hazard if they depend on water from aquifer. Although the problem of aquifer over-exploitation is often exaggerated (Custodio, 1992; Burke & Moench, 2000), variable water regime and low water flows during periods of maximum demands (summer months) can cause stress in many aquatic systems. The problem is much more sensitive when it comes to the area of transboundary concern (Chilton, 2002; Puri & Aureli, 2005).

Table 20.1 Proposal of environmental quality targets and indicators.

Strategic EQO and targets	Partial EQ targets to be achieved in next 15 years (up to 2030)	Related actions	Main environmental impact indicators
Groundwater sustainably used	Created expert teams and completed WR survey in designated areas in Dinaric karst and TBA	WR surveying and assessing	Renewable freshwater resources
	Reduced water spending by some 10% and specific consumption rate fell from ca 300 l/cap/day to 270 l/cap/day	Promoting water savings, gradually increasing water taxes	Renewable freshwater resources in recession (drought) periods
	Designed and implemented 6 pilot projects for aquifer engineering regulation and improved minimum flows	Introducing into water practice aquifer engineering regulation to improve minimum flows (managing aquifer recharge)	'Domicile' (and 'External') freshwater resources
	Repaired and replaced pipelines in problematic waterworks and reduced losses in the region by 5% on average	Increasing efficiency and reducing losses in centralised waterworks;	Renewable GW resources (Dynamic reserves)
	Designed and implemented 4 pilot projects for water recycling and reuse	Controlling fresh GW extraction especially along coastal strips and on islands;	Renewable GW resources (Dynamic reserves) in critical drought periods
	Established proper GW monitoring network by systematic control of water extraction rate and water quality by operators along with created and equipped new GW monitoring sites	Increasing number of connections to waterworks and improving rural water supply.	Water exploitation index
	Regional Consultation and Information Exchange Body (CIE) created and put into practice consultation / intern. reporting body and mechanisms for equitable and sustainable water use		Groundwater exploitation index
			Water demands (availability)
			Drinking water demands
			Water available per capita
			Groundwater depletion
			Water losses

Water and soil quality controlled and improved	Major pollutants eliminated or emission of pollution drastically reduced by putting into practice the principle 'polluter pays'	Controlling and inspecting systematically all registered major pollution emitters	Drinking water quality
	Waste water treatment plants constructed in 30 cities and treatment of communal waste water increased by 30% in comparison with actual status	Building new waste water treatment plants	Industry waste water index
		Constructing environmentally safe and controlled regional and local landfills	Household waste water index
	Constructed 15 new regional landfills and 50 smaller landfills at municipality level. All landfills are sanitary equipped and operated in a way that prevents adverse environmental impacts	Cleaning and remediating illegal and improperly built landfills and dump sites	Specific pollutants index
	Eliminated and cleaned sites of more than 300 illegal small landfills and solid waste dump sites with special attention to sanitation of karst depressions and features (small poljes, uvalas, dolines, potholes)	Removing solid waste to the newly built and secure landfills	Landfill status
		Removing and preventing further depositing of all highly toxic substances at communal landfills	Water reuse
	Increased number of connections to municipal sewage system by 40%	Widely introducing household waste separation and recycling; Sending electronics, electrical equipment and batteries for recycling	Salt water intrusion
	Reduced number of used septic tanks in semi-urban and rural areas by 30% by constructing new sewage systems	Constructing new sewage systems in number of settlements of various size	
	Established functional system for water sanitary control in all waterworks for centralised potable water supply which supplying more than 1300 users	Introducing and promoting advantages of water reuse and recycle	
	Reduced by 30% fresh GW pumping in coastal areas and islands during summer months to prevent salt intrusions	Imposing system for regular water sanitary control (constituents in accordance with WFD, plus specific components putting water at risk locally) and sharing information of all water utilities (National institutions, EIOnet Water, DIKTAS area)	

(Continued)

Table 20.1 (Continued).

Strategic EQO and targets	Partial EQ targets to be achieved in next 15 years (up to 2030)	Related actions	Main environmental impact indicators
	Created and put into practice new harmonised legislation for sources' sanitary protection and zoning by emphasising specificity of karst		
	Established sanitary protection zones and applied sanitary protection measures over all sources for centralised potable water supply for settlements over 1000 inhabitants		
	Constructing additional treatment segments for potable water in ca 20 water utilities (filtration, coagulation, removal of chemical contamination, and if and where necessary, ozonisation versus viruses, faecal coliforms)		
Agricultural, industrial and touristic sectors sustainably developed	Reduced by 50% use of pesticides, fertilisers and herbicides in agriculture	Producing organic food and limiting use of agro-technical protective agents	Fertiliser index Pesticide index Irrigation water demands and use
	Over-grazed pastures reduced by 10%	Controlling and limiting use of compost and sewage sludge that are applied to land for organic food production	
	Forested areas increased by about 10%		
	Reduced use of nitrate based fertilisers in agriculture by 30%	Constructing small waste water treatment unit at all slaughterhouses or removing organic wastes to nearby collecting /recycle centers	
	Decreased concentration of nutrients in rivers, lakes and sea water by 30%		
	Improved quality of littoral sea waters by 15%	Stimulating farming, keeping number of farms, farmers and livestock (large and small ruminants)	
	Fish stock managed and harvested sustainably and legally by applying ecosystem- based approach, so that overfishing is avoided	Conducting various erosion protective measures, especially in areas with cultivated land	
	Economically used water from streams and lakes and to a minimal extent from aquifer systems for irrigation purposes. Groundwater pumping rate generally reduced by about 20% in comparison with current stage	Establishing new forest nurseries. Developing and implementing effective techniques for forest restoration	

	Imposed '"green practice' concerning cleaning/washing in all hotels in tourist resorts, reduced specific consumption for ca 500 l/tourist/day to 350 l/tourist/day	Systematically planting new trees
		Widely introducing water saving technologies in irrigation such as drip or root irrigation
		Fishing to be controlled by law and conducted in sustainable manner
		Building small biological treatment units at resorts, hotels, restaurants and other touristic infrastructure objects
		Promoting 'green concept' in main resorts and informing tourists and visitors on importance of water and energy saving and supporting their environmentally friendly behaviour during their stays and tours
		Organising ecological and geo-heritage tours focusing on karst features, wetlands, specific biodiversity habitats and endemic species
Energy used efficiently and in sustainable manner	Completed planned electro-energetic systems including dams and energy transmission lines	Completing environmental impact assessment studies for designed and newly planned energy objects and implementing all proposed mitigation/monitoring actions	Hydropower water use 'Green' energy contribution
	Installed appropriate filters at all air pollution emitters (such as thermoelectric plants or big industrial factories)	Constructing dams as well as water transfer and energy transmission lines
	Controlled greenhouse gases emission and kept at existing level	Harmonizing and respecting legislation in terms of maximaumgreenhouse gases emission
	Introduced use of renewable vehicle fuels (biofuel)	Controlling sulphur dioxide, nitrogen dioxide and oxides of nitrogen, particulate matter (PM10 and PM 2,5), lead, benzene and carbon monoxide in ambient air
	Treated all mine waters before flowing to nearest recipients	Constructing new hydro-energetic plants and numerous small hydropower plants on small streams
	Constructed numerous solar energy plants and wind energy objects by increasing share of renewable energy by 10% in total energy sector	

(Continued)

Table 20.1 (Continued).

Strategic EQO and targets	Partial EQ targets to be achieved in next 15 years (up to 2030)	Related actions	Main environmental impact indicators
	Increased number of installed heat pumps for sub-geothermal water use by 100% compared to actual status	Building solar and wind energy plants and implementing demonstration projects for sea waves energy use	
	Constructed 2 pilot plants for conversion of sea water waves into energy	Drilling and utilising groundwater for heat/cooling purposes including dublet systems (pumping/re-injecting water)	
		Stimulating use of solar panels for households especially along coastal area and islands	
		Imposing stimulation rates for renewable energy producers	
		Building local capacities and publicly promoting 'green concept' for energy saving and rational use	
Measures to protect nature formulated and applied	Reduced negative impacts of air pollution and noise disturbance of existing eco-systems by 30%	Filtering gases emitted from industrial objects and controlling noise in tourist resorts and in off-roads in mountainous areas	Protected habitat
	Ensured ecological flows for all dependent eco-systems by regulated minimal stream flows and GW discharges at source sites	Managing aquifer recharge and regulating GW discharge and stream flows	Water demands of dependent eco system
	Increased number of protected areas under various protective statuses and expanded pro-tected land surface by 10%	Prohibiting hunting and controlling access to special reservation areas	Specific endemic and endangered species
	Proposed selected karst features for protection as natural monuments and geo-heritage sites due to their importance for humanity and karst science (caves, potholes, uvalas, blind valleys, karrenfelds, etc.)	Establishing criteria and forming teams and conducting surveys for selecting karstic fea-tures of special importance to be protected as geo-heritage sites, and proposing conserva-tion measures	
	Protected habitats and all threatened species from IUCN red list	Establishing system of control and protection of all endangered and endemic species in karst areas and in karst underground	
	Protected and monitored all endemic species that inhabit the region, with special emphasis on those in wetlands and caves		

There are many references and projects related to environmental indicators which cover different components of aquatic systems (including springs, streams, rivers, lakes, wetlands, coastal lagoons and estuaries). Some of the more recent, such of Vrba and Lipponen (2007) or UNECE (2007), pointed to a group of indicators helping to evaluate pressures on water quantity and water quality.

Not all the indicators proposed (Table 20.2) have to be followed continually. Those proposed for observation on an annual basis are as follows: Renewable groundwater resources (item 1 from Table 20.2); Groundwater exploitation index (5); Groundwater depletion (12). Others such as Specific pollutants index (17) and Drinking water quality (by observing selected critical parameters), need more frequent monitoring and the sampling frequency should be at least in accordance with EU Water Framework Directive and European experiences (Jousma & Willems, 1996). However, many others should be observed continuously in an established Groundwater Monitoring Network due to the specific intensive and variable regime of Dinaric karstic aquifer systems, proportional to the complexity of the status assessment of the groundwater body and presence of pollution trends. Most of the monitoring sites should be located in drainage areas i.e. along base levels of erosion and near recharge (ponors) and extraction sites (well fields, intakes).

Harmonisation of national legislation, legal and institutional reforms, creation of a common or unique Water Information System and protocol for data exchange are some of the proposals included as outputs in the Strategic Action Plan.

20.1.5 General Setup for Monitoring Network in selected TBAs

One of the tasks of the DIKTAS project was to prepare a proposal for the creation of a new Groundwater Monitoring Network in designated areas of transboundary concern which will fully respect specific karst behaviour. But, even in large international aquifer systems where a multilateral agreement of equitable water use and protection have been reached, no systematic monitoring of groundwater quantity and quality has taken place. The only exception is Genevese Aquifer, shared by France and Switzerland.

The Nubian Sandstone Aquifer System is one of the largest aquifers in the world and spans approximately 2 million km² across Libya, Egypt, Chad and Sudan. For sustainable utilisation of the Nubian Sandstone aquifer, the four countries represented by their National Coordinators formulated and signed several agreements. Within the framework of the second agreement NSAS countries agreed to update the information by continuous monitoring and sharing of the following information:

1. Yearly extraction in every extraction site, specifying geographical location and number of producing wells and springs in each site.
2. Representative Electrical Conductivity measurements (EC), taken once a year in each extraction site, followed by a complete chemical analysis if drastic changes in salinity is observed.
3. Water level measurements taken twice a year in locations specified on maps and tables.

Table 20.2 Proposed Environmental status indicators for DIKTAS project area and TBAs.

No	Group	Indicator	Expressed as	Unit
I	**Water Resources Availability (Pressures on Water Quantity)**	Renewable freshwater resources	ratio: Total flow of surface and groundwater in the study area vs. Total rainwater in study area (TBA catchment)	mM³/year: mM³/year or%
1a		Renewable freshwater resources in recession (drought) periods	Sub-indicator: As above but in critical drought periods (summer-autumn)	mM³/4 critical months: mM³/4 critical months or%
2		'Domicile' (and 'External') freshwater resources	ratio: Total flow of surface and groundwater generated in the part of TBA inside each country vs. Total flow of surface and groundwater in the entire TBA catchment	mM³/year: mM³/year or%
3		Renewable GW resources (Dynamic reserves)	ratio: Total flow of groundwater in the studied TBA catchment vs. Total rainwater in the studied TBA catchment	mM³/year: mM³/year or%
3a		Renewable GW resources (Dynamic reserves) in critical periods	Sub-indicator: the same as above but in critical drought periods (summer-autumn)	mM³/4 critical months: mM³/4 critical months or%
4		Water exploitation index	ratio: Total water amount utilized for different purposes[1] vs. Total renewable freshwater resources	mM³/year: mM³/year or%
5		Groundwater exploitation index	ratio: Total groundwater utilized for different purposes[2] vs. Total flow of groundwater in the study area	mM³/year: mM³/year or%
6		Water demands (availability)	ratio: Total water demands for different purposes[3] vs. Total renewable freshwater resources	mM³/year: mM³/year or%
7		Drinking water demands	ratio: Total water demands for drinking purpose vs. (1) Total renewable freshwater resour-ces and vs. (2) Total flow of groundwater in the study area	mM³/year: mM³/year or%
8		Water available per capita	Water available (household water access) calculated per capita per year	m³/cap/year
9		Irrigation water demands and use	ratio: Total water used for irrigation purpose vs. Total renewable freshwater resources	mM³/year: mM³/year or%
10		Hydropower water use	ratio: Total water used for HP vs. Total renewable surface water resources	mM³/year: mM³/year or%
11		'Green' energy contribution	ratio: Summary green vs. total energy consumption and per every kind vs total	in millions Mw: millions Mw or%,

12		Groundwater depletion	Annual depletion of groundwater table (av. value) due to over abstraction. Punctually measured at selected points	m/year
13		Losses	ratio:Total water losses (non-utilised)[4] from the systems constructed for different purposes vs. Total tapped renewable freshwater resources	%
14	**Pressures on Water Quality**	Drinking water quality	ratio: Number of samples of raw drinking water (from the sources) with inappropriate quality[5] vs. Total number of the controlled samples	no: no or%
15		Industry waste water index	ratio: Flow of untreated industrial (incl. mining) waste water (returned to recipients) vs. Total flow of waste water generated in study area	mM³/year: mM³/year or%
16		Household waste water index	ratio: Flow of untreated domestic waste water (returned to recipients) vs. Total flow of domestic waste water in study area	mM³/year: mM³/year or%
17		Specific pollutants index	ratio: Concentration (average) of selected component (pollutant) vs. maximal permitted level of the same component (pollutant)[6] in drinking water	expressed in mg/l: mg/l (permitted level) or µg/l: µg/l (permitted level) or % of samples of inappropriate quality of cpec. comp. vs. total samples
18		Fertiliser index	ratio:Amount of mineral or organic fertiliers used per unit of arable land	kg/ha or tonnes /ha
19		Pesticide index	ratio:Amount of pesticide used per unit of arable land	kg/ha
20		Landfill status	ratio: Number of inhabitants in study area without sanitary proper solid waste dumps vs. Total population in study area	.000: .000 or%
21		Water reuse	ratio: Reused or recycled water vs. Total flow of waste water in study area	mM³/year: mM³/year or%
22		Salt water intrusion (in coastal aquifers)	ratio:Total water flow - already salty, brackish or under direct threat of intrusion vs. Total renewable freshwater resources	mM³/year: mM³/year or%
23		Protected habitat	ratio:Total surface of protected area vs. Total surface of study area	km²: km² or%
24		Water demands of dependent eco system	ratio:Total water demands for downstream dependent eco system vs. Total renewable freshwater resources-dynamic, or Total water demands for (WDES) vs. Minimal discharge	mM³/year: mM³/year or%

(Continued)

Table 20.2 (Continued).

No	Group	Indicator	Expressed as	Unit
25		Specific species Sub-indicators: Specific endemic and endangered species (list)	Specific water demands (flow) for endangered species throughout the year (e.g. trout)	Presence of protected endemic species – List

[1]Includes different end-users: Drinking water purpose; Irrigation; Industry; Hydropower; Water dependent eco-systems. The Indicator should be calculated for each consumer separately, but also expressed as (1+2+3) vs. (5)
[2]The same as above
[3]Demands to be calculated for each specific end-user as in the case of items 4 and 5.
[4]Note: Mostly referring to water transport. If water leaked from reservoir and is utilized downstream for another purpose this is not a loss.
[5]No compliance with drinking water standards for whatever reasons (microbiology, chemistry, specific comp.)
[6]Pollutant or specific component in concentration higher than permitted, such as NO_3, P or $PO_{3(4)}$, pesticides, PCB, turbidity, biology indicators, etc. List to be specified in accordance to actual situation within TBAs and in compliance with EU Water Frame Directive for surveillance and operational monitoring

The Guarani aquifer system is the largest groundwater resource in the world, with 45,000 km^3 of water and a surface area of 1.2 million km^2 (Valente, 2002). The transboundary aquifer is shared by Argentina, Brazil, Paraguay and Uruguay. Approximately 24 million people live in the area delimited by the boundaries of the aquifer and a total of 70 million people live in areas that directly or indirectly influenced it. The main use of the aquifer is for drinking water supply, but there are also industrial, agricultural irrigation and thermal tourism uses. In article 12 of their agreement for sustainable utilization of GAS, the parties agreed 'to establish cooperation programmes with the purpose of extending the technical and scientific knowledge on the Guarani Aquifer, promoting the exchange of information and management practices, and developing joint projects', but no specific common monitoring programme with identified critical parameters has been put in place.

The Genevese Aquifer extends over 19 kilometers underneath the southern margin of Lake Geneva and the Rhône River across the border between France and Switzerland. The width of the aquifer varies between 1 and 3.5 km. An average of 15–17 × 10^6 m^3 of water is extracted annually from this aquifer. The two bilateral agreements, in 1978 and 2007 (http://www.internationalwatersgovernance.com/franco-swiss-genevese-aquifer.html), clearly define responsibilities, monitoring procedures (extracted water quantity and water table variations, as well as water quality control) and a reporting mechanism to the Management Commission for this aquifer. (http://www.unece.org/env/water/meetings/legal_board/2010/annexes_groundwater_paper/Arrangement_French_Swiss.pdf.). Although not karstic (the aquifer consists of glacial and fluvio-glacial silt-sand and gravel), this TBA is an excellent example of sustainable aquifer development and agreed proportional water share. By introducing artificial recharge these two countries have also prevented aquifer over-exploitation since the 1970s (Wohlwend, 2002).

Several more recent projects (e.g. *GENESIS project*, Preda *et al.*, 2012) or Commissions for large international watersheds (e.g. The International Sava River Basin Commission 2011) classified the indicator packages or established a list of critical parameters recommended to be monitored.

Finally, DIKTAS Strategic Action Plan includes a proposal to establish a fully equipped pilot monitoring site in Cemi/Cijevna karstic aquifer and also to further expand the monitoring network in the other selected TBAs. The general setup for a Monitoring Network should primarily include the following parameters (Stevanović, 2014): *Rainfall and other climate elements* (air temperature, humidity, wind, evaporation) observed on a daily basis. *Riverflow* observed on a daily basis – limnigraphs for automatic recording or gauging stations installed on major rivers and streams in each country sharing TBA (entrance/exit stations). *Springflow* observed on a daily basis – as above, the limnigraphs for automatic recording or classic gauging stations installed on major springs within TBA. *Groundwater table* observed on a daily basis – automatic data logger for groundwater table recording installed in piezometers properly selected to represent aquifer system in recharge/discharge areas in both countries sharing TBA. In addition, a manual recording of the groundwater table on a daily/weekly basis (depending on wet/dry seasons) should also take place on the piezometers of the 2nd rank. *Water quality control* is to be organised in compliance with EU Water Framework Directive requirements for surveillance and operational monitoring. Sampling frequency and the number of observed parameters (salinity, chemistry, turbidity, biology, specific components and pollutants) are to be adapted to local circumstances and pollution risks. As a minimum in the initial stage (surveillance) a set of the complete analyses is to be organised on major springs, streams and piezometers twice a year (high and low water periods).

To be able to define other environmental impact indicators in addition to the above water parameters, relevant information on surface waters and groundwater regime (quantity and quality) should be collected and provided on a regular basis to the responsible authorities and local water management institutions such as water agencies, hydrometeorological surveys, health and sanitary control centres, and municipalities. Groundwater monitoring and data collection must be the task of all those using groundwater for drinking purposes.

20.2   CONCLUSIONS

To establish a list of essential environmental impact indicators and to organise monitoring of these indicators in addition to water quantity and quality parameters. These are not easy tasks in such a dynamic and vulnerable environment as karst. In the case of the international Dinaric karst, it is necessary to expand the existing groundwater monitoring networks through the inclusion of groundwater users' (water supply systems, industry, agriculture) and to establish many new monitoring sites. Monitoring data are to be used to verify risk assessments and complement human impact assessments. The list of indicators can be long and in the case of Dinaric karst comprise 25 indicators appropriate to assess impacts and pressures on water quality and quantity. The same indicators can also be used to assess the feasibility of proposed environmental quality targets.

Responsibility to collect information, organise surveys to designated areas, establish monitoring sites and conduct measurements and provide information belongs to the national authorities of the countries which share an aquifer system. One commonly established body should be responsible to coordinate these activities and establish appropriate communication channels and specialised bodies for implementing strategy and achieving defined environmental and water quality objectives in the karst region.

REFERENCES

Bonacci O. (1987) *Karst Hydrology; with special reference to the Dinaric Karst.* Springer-Verlag, Berlin

Brundtland Commission (formally the World Commission on Environment and Development,WCED), (1987) Our Common Future, Report. Oxford University Press. Published as Annex to General Assembly document A/42/427, Development and International Co-operation: Environment August 2, 1987.

Burke J.J., Moench H.M. (2000) *Groundwater and society: Resources, tensions and opportunities.* Spec ed. of DESA and ISET, UN public. ST/ESA/265, New York. 170 p.

Cvijić J. (1893) *Das Karstphänomen. Versuch einer morphologischen Monographie.* Geographischen Abhandlung, Wien 5(3), 218–329.

Chilton J. (2002) Preliminary assessment of transboundary groundwaters in South Eastern Europe. UN/ECE Working Group on Monitoring and Assessment, Core Group on Transboundary Groundwaters, INWEB. 15 p.

Convention relative à la protection, à l'utilisation, à la réalimenation et au suivi de la nappe souterraine franco-suisse du Genevois, 18 Dec. 2007, available at http://www.unece.org/env/water/
meetings/legal_board/2010/annexes_groundwater_paper/Arrangement_French_Swiss.pdf.

Custodio E. (1992) Hydrogeological and hydrochemical aspects of aquifer overexploitation.,
Hydrogeology, *Selected Papers of IAH*, Verlag Heinz Heise, Hannover. 3, 3–27.

Dahl A.L. (2012) Achievements and gaps in indicators for sustainability. *Ecological Indicators*
17, 14–19.

Eftimi R. (2010) Hydrogeological characteristics of Albania. *AQUAmundi* 1, 79–92.

EEA (1999) URL: http://ec.europa.eu/regional_policy/archive/innovation/innovating/terra/pdf/
indicators.pdf. Last visited 16/11/2014.

EU Commission staff working document (2006) Impact assessment - Proposal for a Directive of
the European Parliament and of the Council on environmental quality standards in the field
of water policy and amending Directive 2000/60/EC {COM(2006) 397 final} {COM(2006)
398 final} /* SEC/2006/0947, Brussels.

EU Commission (2007) Common implementation strategy for the Water Framework Directive
(2000/60/EC). WFD CIS Guidance Document No. 15, Guidance on Groundwater monitoring, Brussels, 54 p.

European Commission (2015) Ecological flows in the implementation of the Water Framework
Directive. Technical Report – 2015 – 086. Environment. Brussels, 108 p.

FAOLEX (FAO legal database online). Reprinted in: Centre for Environment & Development
for the Arab Region and Europe (CEDARE), Regional Strategy for the Utilisation of the
Nubian Sandstone Aquifer System, Volume IV, Appendix II, Cairo, 2001.

Ford D. (2005) Jovan Cvijić and the founding of karst geomorphology. In: Stevanović Z. &
Mijatović B. (eds.): Cvijić and karst, Board on karst and spel. Serb. Acad. of Sci. and Arts,
Belgrade, 305–321.

International Sava River Basin Commission (2011) Sava River Basin Management Plan.
Background paper no.2: Groundwater bodies in the Sava River Basin, v2.0, Zagreb, pp.37,
www.savacommission.org

International Water Governance – Franco-Swiss Genevese Aquifer (http://www.international-
watersgovernance.com/franco-swiss-genevese-aquifer.html)

Jousma G., Willems J.W. (1996) Groundwater monitoring networks. European Water Pollution
Control, Vol. 6, no. 5.

Komatina M. (1983) Hydrogeologic features of Dinaric karst. In: *Hydrogeology of the Dinaric
Karst*. Mijatović B. (ed.). Spec. ed. Geozavod, Belgrade. 45–58.

Kukurić N. (2011) Assessment of internationally shared Karst aquifers: example of Dinaric
karst aquifer system, In: Polk, J. and North, L, Proceedings of the 2011 International
Conference on Karst Hydrogeology and Ecosystems. Environmental Sustainability
Publications. Book 2.

Levy W. (2002) Water Quality Targets: A Handbook. Community Information Unit Environment
Australia, Canberra (http://www.ea.gov.au/water/quality/targets).

Mijatović B. (1983) Karst poljes in Dinarides. In: Mijatović B. (ed) *Hydrogeology of Dinaric
karst*. Field trip to the Dinaric karst, Yugoslavia, May 15–28, 1983, Geozavod and SITRGMJ,
Belgrade, 69–84.

Mijatović B. (ed.) (1984) *Hydrogeology of the Dinaric Karst.* International Association of
Hydrogeologists, Heise, Hannover. Vol. 4.

Milanović P. (2000) *Geological engineering in karst.* Zebra Publishing Ltd., Belgrade. 347 p.

Milanović P. (2005) Water potential in south-eastern Dinarides. In: Stevanović Z. & Milanović P.
(eds.): *Water Resources and Environmental Problems in Karst* CVIJIĆ 2005, Spec. ed. FMG.
Belgrade, 249–257.

Ministry of Sustainable Development, Sweden (2004) Environmental quality objectives: – a shared
responsibility. Summary of Government Bill 2004/05:150. Article no. M2006.26, Stockholm.

Preda E., Kløve B., Kværner J. *et al.*, (2012) New indicators for assessing groundwater dependent eco systems vulnerability. Delivereable 4.3. GENESIS FP 7 project: Groundwater and dependent eco systems, pp.84, www.thegenesisproject.eu.

Puri S., Aureli A. (2005) Transboundary aquifers: A global program to assess, evaluate, and develop policy. *Ground Water* 43(5), 661–668.

Smeets E., Weterings R. (1999) Environmental indicators: Typology and overview. TNO Centre for Strategy, Technology and Policy, The Netherlands, Eur. Environ. Agency, Copenhagen, 19 p.

Stevanović Z. (2011) Management of groundwater resources (in Serbian). University of Belgrade, Belgrade, 340 p.

Stevanović Z., Kukurić N., Treidel H., Pekaš Z., Jolović B., Radojević D., Pambuku A. (2012) Characterization of TB aquifers in Dinaric karst – A base for sustainable water management at regional and local scale. Proceedings of 39 IAH Congress, Niagara Falls.

Stevanović Z. (2014) Environmental impact indicators in systematic monitoring of karst aquifer – Dinaric karst case example, In: Kukurić N, Stevanović Z, Krešic N (eds) *Proceedings of the DIKTAS Conference: Karst without boundaries*, Trebinje, June 11-15 2014, 80-85.

Stritih J., Qirjo M., Cani E., Myftiu A., Spasojević D., Stavrić V., Marković M., Simić D., Deda S., (2007) Environmental Policy in South-Eastern Europe. UNDP Report prepared for Conference Environment for Europe, FSC. Belgrade. 240 p.

UNECE (UN Economic Commission for Europe) (2003) Environmental monitoring and reporting, Eastern Europe, the Caucasus and Central Asia, United Nations Publications, New York, Geneva, 84 p.

UNECE (UN Economic Commission for Europe (2007a) Environmental indicators and indicators-based assessment reports: Eastern Europe, Caucasus and Central Asia. United Nations Publ. ECE / CEP 140, New York; Geneva. 93 p.

UNECE (UN Economic Commission for Europe) (2007b) Guidelines for developing national strategies to use air and water quality monitoring as environmental policy tools for the countries of Eastern Europe, the Caucasus, Central Asia and South-Eastern Europe, New York; Geneva. p. 46.

Valente M. (2002). South America: MERCOSUR vows to take over huge water reserve, Mercosur Article, Inter Press Service (IPS)/Global Information Network, July 22, 2002, pp. 1-2.

Vrba J., Lipponen A. (2007) *Groundwater resources sustainability indicators*, IHP –VI Series on Groundwater No.14, UNESCO, Paris.

Water Framework Directive of EU, WFD 2000/60, Official Journal of EU, L 327/1, Brussels

Wohlwend J.B. (2002) *An overview of groundwater in international law – A case study: the Franco-Swiss Genevese Aquifer*, Workshop III on Harmonization of Diverging Interests in the Use of Shared Water Resources, 17-19 Dec. 2002, Beirut, 1-24. Available at http://www.bjwconsult.com/The%20Genevese%20Aquifer.pdf.

Chapter 21

Hydrogeological settings for underground dam construction – Four case studies from southwest karst area of China

Jianhua Cao[1,2], Yuchi Jiang[1,2] & Petar Milanović[1,3]
[1]*International Research Center on Karst, UNESCO, Guilin, Guangxi, China P.R.*
[2]*Institute of Karst Geology, CAGS, Guilin, Guangxi, China P.R.*
[3]*National Committee of IAH for Serbia, Belgrade, Serbia*

ABSTRACT

Only 10% of the total groundwater resources in the southwest karst area of China is used. Approaches to exploration, utilisation and management, should be based on good understanding the karst aquifer. The palaeozoic carbonate rocks are characterised as old, hard and weakly porous. Karst hydrogeological media are highly heterogenous with many different lithologies including: limestone, dolomite, mixed limestone/dolomite. These lithologies are often intercalated or in alternation with clastic rocks. Spatial distribution of lithology impacts groundwater flow. This chapter considers four case studies of karst groundwater exploitation by underground dams: damming groundwater flow to form a groundwater reservoir and construction of a grout curtain to stop the flow and transfer the groundwater to a reservoir.

21.1 INTRODUCTION

Carbonate strata are widespread across China and outcrop over an area of 3.44 Mkm² (Datong and Yang, 1983; Datong 1985). The strata range in age from Precambrian to Cenozoic. In southwest China there is over 0.53 Mkm² of exposed carbonate terrain, most concentrated in Guangxi, Guizhou and Yunnan provinces (Figure 21.1) (Jianhua and Daoxian, 2005). The distribution of the carbonate rocks in different provinces is shown in Table 21.1.

The carbonate formations are mainly of Triassic, Permian, Carboniferous, Devonian and Cambrian age. They occur over an area of 13.5, 10.2, 8.3, 7.9 and 6.6 × 10⁴ km² respectively, or 87.5% of the entire exposed carbonate in southwest China (Table 21.1). The carbonate rock formations in the Cambrian, Cretaceous, and Triassic supergroups mainly occur in Hubei, Guizhou and Sichuan provinces, while the carbonates in the Carboniferous and Devonian mainly occur in Hunan, Guangxi and Guangdong provinces (Figure 21.2). The dolomite is, therefore, mainly distributed in Hubei, Guizhou and Sichuan, representing 36.9%, 24.8% and 22.8% of the total exposed carbonate rocks. Limestone is mainly distributed in Hunan, Guangxi, and Guangdong, which occupy 68.6%, 71.6% and 69.4% of the total exposed carbonate rocks (Figure 21.3). Therefore, the carbonate is characterised as old, hard and poorly porous. The karst media is highly heterogeneous with different carbonate rock types

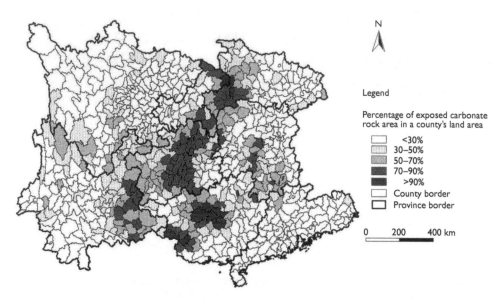

Figure 21.1 Distribution of karst terrains in southwest China (from Jianhua, 2005).

Table 21.1 The exposed area of carbonate rock (10^4 km²) from different geological era and in different provinces in southwest China km².

	K-J	T	P	C	D	O	S	$\in$	Z	Total area
Hubei	0?	1.340	0.710	0.040	0.220	0.450		1.410	1.010	5.180
Hunan	0?	0.340	0.600	1.500	2.360	0.420		0.950	0.190	6.360
Sichuan		3.00	1.050	0.425	0.496	0.430	0.470	0.350	0.810	7.031
Chongqing		1.560	0.457	0.020	0.031	0.404		0.468	0.071	3.011
Yunnan	0.052	3.311	2.700	1.453	1.389	0.371	0.192	0.914	0.447	10.829
Guizhou	0.031	3.177	2.574	1.536	0.584	1.098	0.047	2.515	0.049	11.611
Guangxi		0.760	2.012	2.680	2.640	0.119				8.211
Guangdong		0.031	0.066	0.686	0.226	0.016				1.025
Total area	0.083	13.519	10.169	8.340	7.946	3.308	0.709	6.607	2.577	53.258
%	0.16	25.38	19.09	15.66	14.92	6.21	1.33	12.41	4.84	

K-J - Jurassic, T - Triassic, P - Permian, C - Carboniferous, D- Devonian, O - Ordovician, S - Cambrian

and groups, including massive limestone, massive dolomite, mixed limestone/dolomite carbonate rock intercalated with clastic rocks.

Extensive karstification coupled with tectonic movement creates various types of karst topography in the area, including fault basins, karst plateaus, clustered conduits, high forest plain and stone forest areas. The topography and carbonate rock types constrain the hydrogeological system in each karst watershed. The most significant characteristic in a well-developed karst aquifer is the secondary porosity, fractures coupling many caves, conduits and underground drainage channels (Ford & Williams, 2007; Goldscheider, 2012). Although underground conduit and drainage channel systems occupy only a small part of whole karst hydrogeological system, their

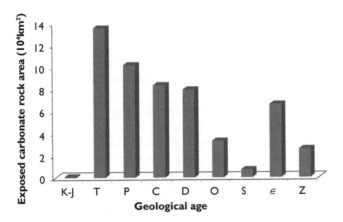

Figure 21.2 Exposed carbonate rock area from different geological age in Southwest China.

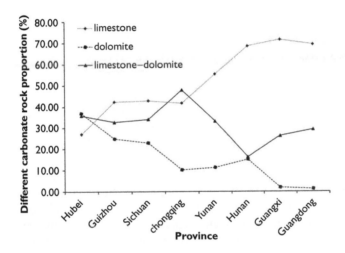

Figure 21.3 The proportion of carbonate rock types in various provinces in Southwest China.

typical high hydraulic conductivity of has a large impact on the entire hydrological system. Groundwater flow in karst medium is significantly different from that in a homogeneous porous medium, where groundwater flow velocity in conduits can be much faster than that in homogeneous porous media. This will lead to faster groundwater transport as well as contaminant dispersion (Daoxian, 1991; Andreo, *et al.*, 2010).

In southwest China a large part of the precipitation is rapidly transferred into the underground system and flows down gradient after the rainfall event. Thus, despite high precipitation and favourable water resources in this region, droughts are often reported, particularly in the dry season. Furthermore, insufficient groundwater exploitation and utilisation, as well as uneven spatial distribution of water, hinders regional economic development (Riyuan *et al.*, 2003). It has been reported by the Guangxi Department of Water Resources that the total annual karst groundwater resource amounts to

48.4 billion m³, and groundwater extraction amounts to 1.274 billion m³, or only 2.6% of the total potential amount (Risheng & Shumin, 2006).

Low utilisation levels of karst groundwater in Guangxi province have constrained domestic water supply, and water for industrial and agricultural development. Global warming and extreme climate events increase the frequency of droughts and floods. Exploitation and management of karst groundwater resources in southwest China is, therefore, imperative.

Despite the difficulties in constructing underground artificial storage systems, successful construction of underground dams in favorable hydrogeological and subsurface geological structures have been reported (Milanović, 2004). Large numbers of underground engineering structures were constructed in the southwest China karst region to achieve better use the groundwater resources (Daoxian, 1991). Construction of underground dams in the karst area requires massive and meticulous geological and hydrogeological investigations, coupled with various geophysical techniques. Four successful cases studies of karst groundwater exploitation by underground dams in the southwest karst of China are presented.

21.2   CASE STUDIES OF WATER USE BY CONSTRUCTION OF UNDERGROUND DAMS

21.2.1   Case 1. Fenfa Cave underground reservoir

Reservoir construction by blocking an underground 'river' outlet is the most common engineering technique for karst groundwater resources utilisation. This requires a retaining dam across the outlet to block the groundwater flow. The structure induces a surface water reservoir in the upstream karst voids.

The Fenfa cave karst underground reservoir is constructed in a tributary of the Yuzhai underground river system, which is located in Dushan County, Guizhou Province (Mingzhang, 2006). The exposed rock formation is mainly from Upper Devonian (D_{3y}) dolomitic limestone. Preliminary speleological investigations show that the Fenfa Cave mainly consists of an upper chamber and passage and a lower chamber and passage, with high heterogeneity and irregular shape. The altitude of the cave entrances is 950 m and 920 m, respectively. Underground river damming to form an underground reservoir was initially constructed by local residents. Due to insufficient understanding of the hydrogeological structure and groundwater flow system, leakage occurred shortly after construction was completed. The local government successively built the second and third dam inside the karst channel (Figure 21.4). These were based on detailed investigations which indicated a fault zone in the outlet section of the Fenfa Cave. The previous failure occurred due to water leaks through the fault and fractures zones. the location for damming was finally selected 55 m from the cave entrance, avoiding the fault zone. The underground dam is constructed in rock with good geotechnical properties.

The dam construction caused the groundwater level to increase to flood the surface and form an artificial reservoir connecting the upstream Hongmei conduits and Powuxi conduits. The water table is 26 m higher than previous water level. The reservoir proved to be technically successful with 2.2×10^5 m³ storage capacity. This supplies irrigation water to an area of 100 ha.

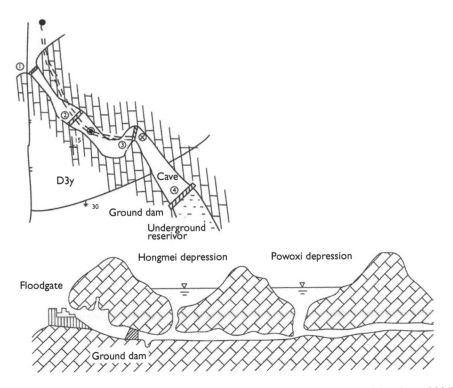

Figure 21.4 Sketch of Fenfa underground reservoir connecting sections (from Mingzhang, 2006).

21.2.2   Case 2. Dalongdong reservoir (DR)

The Dalongdong conduit system is located in Xiyan village, Shanglin County, Guangxi Province (Zhuoxin & Guangyan, 1999). Typical peak cluster topography occurs in this area, and the limestone and dolomite are mainly of Carboniferous and Permian age. The Dalongdong surface reservoir is a natural karst polje 10 km in length and 500–600 m wide. Its total storage capacity is 151×10^6 m³ and effective storage capacity of 109×10^6 m³. The recharge area of Dalongdong Reservoir is 310 km² with average annual runoff 241×10^6 m³ via 6 karst springs flowing to the reservoir. Before the reservoir was constructed, the polje had just the open flow of the Dalongdong conduit system. The flow infiltrates through five sinkholes at the downstream part of polje. In 1957, dams were built using stone and concrete at the five sinkholes to retain water in the polje (Figure 21.5). However, karst collapse occurred after the surface reservoir flooded, and created ten new sinkholes spread along the hillside with maximum water leakage rate of 15 m³/s.

In order to identify the leakage location, detailed investigation was carried out including tracer tests, drilling and electromagnetic CT imaging methods. It was found that the mountain was acting as a dam of solid limestone. Seepage was through numerous large fractures, fissures and multilayer small cave passages along three fault-zones (F1, F2, F3).

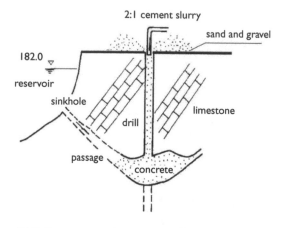

Figure 21.5 Damming to plug the sinkhole (from Zhouxin, 1999).

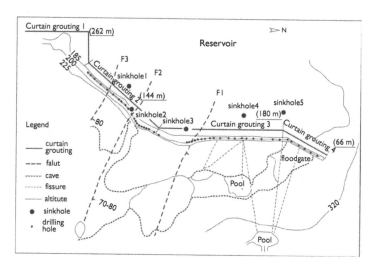

Figure 21.6 Location of sinkholes and curtain grouting in Dalongdong Reservoir (from Zhouxin, 1999).

A curtain grouting method was chosen to block water leakage. The grout curtain extends to a depth at elevation 122.5 m. Four leakage grout curtain segments with total lengths of 652 m were constructed from north to south of the hillside. The individual length of the segments is 262 m, 144 m, 180 m and 66 m, respectively (Figure 21.6).

Presently, Dalongdong Reservoir works well. It provides irrigating water for 1,200 ha of farmland, and the associated hydropower production is 4.48 GWh (Figure 21.7).

21.2.3 Case 3. Suduku underground river damming

The Suduku conduit is located in Zhulin town, Guangnan County, Yunnan Province (Yu *et al.*, 2006). The exposed carbonate rocks in this area are mainly limestone and

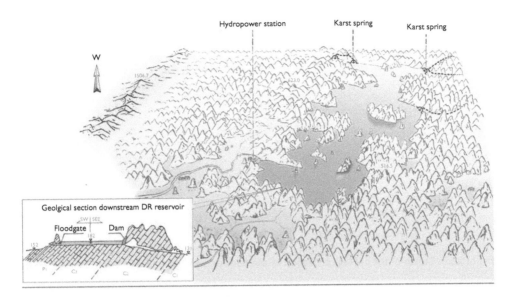

Figure 21.7 Landscape of Dalongdong Reservoir (from Risheng, 2008).

dolomite of Permian and Triassic age. A highly developed conduit system occurs along a fault zone in this area. Most of the water resources are underground water. The drinking water and irrigating water of Suduku village mainly depends on seasonal epikarst springs in the southeast of the village. In the dry season, potable water is insufficient, and sometimes there is a shortage of drinking water.

Hydrogeological investigations were carried out. The water flow path was in a shallow zone of carbonate strata, and frequent crossflow of surface water to groundwater was observed. The depth of the conduit bottom is about 12 m. The irregular karst channel, is 1.8 m high and 2 m wide. The groundwater flow rate of 15.0 l/s.

A dam, 12 m high, was constructed using concrete to increase the groundwater level to intersect the ground surface for local water supply. The Suduku conduit was blocked successfully to form an underground reservoir. The new supply greatly improved the living standard of local residents. It was reported that the captured groundwater flow rate is 1 296 m³/s during the dry season. The volume of the dam is 30 m³ and it cost 50 000 RMB (Figure 21.8).

21.2.4 Case 4. Pijiazhai large spring damming

There are many tectonic karst basins widely distributed in the eastern part of Yunnan province. Due to low precipitation and high evaporation in basins, water shortage often happens in this area where agricultural and industrial activities are also present. There is a water surplus in the surrounding mountains. The Luxi karst basin is typical of the area with a total area of 78.1 km². The Pijiazhai Spring system is located at the northeast edge of Luxi basin. The Pijiazhai Spring are located at E 103°47'20", N 24°32'16", and the altitude of the spring is 1,711 m (Yu, 2008). Water shortage was reported in the cultivated lands located in the east of the basin near Pijiazhai Spring.

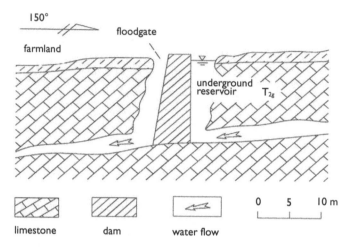

Figure 21.8 Sketch of the geological section of the Suduku underground dam (from Yu *et al.*, 2006).

The elevation of the cultivated lands is much higher than that of the spring outlet. Thus, the spring water could not be used to irrigate this cultivated land. However, flood events happen frequently during the rainy season. In order to optimize groundwater resources a grout curtain structure was used to increase the groundwater level to irrigate the farmland.

Detailed karst hydrogeology investigations were conducted, with application of remote sensing technology, high-density electrical method, groundwater monitoring, hydrogeological boreholes and tracer tests. It was reported that the recharge area of Pijiazhai Spring is 115 km² in area, and 70–100 m higher than that of Pijiazhai Spring outlet. The hydraulic gradient is 1.43–1.62% and the water velocity is 144.82–176.45 m/h. The Pijiazhai Spring water flow has hydro-powerful potential. Therefore, the integrated dam structure was designed for water exploitation and utilisation. It includes an underground anti-seepage grouting curtain, a 'U'shape dam to form a small reservoir to concentrate flow, and pipeline system to connect the water reservoir to a trench by adjusting water pressure and to release water for farmland irrigation (Figure 21.9).

After completion of the Pijiazhai system the groundwater level is now 4.4 m higher than previously. The amount of groundwater which flows through the inverted siphon to irrigate the farmland is 60 000 m³/d. This structure resolved the agricultural water shortage in the downstream farmland, and domestic water needs for potable water in downstream area.

21.3 CONCLUSION

The spatial variation of carbonate rock in the different provinces in southwest China creates varying karst topography. The four successful underground dam projects in the karst areas provide illustrations of groundwater exploitation in southwest China: damming methods including ground reservoir construction by blocking underground

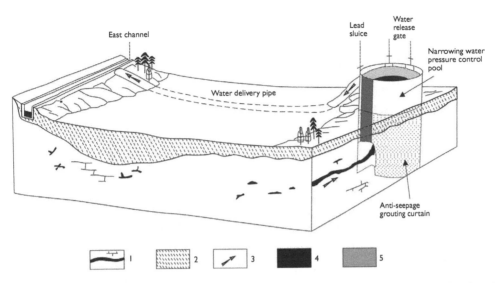

Figure 21.9 Integrated dam engineering in Pijiazhai karst spring (from Yu, 2008). 1. karst condiut 2. soil layer 3. water flow direction 4. previous water level 5. water level after damming.

conduit flow (in Guizhou); construction of grouting curtain to stop water leakage and to transfer the karst flow into the surface reservoir (in Guangxi); damming the underground flow located in the karst window to increase the water level to take water (in Yunnan); and construction of 'U' shaped concrete wall to block groundwater flow to increase water level for irrigation of farmland (in Yunnan).

Successful retaining engineering structures in the karst area are very important for local and regional socio-economic development.

REFERENCES

Andreo B., Carrasco F., Duran J.J., LaMoreaux J.W. (2010) *Advances in Research in Karst Media*. Springer-Verlag Berlin Heidelberg.

Daoxian Y. (1991) *Karst of China*. China Geological Publishing House.

Datong L., Yang L. (1983) Measurement of Carbonate Rocks Distribution Area in China. *Carsologica Sinica* 2(2), 147–150 (in Chinese with English Abstract).

Datong L. (1985) *Explanation of the map of soluble rock types in China*. Cartographic Publishing House (in Chinese).

Ford D., Williams P. (2007) *Karst hydrogeology and geomorphology*. Wiley, Chichester.

Goldscheider N. (2012) A holistic approach to groundwater protection and ecosystem services in karst terrains. *AQUA Mundi* 2, 117–124.

Jianhua C., Daoxian Y. (2005) *Karst Ecosystem of Southwest China constrained by Geological Setting*. China Geological Publishing House. (in Chinese).

Milanović P. (2004) *Water Resources Engineering in Karst*. CRC, Boca Raton.

Mingzhang W. (2006) Exploitation and Utilization of underground water resources in Guizhou Karst Area. In: China Geological Survey, Institute of Karst Geology, CAGS (editors) *Exploitation and Utilization of Karst Groundwater Resources in Southwest China*. China Geological Publishing House. (in Chinese).

Risheng M., Shumin Q. (2006) Exploitation and Utilization of underground water resources in Guangxi Karst Area. In: China Geological Survey, Institute of Karst Geology, CAGS (editors) *Exploitation and Utilization of Karst Groundwater Resources in Southwest China*. China Geological Publishing House (in Chinese).

Riyuan X., Jiansheng T., Yuanfeng Z. (2003) Sustainable utilization measures of groundwater resources in karst areas of Southwest China. *China Population, Resources and Environment* 13(1), 81–85 (in Chinese with English Abstract).

Yu W., Gui Z., Aihua L. (2006) Hydrogeological Setting for karst groundwater resources exploitation and utilization in Yunan and some successful cases. In: China Geological Survey, Institute of Karst Geology, CAGS (editors) *Exploitation and Utilization of Karst Groundwater Resources in Southwest China*. China Geological Publishing House (in Chinese).

Yu W. (2008) Water resource exploitation from big karst spring by flow narrowing-water pressure adjusting-water table raising – A case in Pijiazhai spring, Luxi county, Yunnan, *Carsologica Sinica* 17(1), 1–6 (in Chinese with English Abstract).

Zhuoxin W., Guangyan S. (1999) Treatments against the seepage of Dalongdong Karst Reservoir in Shanglin County, Guangxi. *Pearl River* 48–52 (in Chinese).

Chapter 22

3D Conduit modelling of leakage below a dam situated in highly karstified rocks

Saša Milanović & Ljiljana Vasić

Faculty of Mining and Geology, Centre for Karst Hydrogeology of the Department of Hydrogeology, University of Belgrade, Belgrade, Serbia

ABSTRACT

One of the basic problems of dam and reservoir construction in karst is water tightness. As a consequence of the nature of karst, and the associated insecurity of water storage due to leakage, choosing optimal dam sites can be a highly complex task. Problems associated with leakage are exemplified by the Višegrad dam site located in Bosnia & Herzegovina. During the first year of the operation of this dam, the occurrence of submerged downstream springs was noticed. Flow measurements of these springs established that the discharge increased from ~1.4 m³/s in 1990, to ~14.7 m³/s by 2013. This chapter presents results from field investigations and modelling undertaken to establish the positions, geometry and flow through karst conduits beneath the dam site. The modelling process consisted of formulating a 3D spatial model, and subsequently a mathematical model of the groundwater flows observed in the system. Data collected during subsequent remedial engineering works provided evidence that the method applied for the construction of a 3D spatial model and subsequent parametric model of the karst aquifer, aided by an incomplete data series, is feasible.

22.1 INTRODUCTION

In a number of karst areas the only natural resource of sufficient magnitude to enable regional socio-economic development is the hydro-power potential. However, the natural flow regime in karst presents a great variety of risks associated with the development of this potential. Understanding modifications of surface and groundwater regimes are the key requirements for proper planning of such water resource systems. Successful solutions require serious and complex investigations and close co-operation between a wide spectrum of scientists and engineers, including geologists, civil engineers, biologists, chemists, geophysicists, sociologists and many others.

In karst areas a special approach has to be applied to understanding and mitigating water leakage through the banks of such reservoirs and beneath the dams. An appropriate programme of exploration prior to dam construction can significantly reduce the risks of water losses, or at least minimise them to acceptable levels (Therond, 1972; Zogović, 1980; Milanović P. 2000; Bruce, 2003; Turkmen, 2003; Ford & Williams, 2007; Fazeli, 2007; Bonacci O. & Bonacci R., 2008). Due to of very thorough and complex investigation programmes, including all available investigation methods, whilst leakage can be minimised to an acceptable level, it can never be completely eliminated.

Many analyses show that once a reservoir is filled, groundwater flow that was previously oriented towards the reservoir site can reactivate previously unsaturated (fossil) conduits and pathways, and form a discharge outside the reservoir area. In response to these findings, it is of paramount importance that, at an early stage of the site investigation, the main emphasis must be on understanding the complex conditions of karst groundwater circulation (Milanović, 2015). Some inadequately explored dam sites and reservoirs constructed in karst have never fully filled with water, for example the Hales Bar dam in the USA or the Vrtac dam in Montenegro. Indeed, some have been completely abandoned after unsuccessful attempts to reduce water losses, such as at the Montejaque dam in Spain. Some other dams have had sudden water losses even after years of successful operation, or increases in leakage over time. Examples of this type include Višegrad in Bosnia and Herzegovina; the Great Falls dam in Tennessee, USA; the Mavrovo dam in FRY Macedonia; the Slano dam in Montenegro; the Hammam Grouz dam in Algeria and others (Milanović, 2000).

Romanov and Gabrovšek (2003) state that karst conduits under natural conditions are created on time scales of thousands to hundreds of thousands to millions of years, as natural hydraulic heads drive water through narrow fractures which are gradually widened by solution. By contrast, under reservoir condition where water pressure of 5 to 10 or more Bar can accelerate conduit enlargement by more than ten or one hundred times. For example, Bauer *et al.* (2005) demonstrated, through simulation of a dual porosity system, that small conduits with initial diameters of 4×10^{-4} m can be enlarged by solution within tens of years and may cause serious water losses.

A study of karst conduit genesis and formation of a three dimensional spatial model of the main karst pathways underlying the Višegrad dam site has been conducted with the aim of better understanding and preventing leakage beneath the dam site. The Višegrad hydropower plant is situated on the River Drina, 2.7 km upstream from the town of Višegrad. It was built between 1985 and 1989. The dam of the Višegrad Hydro-Power Plant is a concrete gravity dam. An integral part of the dam is the 594 m long grouting curtain (325 m beneath the dam structure and 65 m in the left abutment and 204 m in the right abutment) and 50 to 130 m deep. In order to define the positions of karst conduits along which groundwater circulates under the dam site, special-purpose investigations and remedial works were undertaken in 2009–2010 and again in 2013–2014. Remedial works are still in progress at the time of writing (2015).

Geological investigations of the karst setting focused on a rather narrow area containing a recharging 'sinking' zone and a drainage 'discharge' zone (Figure 22.1). The initial problem was how to perform a quality analysis to characterise the problem sufficiently in order to enable effective remedial works to prevent leakage. The problem was approached theoretically, which initially played a major role and provided guidelines for field activities; followed by detailed and complex field investigations; a spatial 3D modeling of karst conduits; an empirical approach and later also a mathematical approach aimed at producing the final form of the model.

22.2   MAIN GEOLOGICAL AND HYDROGEOLOGICAL CHARACTERISTICS

The investigation area belongs to the Dinaric–ophiolitic zone, one of the most outstanding and complex geological regions of the Dinaric karst. The dam site is located

Figure 22. I Visegrad dam; Photo above left – submerged springs below the dam site during tracer test (discharge zone); Photo above right – reservoir (sinking zone); Photo below – panoramic view of Visegrad dam.

within an asymmetric tectonic trench formed by gravitational activation of the River Drina fault. This is the contact zone of two regional structural-tectonic mega-blocks: the autochthonous East Bosnia-Durmitor unit, and the overthrust autochthonous Dinaric–ophiolitic unit. These units are highly heterogeneous and comprise Triassic, Jurassic and Quaternary formations. The dam is located on top of the Visegrad tectonised Middle Triassic carbonate sediments, which are on the south – east side less tectonically altered limestones with subordinate dolomites (Milanović, 2000). The central part of the dam, located on the Drina river, is founded on Middle Triassic dolomitic limestones, dolomites and cavernous dolomites with lenses of cherts and sandstones, in tectonic contact with ophiolitic melange.

Under such complex conditions in the carbonate rock mass, which is divided into blocks by fault structures and exhibits variable levels of karstification, dramatic changes in boundary conditions in the existing karst aquifer occurred following the formation of the reservoir.

With the change in the hydraulic gradient along the fault structures beneath the dam, karstification processes have led to the formation of underground channels. The results of water pressure tests indicate a generally permeable rock mass. The central area (beneath the dam) is underlain by locally karstified dolomites. The zones of high permeability are related to fault zones. The right bank of the reservoir consists of karstified dolomites. A number of large (from 0.2 up to 1.2 m in diameter) karstic channels, situated at several different levels, were discovered during drilling and excavation for the foundation of the dam. These caves are mostly empty, although some are filled with calcite and terra-rossa. Due to high-pressure washing out of fine-grained, incohesive sediments from the filled faults, the percolation process became more intensive through time. The constant increase in the measured discharges downstream from the dam, as well as the intermittent increased turbidity (from 5 to 65 NTU) of the water in the springs, indicates that the karstification process is intensified. According to inspection by divers, heavily clay-laden groundwater discharge was emanating from the openings in the river bed downstream of the dam. Abrupt discharge pulsation and pulsation of the quantities of mud in the discharging water was also observed. The number of springs in the downstream river bed increased alongside the discharge of water leaking beneath the dam. In the first five years of dam operation the leakage increased from approximately 1.4 m³/s to 6.5 m³/s, and subsequently to 14.68 m³/s by 2014. The number of observed springs in the downstream bed of the river increased from 3 to 18.

It is important to highlight that during the first filling of the reservoir, water (discharging from the river-bed springs) was clear, but its temperature was higher than average (14.5°C–18.7°C), which indicates a process of forced circulation of hypogenic water within the karst aquifer. Over time and with the increasing discharge capacity of the karst system beneath the dam, the temperature of the downstream river bed springs decreased to match the temperature of water at the bottom of the reservoir.

22.3   OVERVIEW OF THE INVESTIGATIONS PERFORMED

Geological investigations were carried out during all phases of design and construction of the dam for the Višegrad Hydro-Power Plant. The initial period of investigation had the primary goal of choosing the optimum dam site as well as definition of possible seepages. These investigations have yielded a large body of results, a part of which are relevant to the solution of the leakage problem under the present conditions below the dam site.

Geological mapping of the catchment area of the Višegrad reservoir and dam site was done at a scale of 1:10 000, while an engineering geological map of the dam site was produced at a scale of 1:500. More detailed geological mapping of the dam site at scales of 1:50 to 1:1000 were also undertaken. More than 4470 m of drilling was undertaken during the investigation, together with water pressure tests. Groundwater tracer tests were carried out in the phase of choosing the optimal dam site. Systematic monitoring has been undertaken from the stage of initial investigation and construction design to the present day, and leakage beneath the dam has been monitored from 1991 to the present day (Figure 22.2) (IWD, Jaroslav Černi, 2009).

In 1993 an emergency grouting programme was carried out to a depth of 110 m because of excessively large leakage flow. An initial attempt using polyurethane mixed with cement and sand was unsuccessful. A second attempt involved injecting fast-setting thixotropic cement containing 3–10 mm and 8–16 mm sand fractions and which also included a proportion of fine cut plastic sponge. Generally this was unsuccessful at reducing leakage, and leakage was still occurring through reactivated karstic conduits deeper than 110 m (Milanović, 2004). During this period of remedial works groundwater levels were monitored in the piezometers located in dam-abutments and abutment injection galleries, and subsequently on a continuous basis throughout the

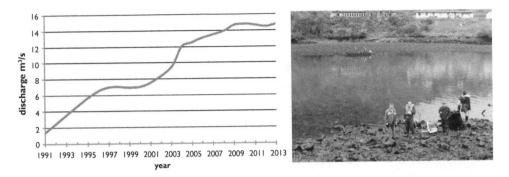

Figure 22.2 Diagram of water discharge (leakage) increasing through time (Milanović, 2015).

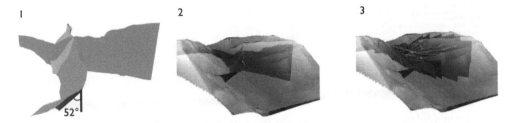

Figure 22.3 Creation of a 3D tectonic model of the Visegrad dam area as one of the main input for the construction of a karst conduit model network. 1. Fault – detail with 3D position and real inclination due to upper and lower layers 2. 3D model of two fault zone detail sketch 3. Fault network in investigation zone with non-permeable layer.

years 1993, 1994 and 1995. At present the groundwater level regime is monitored twice a month in 58 piezometers.

A new period of investigation of the Višegrad dam began in 2009 specifically to define the leakage flowpaths beneath the dam site. The first step was a detailed investigation of the narrower area around the dam site involving a geodetical survey of the dam and the appurtenant structures on the scale of 1:1000 (IWD, Jaroslav Černi, 2009). The second step was remote sensing investigations, based on the analysis of satellite images and aerial photographic images from periods both before and after construction of the reservoir. The main goal of these investigations was to determine the fracture locations and morpho-structural texture of the rocks as the input data for 3D modeling of faults, including their dominant orientation, karstification and control on water circulation (Figure 22.3). By combining the geodetic and remote sensing data with existing results from geological mapping and coring, as well as the mapping of the dam foundation and results obtained during the grouting program, more than 40 faults in the area of the dam were detected and classified.

The fault zones were assigned numerical designations to facilitate the definition of their characteristics. The classification included the separation of regional and local as well as less important structures. Structural blocks were identified based on the positions and lengths of the structures, as well as fault widths and the positions and thicknesses of lithological units mapped from boreholes. Larger structural blocks are restricted by

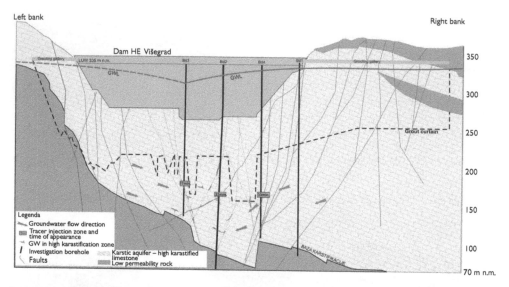

Figure 22. 4 Hydrogeological cross section of Višegrad dam site with position of grout curtain and investigation boreholes.

significant faults, and these large faults have widths such that cavities can readily be enlarged due to flushing and chemical corrosion (i.e. karstified rock). Identified cavities were classified based on derived tectonic characteristics (i.e., all faults and dominant fractures within the dam area were classified based on their importance and size).

A second group of faults comprised of faults within larger structural blocks were divided into two sub-groups – major faults with wide zones, and structures with smaller fault zones. The analysis was undertaken with satellite imagery analysis with the main task being to identify the largest faults, primarily those located within the area of the Višegrad Hydro Power Plant reservoir.

Geological investigations of the dam site started with new, detailed geological mapping of the terrain (Figure 22.4). Geological mapping of the site was the basis of the investigation process, which included verification of previous data and additional mapping. These materials were used to generate a detailed cross section of the dam area (Figure 22.4).

Cross-hole geoelectrical tomography scanning in the left and right dam abutment and over the reservoir was carried out, as well as reflective seismic investigations in dam galleries. The electrical tomography method was applied in two stages: first between the boreholes BD-1 and BD-2 and then at the profiles between the boreholes BD-1 and BD-4, BD-4 and BD-2 and BD-2 and BD-3. Following the completion of both test stages (i.e., after geo-electrical tomography measurements at all four bore-holes), a 2D model was created based on a finite element analysis, for the BD-3 – BD-2 – BD-4 – BD-1 profile as a base for further 3D model forming.

Self-potential measurements were performed simultaneously in the reservoir zone. Survey in the reservoir zone led to the detection of anomalous locations that indicated possible water percolation. Two profiles were located in dam abutments (left and right) and one located partly across the reservoir and partly across the left dam abutment. The depth coverage was from 80 to 130–140 m. Geo-electrical scanning performed along dam abutments and the reservoir also indicated the positions of anomalous geo-electrical environments with low electrical resistance values (under 60 Ohm · m) and which were interpreted as fault structures through which water circulation and water losses most likely occur.

At the locations of the anomalies, detailed diving investigations and underwater video camera recordings were undertaken. This led to the discovery of a large sinkhole at a depth of 50 m in the reservoir which was measured and at which the water inlet velocities (0,5 m/s average) were determined (Figure 22.5). Investigation of the large sinkhole was made using underwater robots and other special hydrogeological underwater equipment constructed by the Centre for Karst Hydrogeology, Faculty of Mining and Geology, University of Belgrade. The sinkhole was also sed to investogate spatial groundwater flows upstream from the dam by the 'misse a la masse' method. Downstream from the dam, in the part of the riverbed near the dam, a bathymetric survey was performed, as well as further diving investigations and measurements of submerged spring discharge. During this investigations divers explored 8 large and 12 small springs with discharges from a few l/s up to a few m³/s ($Q_{sum} = 14.91$ m³/s) (Figure 22.5).

Exploratory drilling and corresponding investigations in the boreholes (downhole-video, geophysical logging etc.) were carried out. The locations of the boreholes was determined from the results of all previous investigations. Video recording of boreholes was one of the important contributors to the 3D spatial model. More than 350 individual karst features were recorded and correlated with previous geological, tectonic and hydrogeological data (Figure 22.6). The change in chemical composition of the water in the piezometers drilled at various positions in the reservoir and downstream was more pronounced in the deepest levels 180–230 m than in zone of grout curtain 50–80 m. This indicated that water flowing under the grout curtain was

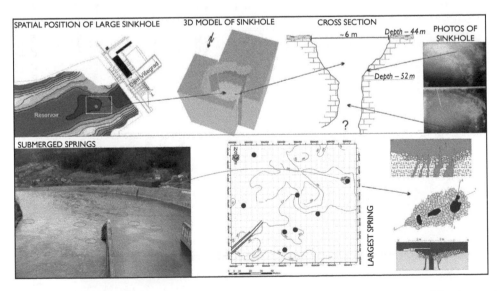

Figure 22. 5 Spatial position of sinkhole in the reservoir area and springs downstream of the dam and some of underwater investigation results.

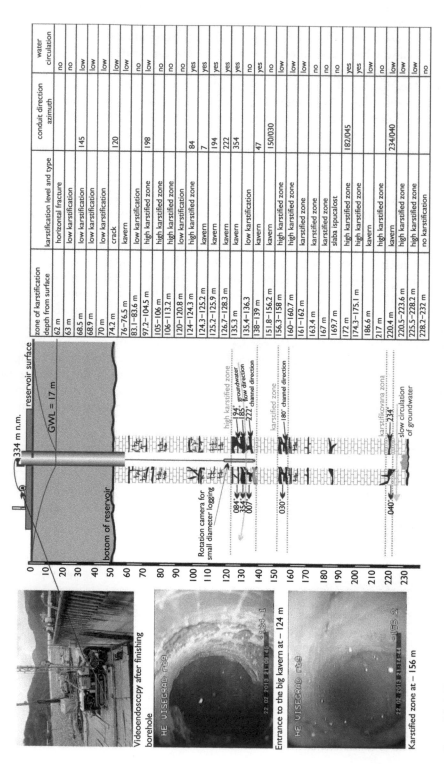

zone of karstification depth from surface	karstification level and type	conduit direction azimuth	water circulation
62 m	horizontal fracture		no
63 m	low karstification		no
68.5 m	low karstification	145	low
68.9 m	low karstification		low
70 m	low karstification		low
74.2 m	crack	120	low
76–76.5 m	kavern		low
83.1–83.6 m	low karstification		no
97.2–104.5 m	high karstified zone	198	no
105–106 m	high karstified zone		no
106–113.2 m	high karstified zone		no
120–120.8 m	low karstification		no
124–124.3 m	high karstified zone		no
124.3–125.2 m	kavern	84	yes
125.2–125.9 m	kavern	7	yes
126.7–128.3 m	kavern	194	yes
135.3 m	kavern	222	yes
135.4–136.3	low karstification	354	yes
138–139 m	kavern	47	yes
151.8–156.2 m	kavern	150/030	no
156.3–158 m	high karstified zone		low
160–160.7 m	high karstified zone		low
161–162 m	karstified zone		low
163.4 m	karstified zone		no
167 m	karstified zone		no
169.7 m	slaba ispucalost		no
172 m	high karstified zone	182/045	yes
174.3–175.1 m	high karstified zone		yes
186.6 m	kavern		low
217 m	high karstified zone		no
220.4 m	kavern	234/040	no
220.5–223.6 m	high karstified zone		low
225.5–228.2 m	high karstified zone		low
228.2–232 m	no karstification		no

Videoendoscopy after finishing borehole

Entrance to the big kavern at – 124 m

Karstified zone at – 156 m

Figure 22.6 Photos of drilling rig position at the surface of the reservoir (top left); characteristic caverns identified in boreholes (middle and bottom left); borehole log (centre) and summary data table of borehole RB 9 CCTV survey (right).

flowing faster than it was through the curtain itself. Similar results were established during investigations of a leakage problem at Ataturk dam (Unal *et al.*, 2007).

More than 40 tracer tests were carried out, employing in total more than 80 kg of sodium fluoresceine and more than 500 kg of sodium chloride. Tracers were injected in the existing piezometers upstream from the dam, then in known sinkholes, as well as into the newly drilled boreholes. Importantly, tracer tests were repeated several times in the large sinkhole and in some investigative boreholes as a basis for model testing and calibrating.

The results of the tracer tests were generally indicative of possible directions and lengths of karst conduits. The results of the tracer tests conducted in the large sinkhole; boreholes BD-2, BD-3 and BD-4; and piezometers UD-42k and UD-45k, demonstrated a complex network of rapid flow connections in the conduit system beneath the grout curtain, connecting the large sinkhole with the springs in the River Drina riverbed downstream of the dam (Figure 22.7).

Results from the tracer tests in the large sinkhole at ~50 m depth showed that, in every test, the first arrival of tracer was recorded after 29 minutes at the spring located in the left portion of the riverbed (looking downstream, No. 17b) and then at springs in the central portion of the riverbed after 31 to 34 minutes (in the following order: springs 8, 7, 5, 3b and 4). Somewhat later, but with the longest duration of the emergence wave, the tracer emerged at the largest spring (No. 1) after 36 minutes. The final springs at which the tracer emerged (after 40 and 41 minutes) are located further downstream, towards the right bank (13b and 18b).

Tracer tests conducted in the boreholes BD-4, BD-2 and BD-3 (located between the right and central portions of the dam profile) and at piezometers UD-42k and UD-45k, also confirmed the underground connection with all springs downstream from the dam. In all tests with sodium chloride, as well as with sodium fluorescein, the tracer discharged in the same order as in the case of the sinkhole tracer tests. The main findings from the tracer tests are shown on Figure 22.7 and Table 22.1.

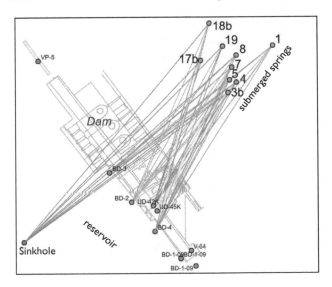

Figure 22.7 Map of tracer tests results.

Table 22.1 Summary overview of tracer tests.

Location of tracer injection	UD-42k	UD-45k	BD-2	BD-3	BD-4	Sinkhole	
Observation points/spring	Tracer emergence time (min.)						Fictitious velocity from sinkhole (m/s)
	Straight line distance (m)						
17b	24	17	22	17	34	29	0.189
	199	204	231	205	188	329	
8	25	18	23	17	35	30	0.204
	225	229	235	225	254	368	
7	26	19	24	18	36	31	0.189
	208	212	218	209	236	352	
5	26	20	25	18	37	34	0.167
	195	197	204	197	220	340	
3b	26	20	25	20	38	34	0.161
	178	182	190	186	205	328	
4	26	20	25	20	38	34	0.170
	195	220	208	203	223	346	
1	30	22	28	22	41	36	0.191
	262	265	275	270	288	412	
18b	38	27	37		50	40	0.155
	248	254	253		280	372	
19	29		28		40	41	0.147
	266		234		257	362	

Tracer tests conducted from the right-hand bank of the reservoir resulted in low tracer concentrations and long emergence times, suggesting a lower degree of karstification in this area.

It was considered appropriate to assume either that interactive work and integrated use of known 2D and partly-defined 3D parameters was sufficient to produce an output of a three-dimensional conduit defined within a 3D physical model. A conceptual model based on the results of all of the investigations undertake is presented schematically in Figure 22.8.

22.4 3D MODELING OF KARST CONDUIT BELOW VISEGRAD DAM SITE

Analysis of the geometry of the main karst conduits in the saturated zone below the dam site and grout curtain connecting the reservoir with the discharge zone has enabled the creation of a 3D model of karst channels. Analysis of the parameters obtained through the quantitative and qualitative monitoring of groundwater characteristics, and their analysis for the physical model, provided data on the relationship between the recharge and discharge zones. Such a model was used for an analysis of speleogenesis and the detailed hydrogeology of the dam area.

The basic problem in the determination of this methodology was how to develop a model to further analyse the spatial position of karst conduits, integrating

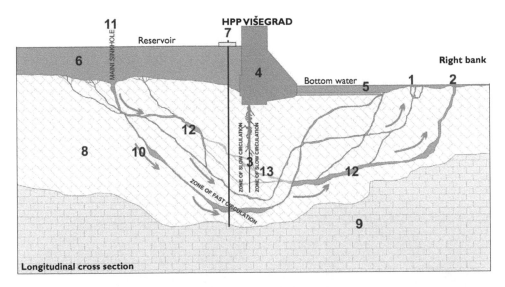

Figure 22.8 Schematic longitudinal cross sections; 1 – Location of first tracer appearance, 2 – Location of last tracer appearance, 3 – Grout curtain, 4 – Visegrad Dam, 5 – Bottom Water, 6 – Visegrad Reservoir, 7 – Investigation borehole, 8 – Karstified rock, 9 – Limestone with low karstification, 10 – Groundwater flow direction, 11 – Main ponor (sinkhole), 12 – Channel with active water circulation, 13 – Channel with slow water circulation due to injection works.

hydrogeological laws and geological characteristics. So far, only a few techniques have been developed to structure a karst conduit network model in a 3D environment (Stevanović *et al.*, 2010). Some of them are based on modelling the physics and chemistry of the speleogenesis processes, as well as simulating the geometry of karstic conduits on a regional scale, as constrained by the knowledge of the regional geology, hydrology and hydrogeology, and by a conceptual knowledge of karst genetic factors (Borghi *et al.*, 2012; Filipponi, 2009).

The 3D modeling was approached from three parallel directions:

- Theoretical approach, which initially played a major role and provided guidelines for field activities.
- Detailed field investigations.
- Development of a basic input 3D model, then an empirical approach and later also a mathematical approach, aimed at producing the final form of the model.

The 3D geological model was developed for the purposes of generating a network of potential karst conduits running from the identified infiltration zone to the accurately defined karst discharge zone. The 3D geological model was developed using ArcGIS software and the 3D Analyst, Spatial Analyst and Network Analyst extensions (Milanović, 2010). All spatial data, such as geological maps and profiles, as well as the positions of the dam, grout curtain, grout galleries and piezometers, were converted into digital format, and each spatial unit was defined by its x, y and z coordinates.

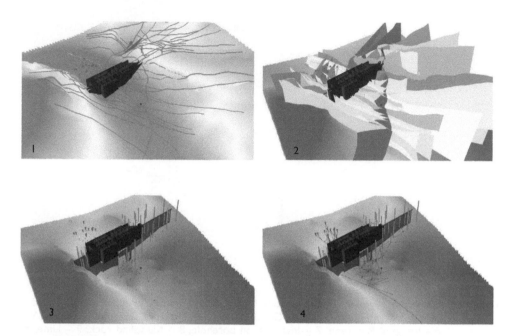

Figure 22. 9 Steps of 3D spatial network generation; 1. 2D data layer with all collected data, 2. 3D modeling of faults, 3. 3D modeling of all other data (boreholes, diving, tracer tests, geophysics etc.), 4. Final formulation of 3D conduit network.

Detailed development of separate inputs, the components of the basic model (Figure 22.9), were separated into three categories:

1. Individual fault digitisation, i.e. the creation of a series of 3D fault planes which, in addition to the basic extent shown in the 2D presentation (plan view), are also characterised by vertical attributes relating to depth and the angle of dip.
2. Generation of DEMs of the land surface and the base of karstification, as well as 3D models of faults whose outputs (nodes and lines) provide the starting point for 3D analysis.
3. Generation of an interactive node and line network based on faults, DEMs and boreholes logs, as well as other directly quantifiable investigative data which were integrated in a 3D spatial model of the karst conduit network.

The karst conduit network was generated by linking the most probable flow directions, using tectonics as the safest parameter for the initial phase of evolution of karst conduits over a certain time interval. A 3D network of potential karst conduit pathways (Figure 22.9) was constructed based on the results of GIS data processing. The 3D model described the network of directions of potential karst conduits, from the sinking zone to the discharge zone (Figure 22.10).

As required by the numerical algorithms, this network was described by means of model nodes and elements. They made up the topology of the numerical model and spatially corresponded to the geological model. Nodes were represented via x, y and

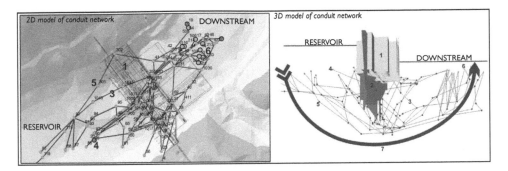

Figure 22.10 Left – 2D conduit network of possible leakage pathways below the dam site (model result) Right – 3D conduit network of possible leakage pathways below dam site (Milanovic, 2015); 1 – Dam, 2 – Grout curtain, 3 – Karst conduit, 4 – Node at the sinking zone, 5 – Node at the intersection of fault (or on fault), 6 – Node at the drainage zone, 7 – General groundwater direction below dam site.

z coordinates, where potentials were computed. 1D elements were used for hydraulic calculations based on the finite element method. The lengths of these elements were defined by the corresponding nodes, and spatial mathematical calculations were performed along them (Milanović, 2015).

The network comprised 177 nodes linked by 226 elements (Figure 22.10). Using an optimisation algorithm, the number of nodes, elements and free parameters is indicative of the complexity of the problem. As a result, 1130 parameters were determined and at the same time more than 100 computed values were compared with measured values. This optimisation required substantial processor time and a parallel genetic algorithm was, therefore, used for the final calculation.

22.5   RESULT OF CONDUIT MODELLING

The principal objective of the mathematical model was to determine the spatial layout of the principal karst conduits and their physical characteristics. Taking into account the satisfactory degree of congruence with the real system, the model defines all the main groundwater flows, with the corresponding parameters: dimensions, resistances, potentials, velocities and discharges. This model, with its results, represents a foundation for the interpretation of geological data and the determination of the direction of development of the dominant karst channels, i.e. water circulation paths. The mathematical model was based on known physical dependencies which approximately define the behaviour of the real system. Some model parameters, which affect model performance, were unknown and their values were assumed from a set of possible values.

By simulating the adopted topology and flow characteristics in the hydraulic model using all of the hydraulic parameters that had been obtained, an analysis and definition of the dominant directions of water flow beneath the dam body can be undertaken (Figure 22.11).

The results of these investigations show that the greatest amount of water flows from the reservoir underground through the large sinkhole which is best modelled

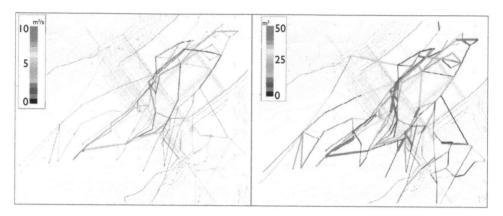

Figure 22.11 Left – calculation values of discharges along modeled karst conduits (m³/s). Right – calculation dimensions of underground flows represented as the areas of a transverse section in m² (IWD, Jaroslav Černi, 2009).

with length ~60 m and average cross-sectional area of 7.1 m². According to the measurement results ~8.15 m³/s infiltrates into the groundwater system through this feature, with another ~6.5 m³/s infiltrating elsewhere in the reservoir bed. From the large sinkhole are formed two karst conduits through which groundwater flows.

Conduits on the left side, through which the water flows downwards from an elevation of 260 m asl to an elevation of 235 m asl, have an average transversal section area of 19.8 m². Through these water flows with an average velocity of ~0.2 m/s, and the total discharge through this channel is ~4.1 m³/s. Along conduits with a spatial position in the middle of the conduit net direction, flows underground are 5.5 m³/s. The area of the transverse sections of these channels amounts to on average 23.9 m². The erosion process along the channel of diameter 14.4 m², up to an elevation of 228 m asl, the underground flow bifurcates into two arms. The average area of the transverse section near the borehole profiles amounts to ~10 m², with an average water flow velocity of ~0.3 m/s. At the elevation of 147 m asl this profile bifurcates, and the major portion of the water flows towards the middle part of the riverbed toward springs 17, 8 and 1, where, at an elevation of 130 m asl, it joins the smaller channel to the right. Flows ~2.5 m³/s occur in the channel that runs from the left abutment, when the total discharge amounts to 3.8 m³/s, established through 3D modelling and mathematical modelling of the karst conduit system (IWD, Jaroslav Černi, 2009).

22.6   CONCLUSION

The investigations defined the underground water courses and parameters necessary for development of the technical solution for the permanent rehabilitation of seepage. However, the sealing rehabilitation works under such complex conditions are complex, and must be followed by detailed monitoring.

The results of investigations show that the largest amount of water flows from the reservoir underground through the bottom and through the main sinkhole with area of 7.1 m². The 'left' direction of underground flow, i.e. the karst channel, was formed along the dominant longitudinal fault structures. Along this direction is a system of karst channels with two courses, that generally run towards the boundary of the dam blocks 8 and 9. The 'right' direction of the underground flow is along the fault structure and water flows through it generally towards block 5 of the Višegrad Dam. This direction upstream from the dam bifurcates into two courses; one of them, the more important one, passes under block 5, while the other turns towards the central part of the riverbed. These two courses are joined downstream from the dam axis. In the region downstream from the dam, underground courses are mainly developed along the directions of the longitudinal faults closer to the left abutment. The connection with the springs in the River Drina riverbed is along the axis of the diagonal faults.

The 3D and mathematical modelling of the spatial position of conduits and flows will serve as a foundation for the simulation of technological processes in the process of rehabilitation regarding water seepage.

The 3D position of karst conduits and parameters necessary for the development of technical solutions for the permanent rehabilitation of seepage, were defined by the investigations. However, the sealing rehabilitation works are complex, and detailed post-remediation monitoring is required. This is necessary in order to make correct decisions during and after remediation.

The new data collected during grouting works (2012–2014) showed that the method which was applied for the construction of a 3D spatial (physical) model and the model of the karst aquifer, aided by an incomplete data series, was useful. The underground cavity is now filled with more than 37 000 m³ of inert material (sand fractions from 4 to 32 mm) and leakage has been reduced by up to 65%.

REFERENCES

Bruce D.A. (2003) Sealing of massive water inflows through karst by grouting: principles and practice. In: Back B (ed), Sinkholes and the Engineering and Environmental Impacts of Karst. ASCE Geotech. Spec. publ. No.122.

Bauer S., Birk S., Liedl R., Sauter M. (2005) Simulation of karst aquifer genesis using a double permeable approach – Investigations for confined and unconfined settings, Processes of Speleogenesis: A modeling approach, Editor Gabrovsek F., ZRC SAZU, Postojna.

Bonacci O., Roje-Bonacci T. (2008) Water losses from the Ričice reservoir built in the Dinaric karst, *Engineering Geology* 99 (2008) Elsevier, pp 121–127.

Borghi A., Renard P., Jenni S. (2012) A pseudo-genetic stochastic model to generate karstic networks. *Journal of Hydrology* 414–415, 516–529.

Fazeli M. A. (2007) Construction of grout curtain in karstic environment case study: Salman Farsi dam. *Environmental Geology* 51.

Filipponi M. (2009) Spatial Analysis of Karst Conduit Networks and Determination of Parameters Controlling the Speleogenesis along Preferential Lithostratigraphic, Horizons, thèse no 4376, Suisse.

Ford D., Williams P. (2007) Karst hydrogeology and geomorphology. Wiley.

Institute for Developing of Water Resources (IWD) 'Jaroslav Černi', 2009., Design on rehabilitation regarding water seepage beneath the dam of the Višegrad hydropower plant, Summary report on performed investigations, Belgrade 2009.

Milanović P. (2000) *Geological Engineering in Karst.* Monograph, Zebra Publ. Ltd, Belgrade.

Milanović, P. (2004) Water resources engineering in karst, CRC Press, 143 p.

Milanović S. (2007) Results of recording of karstic features – Ourkiss dam project. Report. Hidrotehnika, Belgrade.

Milanović S. (2009) Report on special investigation on dam site Višegrad, Inst. for Develop. of Water Resources 'Jaroslav Černi', Belgrade.

Milanović S. (2010) Creation of physical model of karstic aquifer on example of Beljanica mt. (eastern Serbia), Doc. dissert, FMG, University of Belgrade, Beograd.

Milanović S. (2015) Choosing optimal dam sites and preventing leakage from reservoirs, In: Karst Aquifers - Characterisation and Engineering. (Ed. Z. Stevanović), Springer, Professional Practice in Earth Sciences, pp. 531–549.

Romanov D., Gabrovšek F., Dreybrot W. (2003) Dam sites in soluble rocks: a model of increasing leakage by dissolutional widening of fractures beneath a dam. *Engineering Geology* 70, 17–35.

Romanov D., Kaufmann G., Hiller T. (2010) Karstification of aquifers interspersed with non-soluble rocks: From basic principles towards case studies, *Engineering Geology* 116, 261–273, Elsevier.

Stevanović Z., Milanović S., Ristić V. (2010) Supportive methods for assessing effective porosity and regulating karst aquifers. *Acta Carsologica*, 39(2), 313–329.

Therond R. (1972) *Recherche sur l'etancheite des lacs de barrage en pays karstique.* Eyrolles, Paris.

Turkmen S. (2003) Treatment of the seepage problems at the Kalecik Dam (Turkey). *Engineering Geology* 68, 159–169.

Unal B., ErenM., Yalcin G. (2007) Investigation of leakage at Ataturk dam and hydroelectric power plant by means of hydrometric measurements, *Engineering Geology* 93, 45–63.

Zogović D. (1980) Some methodological aspects of hydrogeological analysis related to dam and reservoir construction in karst. Proceedings of 6th Yugoslav Symposium for Hydrogeology and Engineering Geology, Portoroz.

Chapter 23

Reactivation of karst springs after regional mine dewatering in the Tata area, Hungary

Attila Kovács & Teodóra Szőcs
Geological and Geophysical Institute of Hungary, Budapest, Hungary

ABSTRACT

Mine dewatering operations in the Transdanubian Mountains, Hungary between 1960 and 1990 caused significant groundwater depressurisation and drying up of several karst springs in the city of Tata. Following the termination of mining operations, the groundwater flow system started to recover. Many of the former springs reactivated and further springs are expected to reappear in the future, causing environmental issues. Spring locations in Tata are aligned with deep tectonic structures both in uncovered and confined karst areas. The analysis of well hydrographs indicates that there is no hydraulic connection between shallow and karst groundwater bodies. The prediction of karst water levels based on physical curve fitting suggests that equilibrium karst water level will be reached around the year 2018 at approximately 140 m asl. The chemical composition of karst waters in the Tata region confirms that they come from a dolomitic aquifer. While the karst waters of the region show a uniform composition, shallow groundwater shows a variable composition. The chemical composition of most reactivating springs indicates karst water origin. Some springs discharge shallow groundwater and show signs of local pollution.

23.1 INTRODUCTION

The area of interest is located on the northern edge of the Transdanubian Mountains, Hungary (Figure 23.1). The Tata Springs represent the natural outlet of the Transdanubian karst aquifer. The aquifer is situated in Triassic limestones and dolomites. The Transdanubian karst system was strongly affected by mine dewatering related to bauxite and coal mining from the beginning of the 1950s (Figure 23.2). The total groundwater abstraction rate reached 12 m³/s during the period between the late 1960s and the late 1980s. The intense karst water abstraction caused regional groundwater recession (VITUKI, 2000; Csepregi, 2007), and as a consequence, several springs in the Tata area disappeared during this period. Following the termination of mining operations in the early 1990s, mining related groundwater abstraction significantly decreased, and total abstraction rates had dropped to 3.5 m³/s by the late 2000s . As a consequence the flow system started to recover. Since the late 1990s the karst water table has risen by more than 40 m in the Tata area. As a result, some of the former springs reactivated and further springs are expected to reappear (Maller & Hajnal, 2013).

During the 1970s significant developments took place in areas that were previously used as agricultural land. Currently 30% of the population of Tata lives in

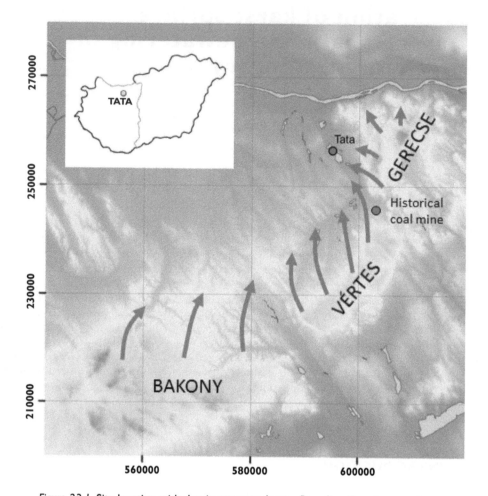

Figure 23.1 Site location with dominant groundwater flow directions in natural state.

this area. The reactivating springs cause significant environmental problems related to sewerage and water quality.

The aim of the study was to understand the hydraulic and hydrogeochemical behaviour of the recovering flow system, to delineate affected areas and to provide predictions on the location and timing of spring reactivation.

23.2 HYDROGEOLOGICAL CONDITIONS

The karst springs at Tata represent one of the main natural outlets of the Transdanubian carbonate aquifer system. The regional erosion base is the Által-ér creek. The recharge areas of the springs are located in the north western uncovered carbonate aquifers of the Gerecse and Vértes Mountains, and in the north eastern karst areas of the Bakony Mountains (Csepregi, 2002) (Figure 23.1).

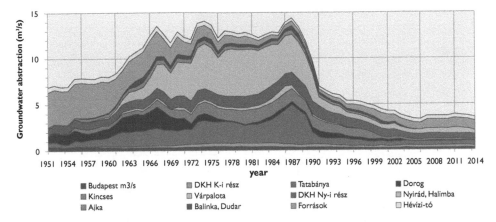

Figure 23.2 Total groundwater extracted by minng operations in the Transdanubian Mountains after (Csepregi, 2007),

The karst springs in the Tata area are located along a chessboard-like fault system which developed during tectonic movements between the Eocene and Pliocene periods (Figure 23.3). Before the beginning of mine dewatering operations, there were several active springs in the area, at topographic elevations between 118–141 m asl with yields between 0.00001 and 1 m³/s (Horusitzky, 1923). Spring locations are aligned with deep tectonic structures both in unconfined and confined karst areas. Although there is little evidence of karstification, concentrated groundwater flow takes place along tectonic structures representing preferential flowpaths. The confining marls in the city of Tata do not block groundwater discharge to the surface along tectonic features.

Potentiometric data in the monitoring wells indicate a very flat karst water surface. The hydraulic gradient is in the range of 0.001–0.0005. The natural regional flow direction in the carbonate basement was from southwest to northeast. As a consequence of mine dewatering, the natural flow directions were altered because of a large depression located south east of the Tata area.

The karst water level in the vicinity of Tata was around 136 m asl (in the observation wells Té-2: 135.6 m, Té-3: 135.5 m, Lo Presti: 136.0 m, Tükör: 136.2 m, Fényes-1: 138.3 m; locations shown on Figure 23.3.) based on data from January 2014. According to the low hydraulic gradients throughout the investigation area the karst water level can be approximated with a uniform value of 136 m asl at the beginning of 2014.

The original, undisturbed karstic water level data differs between various references, and vary between 140 and 143.5 m asl (Horusitzky, 1923; Fogarasi, 2001). A modelled water level was calculated at 138 m asl for 2020 by Csepregi (2002). The natural-state water level data involves uncertainties. It is assumed that the karstic level was about 140 m asl near Tata based on historical spring levels, and expect a similar water level by the end of water level recovery, knowing that this also depends on water extraction and on future climate conditions.

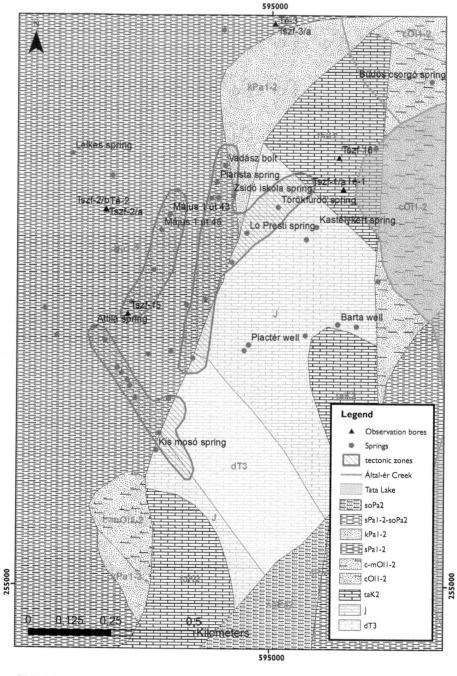

Figure 23.3 Geological settings and locations of springs and monitoring bores. Pa1-2 and Ol1-2 indicate confining Pannonian (Upper Miocene – Pliocene) and Oligocene sediments. T3J and K2 indicate the unconfined zones of the Upper Triassic, Jurassic and Upper Cretaceous carbonate aquifers.

23.3 SHALLOW GROUNDWATER

The Mesozoic carbonate aquifer is overlain by a series of younger sediments which contain shallow groundwater bodies. The main shallow aquifer is unconfined and is a few meters thick. The groundwater table follows the ground surface at an average of 1–3 m depth but is deeper in the hilly areas. The flow direction of the shallow ground-water is generally in the direction of the Által-ér creek.

The comparison between time series of monitoring wells screened in the shallow aquifer and those installed in the deep karst aquifer suggest that there is no hydraulic connection between shallow and karst aquifers. The effect of karst water recovery is not seen in hydrograph data for shallow bores.

A local confined shallow aquifer can be found in Upper Pannonian (Late Miocene) sediments covered by a marl aquitard. The potentiometric levels in the confined shallow groundwater show a damped effect of karst water level recovery, suggesting a leaky connection to the karst aquifer which is indicated by monitoring well Tszf-2/a (Figure 23.4).

23.4 PIEZOMETRIC MONITORING DATA

A groundwater level monitoring system has been operated by the regional Water Directorate for several decades, and this has been expanded with new auxiliary karstic and shallow groundwater level monitoring wells and contamination monitoring wells during the implementation of the safety regulations for drinking water protection areas in 2002 (Csepregi, 2002).

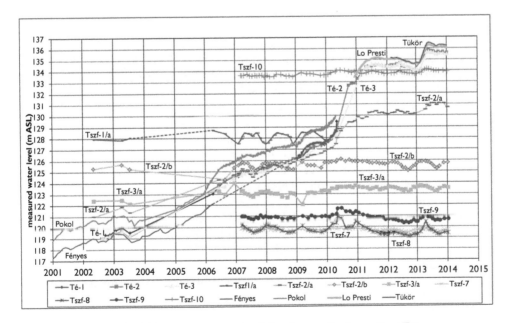

Figure 23.4 Time series water level data for monitoring wells at Tata.

The changes in groundwater levels in the monitoring wells between 2001 and 2014 are shown in Figure 23.4. Wells Té-1, Té-2 and Té-3 were sited to monitor the changes in groundwater level and groundwater chemistry in the main karst aquifer. Wells Tszf-1 to Tszf-18 were installed to characterise the changes in groundwater level and groundwater chemistry of shallow groundwater, and to characterise the hydraulic connection between shallow groundwater and karst water.

23.5   WATER LEVEL PREDICTION

One of the practical goals of the study was to predict when the flow system reaches natural equilibrium. The karst water level prediction was based on time series of several karst water monitoring wells. A logarithmic trend line was fitted on the time series of wells Té-1, Té-2 and Té-3. The application of a logarithmic function for curve fitting was based on the assumption that the recovery follows the Cooper and Jacob (1946) well function. The Cooper and Jacob solution is an approximation of the Theis (1935) non-equilibrium method.

Curve fitting was performed for the 2001–2009 period (Figure 23.5), since the extremely high precipitation in 2010 broke the trend of previous years, causing more than 4 m rise in karst water levels. The trend-line was shifted by 4 m in 2010 to represent the trend characteristic of the following years. The fitted equation and the karst water prognosis up to 2020 are shown in Figure 23.5. The prediction based on curve fitting suggests that the equilibrium level of the karst water table in the study area will be reached around 2018 at approximately 140 m asl. This prediction is based on the assumption that water abstraction rates and climatic conditions recorded between 2000 and 2009 remain constant during the following years. Any significant change in these parameters might influence the recovery process and thus the time of spring reactivation.

The horizontal dashed lines in Figure 23.5 show the topographic levels of the springs. The intersection of these topographic levels with the fitted trend line determines the predicted date of reactivation of the surveyed springs, and also provides information about expected karst water level changes until 2020.

While most springs have already reactivated during the past years, the Lo Presti, Kismosó, Pokol and Tükör springs are expected to reactivate until 2018. The water level in the wells located at higher topographic levels (Piarista well, Barta well, Piactéri well) are predicted to rise by a further 4–5 m by 2018.

After the end of karst water recovery any oscillations in water levels will be determined by climatic conditions. These water level variations are expected to range between 0 and 2 m.

Locations of natural springs and dug wells are indicated in Figure 23.6. The 136 m asl topographic level corresponds to the current karst water level, while the predicted equilibrium karst water level is around 140 m asl. The grey area in between the two isolines is subject to the possible reappearance of the springs in later years. The striped areas indicate tectonic zones assumed to represent preferential pathways for groundwater flow. The outcrop line of the Mesozoic carbonate aquifer is indicated with a grey dashed line.

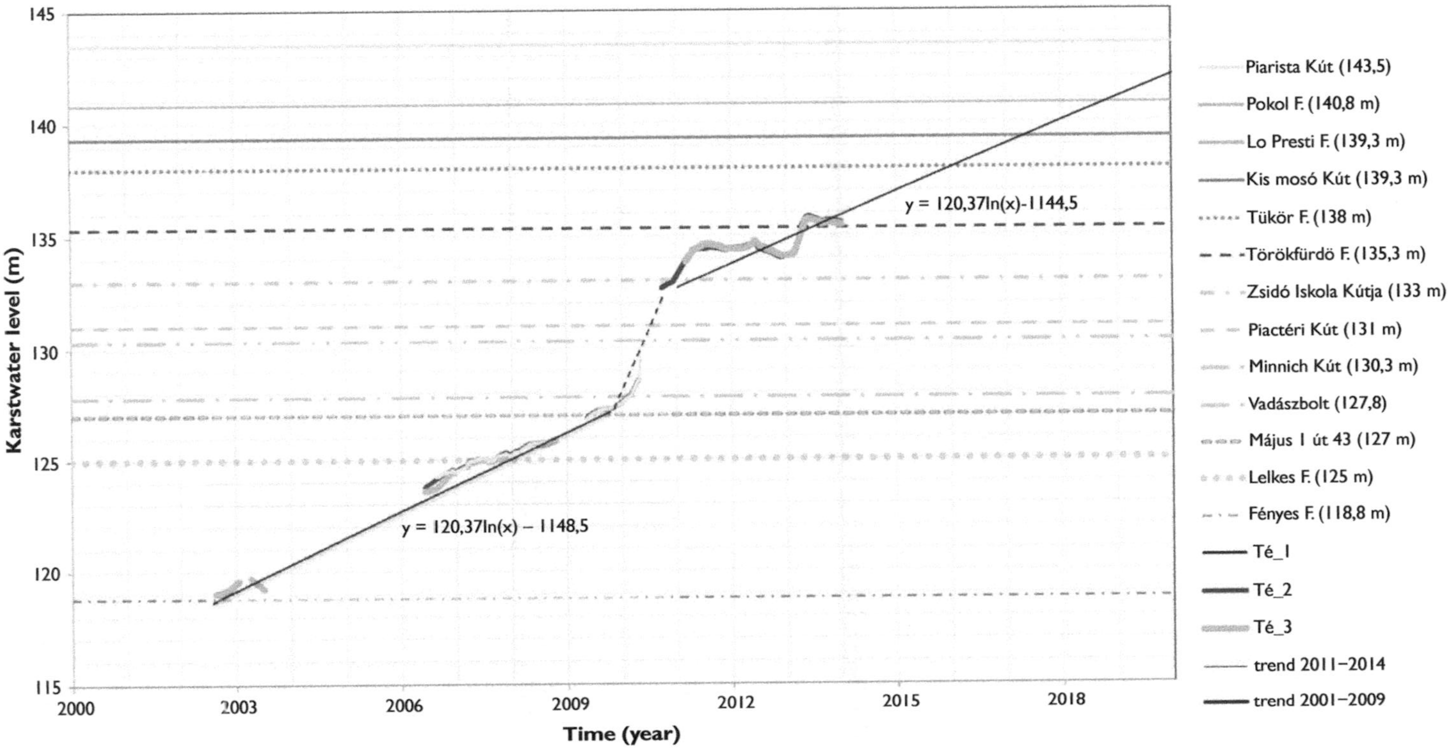

145
140
135
130
125
120
115
Karstwater level (m)
2000
2003
2006
2009
2012
2015
2018
Time (year)
y = 120,37ln(x)-1144,5
y = 120,37ln(x) – 1148,5
Piarista Kút (143,5)
Pokol F. (140,8 m)
Lo Presti F. (139,3 m)
Kis mosó Kút (139,3 m)
Tükör F. (138 m)
Törökfürdö F. (135,3 m)
Zsidó Iskola Kútja (133 m)
Piactéri Kút (131 m)
Minnich Kút (130,3 m)
Vadászbolt (127,8)
Május 1 út 43 (127 m)
Lelkes F. (125 m)
Fényes F. (118,8 m)
Té_1
Té_2
Té_3
trend 2011–2014
trend 2001–2009

Figure 23.5 Karst water level prognosis based on time series of the karst water monitoring wells in the Tata region and the topographic levels of main springs.

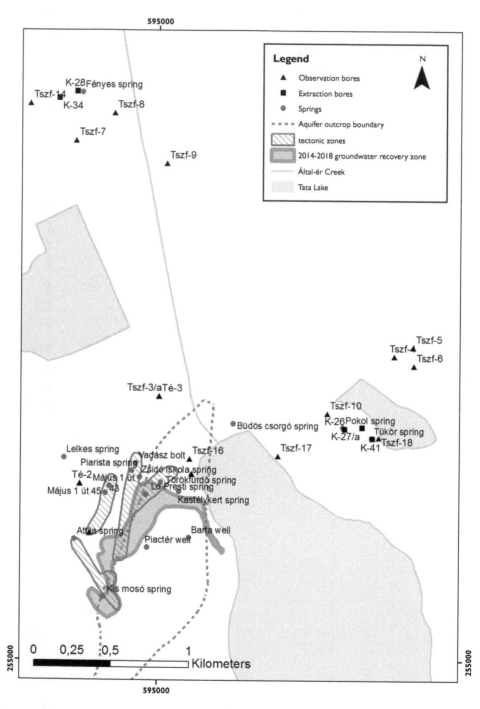

Figure 23.6 Locations of natural springs and dug wells together with topographic elevation contours. Shaded area indicates the future zone of groundwater heads rising above ground surface.

23.6   GENERAL HYDROGEOCHEMICAL CHARACTERISTICS

The evaluation of the hydrogeochemical conditions was based on historical data, on past surveys (Csepregi, 2002) and on data collected by the Tata Municipality and the Geological and Geophysical Institute of Hungary. New groundwater samples were collected during early spring and late autumn in 2014. The study gives an overview of the chemical composition and isotope data distribution of the springs and seepages, and shallow and deep wells. Although the analyses of the different data sources were carried out by different methods and in some cases with different detection limits, or they were limited to just a few parameters, they can be considered reliable for following the changes in the water chemistry due to karst water abstraction followed by karst water table rise after the mine closures.

The karst waters in the Tata region are CaMg-HCO$_3$ type, indicting they come from a dolomitic aquifer. In order to present the chemical characteristics of groundwater the data were plotted on Piper diagrams. Figure 23.7 shows the data for karst water wells (Tata 26, 27/A, 28, 34, 41 and Karst 'recent' representing Tata 28 and Tata 34 data for samples collected in 2014), springs (Fényes, Lo Presti, Pokol, Törökfürdő, Kastélykert, Büdös csorgó, Zsidó iskola), seepages (43 Május 1 street, 45 Május 1 street, Attila, Lelkes, Vadászbolt), a stream (Kismosó) and a dug well (Kismosó well) next to the stream. At those sites where three or more data sets were available, the median values were applied. In the case of two karst water wells (Tata 28, Tata 34), the time series were grouped in two parts (data before 1983 and after

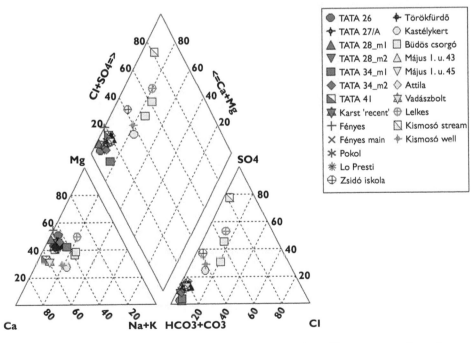

Figure 23.7 Chemical characteristics of groundwater based on data of karst water wells, springs, seepages, a stream and a dug well.

1983) based on the changes in the chemical composition due to the effects of mining activities. Medians were calculated for both data groups.

The chemical composition of the surveyed wells, the Fényes, Lo Presti, Törökfürdő, Pokol springs and the new Attila seepage and seepages at Május 1.u 43 and 45 show karst water composition with CaMg-HCO$_3$ or MgCa-HCO$_3$ water types.

The chemical composition of Kastélykert spring and of the dug well at Kismosó stream are similar, both having a CaMg-HCO$_3$SO$_4$ water type. This shows the effect of mixing with shallow groundwater and potentially also of local pollution. The high nitrate concentration in the Kastélykert spring also suggests pollution from an anthropogenic source.

Sampling sites Büdös csorgó and Lelkes seepages have a distinct chemical composition of the MgCa-SO$_4$HCO$_3$-MgCaNa-SO$_4$HCO$_3$ water type.

Kismosó stream as a local discharge area, does not show any connection with the karst water, is of the CaMg-SO$_4$ water type, and is badly polluted.

Figure 23.8 illustrates the data obtained from chemical analyses for groundwater from shallow bores. The chemistry of shallow groundwater (TSZF wells) is typically of CaMg-HCO$_3$SO$_4$ type, but some of them are of NaCaMg-HCO$_3$ or CaMg-SO$_4$ type. Their water composition covers a much larger range than the karst waters, reflecting the effect of the local near surface geology and hydrogeology, which is probably influenced by local tectonics. While the karst waters of the region show a uniform composition, the shallow groundwater shows a variable composition. Many of the shallow groundwater wells, not just the local dug wells, are polluted with nitrate. This pollution is local, and usually cannot be found in the adjacent wells or springs suggesting that nitrate might originate from septic tanks or a leaking sewage system.

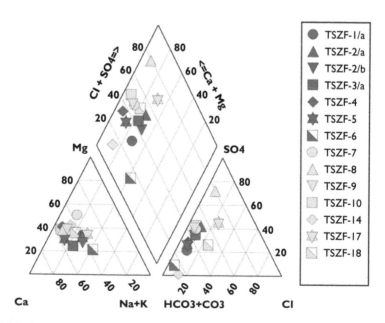

Figure 23.8 Chemical characteristics of groundwater based on data from shallow boreholes.

23.7   EFFECTS OF MINE DEWATERING ON GROUNDWATER CHEMISTRY

The changes in chemical composition through time are shown for some parameters in Figures 23.9–23.11. The hydrogen-bicarbonate content of karstic waters is typically 400 to 500 mg/l (Figure 23.9), while in the shallow groundwater (TSZF wells) varies in the range of 100 to 800 mg/l (not shown in figure). The effect of mining activities on the karst water composition is shown in Figure 23.9 where the hydrogen-bicarbonate data of some representative karst wells and springs are plotted against time. It can be seen that the hydrogen-bicarbonate content dropped at the beginning of 1983, with median values decreasing from 476 mg/l to 458 mg/l. This might indicate a hydrochemical response to aquifer dewatering which started in the early 1950s and intensified around 1972 to 1973 in the Tatabánya region.

The recent data from new karst water seepages (43 Május 1 street. 45, Május 1 street) also show these decreased concentrations, but samples collected from Lo Presti karst spring and the two drinking water supply wells (Tata 28, Tata 34) in the last quarter of 2014 and in 2015 show slightly increasing concentrations (median 3 = 468 mg/l). These minor concentration changes are within measurement error limits, but the general pattern, including concentration changes in other anions, suggests a recent change in karst water composition starting probably from the mid-2000s, and increased since 2010.

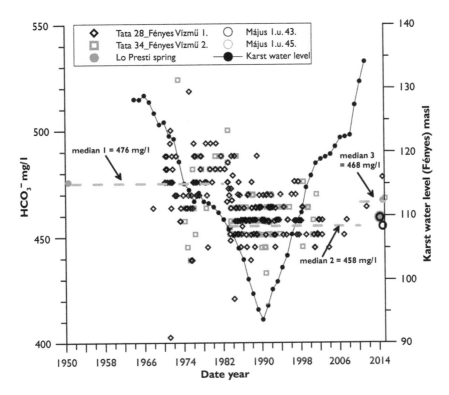

Figure 23.9 Changes with time in the hydrogen-bicarbonate concentration based on some representative karst water wells, spring and seepages.

The decrease in the hydrogen-bicarbonate content was accompanied by an increase in sulphate content from 1977 (Figure 23.10), which supports changes in groundwater chemistry in response to mine dewatering. The initial 10 mg/l sulphate median values rose to 59–67 mg/l by 1986–1987. The data for the last few years show a stable or even slightly lower sulphate value (median 3 = 54 mg/l) which might be a sign of the start of hydrochemical regeneration of the karst water flow system.

No data were available on anion concentrations before 1967. From the beginning of the available data series chloride concentrations remained stable until about the mid-2000s. Recent data show a clear decrease in chloride concentrations. This may indicate a hydrochemical response to the recovery of the karst water flow system.

No cation data were available before the 1980s, so any changes in cation concentrations could not be analysed. The calcium data of the last 30 years are variable, but a slight increase in calcium concentration can be identified.

Groundwater lowering in Tatabánya started in the early 1950s with increasing abstraction rates throughout the following years. The maximum amount of abstracted karst water reached its peak in 1988 with 2 m³/s, after an almost 25 years long period with a similarly high abstraction level. In the following years, mostly between 1989 and 1992, with the progressive mine closures, the karst water abstraction rates

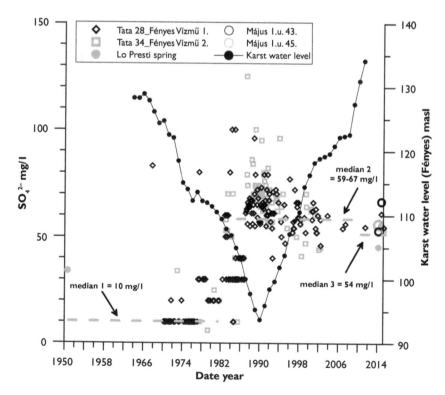

Figure 23.10 Changes with time in the sulphate concentration based on some representative karst water wells, spring and seepages.

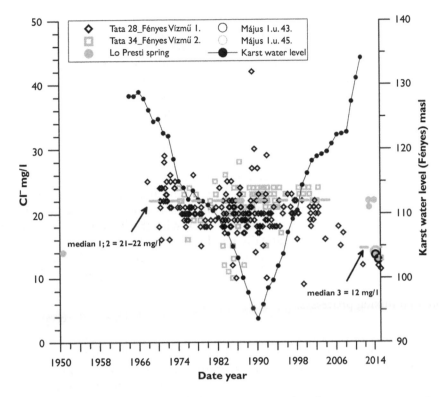

Figure 23.11 Changes with time in the chloride concentration based on some representative karst water wells, spring and seepages.

dropped to about 0.5 m³/s (Figure 23.2). Figures 23.9–23.11 show the deepest point in karst water level that was reached in 1990.

The remarkable changes in the concentration of hydrogen-bicarbonate around 1983–84 and of sulphate between 1977 and 1986–1987 are assumed to indicate the hydrochemical response of the groundwater system due to large-scale aquifer dewatering. As no sufficient data on hydrochemistry was available at the early stages of the mining activity, nor on extraction rates, the lag between dewatering and chemical response cannot be estimated confidently.

Recent hydrochemical data indicate that the concentration of the main water components started to rebound towards their original values around the mid-2000s, emphasised from 2010, presumably indicating the hydrochemical recovery of the groundwater system. The comparison between abstraction rates and chemical data suggests a delay of 10–20 years between groundwater chemistry and groundwater flow conditions.

Although further studies are required to support this conclusion, it can be assumed that the hydrochemical changes were caused by the reversal of hydraulic gradients and the subsequent changes in regional flow directions between the natural north easterly flow and a modified south–westerly groundwater flow.

23.8 MIXING PROCESSES

One of the major questions is whether the regenerating karst springs or new seepages contain pure karst water, or are a mixture of karst and shallow groundwater, or are seepages of shallow groundwater. Since very few samples could be analysed for isotopes, a conservative anion in the groundwater flow system, namely chloride, was used to study mixing processes. Most of the dug wells in the city of Tata contain very high nitrate concentrations, and sometimes ammonium, reflecting a strong anthropogenic influence. Therefore, data from shallow boreholes were used (Figure 23.8). 16 shallow boreholes were drilled in the drinking water protection area safety survey (Csepregi, 2002). Some of them provided information on the connection between shallow groundwater and karst water, others enabled an evaluation of the effects of former pollution or potential pollution. The chemical composition of the water in these shallow wells varies greatly, but for different reasons. The deeper wells show connections either with the deeper groundwater, with the karst water, or one of the wells in contact with the systematically regulated water level of Lake Tata. The variable shallow groundwater composition is probably controlled by the fractures in the area. Well TSZF-2/b shows a typical, shallow groundwater composition. This median data for this well was used as an end member for shallow groundwater in the study of potential mixing processes.

The concentration ranges with whiskers representing the 10 and 90% percentile values of the main anions and their median values are shown in Figures 23.12–23.15. In addition to the previously shown hydrogen-bicarbonate, sulphate and chloride concentration time series of some representative karst water wells, springs and seepages (Figures 23.9–23.11), these figures illustrate not just the typical concentration ranges based on karst water wells and 'recent' (Tata 28 and Tata 34 data of samples collected in 2014 and 2015) karst water data (left side), but also the data of characteristic shallow groundwater (TSZF-2/b), karst springs (Fényes, Fényes main, Pokol, Lo Presti, Zsidó iskola, Törökfürdő) and new seepages (Május 1.u. 43, Május 1.u. 45, Attila, Vadászbolt). It can be seen that the karst springs have similar concentration values as the karst water wells. While the CaMg-HCO_3-water type of the new seepages show a typical karst water composition, the higher Cl^-, NO_3^- concentrations (first samples) at 45 Május 1 street also show an anthropogenic influence and potential mixture with shallow groundwater. At the Attila and Vadászbolt seepages a very small percentage of shallow groundwater mixing can also be seen, based on the slightly higher hydrogen-bicarbonate and nitrate concentrations. The appearance of nitrate in the karst springs and seepages show some local, minor anthropogenic contamination of the shallow aquifer which through mixing with karst water results in elevated concentrations of these parameters in certain springs. The mixing between karst water and shallow groundwater primarily takes place along deep fractures where the Mesozoic carbonate reservoir discharges to the surface.

23.8.1 Isotope geochemical characteristics

Some of the samples were analyses for δD-$\delta^{18}O$, $\delta^{13}C$, ^3H and ^{14}C in order to get a better understanding of the karst water system, and the origin, as well as the relative or absolute ages of water from the springs and seepages.

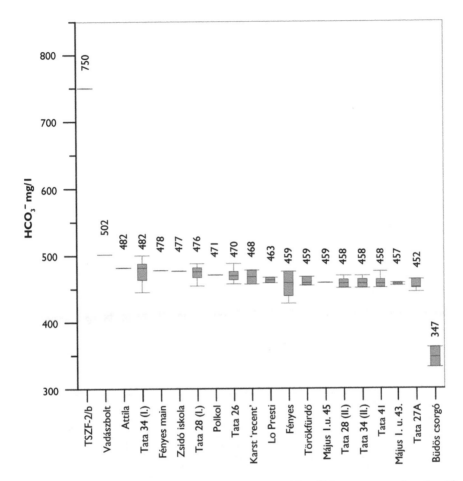

Figure 23.12 Hydrogen-bicarbonate concentration ranges with whiskers representing the 10 and 90% percentile values and their median values (left: karst water wells; right: shallow groundwater (TSZF-2/b), karst springs and new seepages).

The tritium content of the sampled karst waters is below the detection limit (0.059 Bq/l), which suggests that there is no precipitation component younger than 50 years in the karst waters (Figure 23.16). This also means that shallow groundwater does not mix with karst waters at the surveyed sites.

Based on previous studies (Deák, 2006) groundwater in Hungarian aquifers with $\delta^{18}O$ values higher than $-10‰$ are generally considered to be of Holocene origin. However, where mixing has occurred, especially with deep formation waters, higher values are also recorded. The $\delta^{18}O$ data of the karst wells cluster very close to each other with values of $-10.9‰$, while the δD data range between -75.1 and $-76.2‰$ respectively (Figure 23.17). The data ($-6.6‰$ $\delta^{18}O$ and $-52.2‰$ δD) of the Büdös csorgó spring which is shifted from the Carpathian Basin Meteoric Water Line (CBMWL) $\delta D = 7.8 * \delta^{18}O + 6$ defined by Deák (1995) shows the effect of evaporation (Clark &

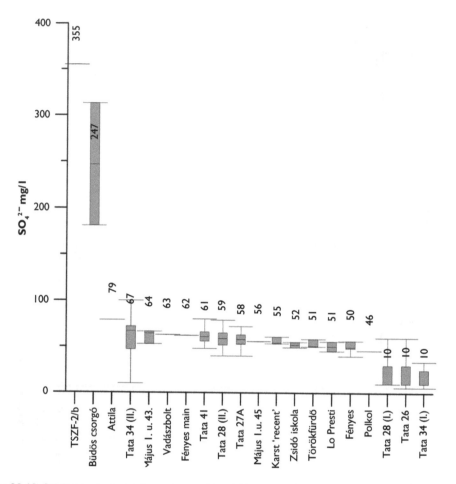

Figure 23.13 Sulphate concentration ranges with whiskers representing the 10 and 90% percentile values and their median values (left: karst water wells; right: shallow groundwater (TSZF-2/b), karst springs and new seepages).

Fritz, 1997). The CBMWL is very close to the Global Meteoric Water Line defined by Craig (1961).

Age calculations based on Carbon-14 measurements were done by the application of a Carbon-13 correction (Pearson, 1965). Carbon-14 data vary between 8.5–9.7% pmC, while $\delta^{13}C$ data vary between –8 and –9‰. The radiocarbon ages calculated by the widely used Carbon-13 correction method vary between 11 800 years and 13 100 years.

The isotope compositions of the sampled karst waters clearly show Pleistocene recharge except for well K-28. It can, therefore, be assumed that the karst waters in the Tata region are older than 10 000 years. Importantly, despite a rise in the karst water level, no sign of recent (young) infiltration could be detected in the karst water. Büdös csorgó spring and a dug well at the Kismosó stream have a significant or wholly recent infiltration origin. Based on tritium data, mixing with karst water cannot be

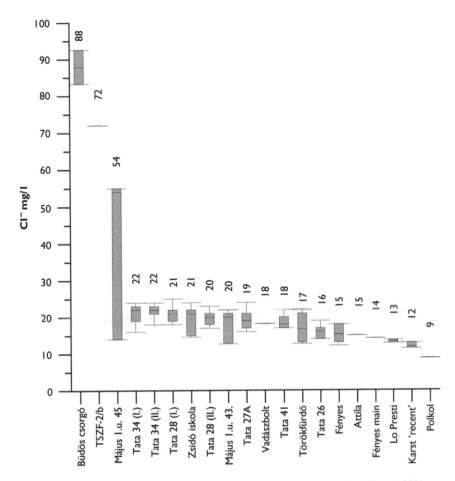

Figure 23.14 Chloride concentration ranges with whiskers representing the 10 and 90% percentile values and their median values (left: karst water wells; right: shallow groundwater (TSZF-2/b), karst springs and new seepages).

completely excluded, but neither the main nor the trace element data support the possibility of mixing. The high chloride concentration (Figure 23.14) shows mixing with shallow groundwater. More information on mixing could be gained by using δD–$\delta^{18}O$ data. Figure 23.18 shows that with an increase of the recent infiltration component, shown by an increase in the tritium content, there is a significant increase in the chloride concentration which suggests mixing with shallow groundwater at these sites.

23.9   CONCLUSIONS

A hydrogeological and hydrochemical study was implemented to investigate hydrogeological settings and the effects of mine dewatering on the historical Tata springs.

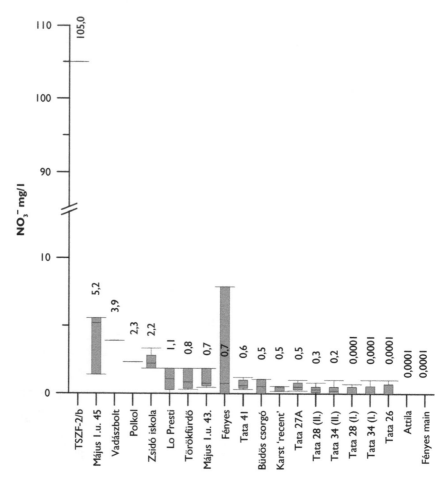

Figure 23.15 Nitrate concentration ranges with whiskers representing the 10 and 90% percentile values and their median values (left: karst water wells; right: shallow groundwater (TSZF-2/b), karst springs and new seepages).

Spring locations are aligned with deep tectonic structures both in uncovered and confined karst areas. This indicates that concentrated groundwater flow takes place along tectonic structures and the confining marls do not prevent groundwater discharge to the surface.

The comparison between time series of monitoring wells screened in the shallow aquifer and those installed in the deep karst aquifer suggest that there is no hydraulic connection between shallow and karst aquifers.

The prediction of karst water levels based on physical curve fitting suggests that equilibrium level of the karst water table will be reached around 2018 at approximately 140 m asl. While most springs have already reactivated in recent years, the Lo Presti, Kismosó, Pokol and Tükör springs are expected to reactivate until 2018. The water level is predicted to rise by a further 4 m.

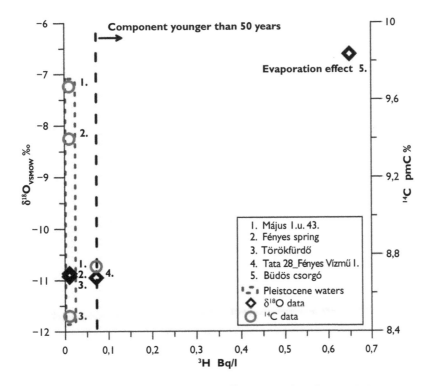

Figure 23.16 Tritium, oxygen-18 and ^{14}C isotope data characteristics.

The karst waters in the Tata region are CaMg-HCO$_3$ type, indicating that they come from a dolomitic aquifer. While the karst waters of the region show a uniform composition, the shallow groundwater shows a variable composition. The chemical composition of the surveyed wells, the Fényes, Lo Presti, Törökfürdő, Pokol springs, and the new Attila seepage and seepages at 43 and 45 Május 1 street show karstic water composition. The chemical composition of Kastélykert spring and the dug well at Kismosó stream are similar, and show the effect of mixing with shallow ground-water and potentially also local pollution. Kismosó stream does not show any connection with the karst water and is badly polluted.

Geochemical data indicate significant changes in karst water chemistry in response to groundwater depressurisation and later recovery. While bicarbonate concentrations decreased, sulphate concentrations increased during the mine dewatering operations. Recent hydrochemical data indicates that the concentration of the main water components started to recover towards their original values around the mid-2000s, increasingly since 2010. This presumably indicates the geochemical recovery of the groundwater system. The available data suggest a delay of approximately 10–20 years between the changes in dewatering rates and the subsequent hydrochemical reactions. The hydrochemical changes were caused by the reversal of hydraulic gradients and regional flow directions between the natural north easterly flow and a modified south westerly groundwater flow direction.

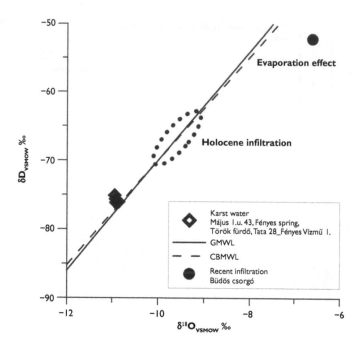

Figure 23.17 δ¹⁸O versus δD.

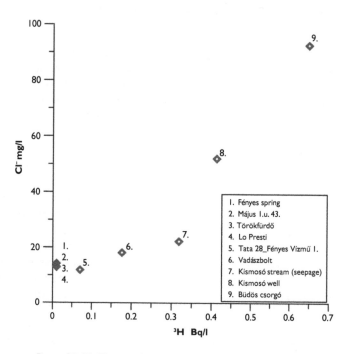

Figure 23.18 Tritium data versus chloride concentration.

Chloride as a conservative anion was used to study the mixing processes. Most springs have similar concentration values as karst water wells. While new seepages show a typical karst water composition, the higher Cl^-, NO_3^- concentrations at 45 Május 1 street also show an anthropogenic influence and potential mixing with shallow groundwater. At the Attila and Vadászbolt seepages a very small percentage of shallow groundwater mixing can also be seen.

The tritium content of karst waters is below detection limit, which means there is no precipitation component younger than 50 years in the karst waters. The isotope compositions of the sampled karst waters clearly show Pleistocene recharge except for well K-28. Based on this it can be assumed that the karst waters in the Tata region are older than 10 000 years. Importantly, despite recent rising water levels no signs of recent infiltration could be detected in the karst water discharges. Büdös csorgó spring and a dug well at the Kismosó stream have a significant or wholly recent infiltration origin.

ACKNOWLEDGMENTS

We would like to thank the City Council of Tata for providing support and making their data available for this research. We also acknowledge the North Transdanubian Water Directorate, the Északdunántúli Vízmű Zrt., Márton Maller, Géza Hajnal and András Csepregi who provided data for this study, and Fényes Spa for providing local knowledge. We would like to thank Zsolt Galgóczy and András Rehák for their field help.

REFERENCES

Clark I., Fritz P. (1997) *Environmental isotopes in Hydrogeology.* Lewis Publishers, USA.

Cooper H.H., Jacob C.E. (1946) A generalized graphical method for evaluating formation constants and summarizing well field history, *American Geophysical Union Transactions* 27, 526–534.

Craig H. (1961) Isotopic Variations in Meteoric Waters. Science, 133, pp. 1702–1703.

Csepregi A. (2002) A tatai vízbázis biztonságba helyezési terve. Report of Csepregi Ltd.

Csepregi A. (2007) A karsztvíztermelés hatása a Dunántúli-középhegység vízháztartására. In: Bányászati karsztvízszint süllyesztés a Dunántúli-középhegységben. (Ed.: Alföldi, L, Kapolyi L.) pp. 77–106.

Deák J. (1995): A felszín alatti vizek utánpótlódásának meghatározása izotópos módszerekkel az Alföldön. VITUKI report. Budapest.

Deák J. (2006): A Duna-Tisza köze rétegvíz áramlási rendszerének izotóp-hidrológiai vizsgálata. PhD thesis. Budapest.

Fogarasi S. (2001) Visszatérnek-e a tatai források? Földrajzi Konferencia, Szeged. Report. 15.p.

Horusitzky, H. (1923) Tata és Tóváros hévforrásainak hidrogeológiája és közgazdaságijövője – A Magyar Királyi Földtani Intézet Évkönyve XXV. kötet 3. füzet, Budapest. pp. 37–83.

Maller M., Hajnal G. (2013) A tatai források hidrogeológiai vizsgálata. Mérnökgeológia-Kőzetmechanika (Ed.: Török Á., Görög P. & Vásárhelyi B.), pp. 7–18.

Pearson F. J., Jr. (1965) Use of 13C/12C ratios to correct radiocarbon ages of materials initially diluted by limestones, In: Chatters, R M and Olson, E A, (Eds.), Intern. conf. on 14C and tritium dating, 6th, Proc: Clearinghouse Fed Sci Tech Inf, NBS, Washington, DC, p 357–366.

Theis C.V. (1935) The relation between the lowering of the piezometric surface and the rate and duration of discharge of a well using groundwater storage, *American Geophysical Union Transactions* 16, 519–524.

VITUKI Rt. (2000) Karsztvízvédelem a Közép-Dunántúli Régióban, Budapest. Report. (T.: 721/1/4846-1)

Subject index

Geographic names (Regional)

Series IAH-selected papers

14. Advances in Subsurface Pollution of Porous Media: Indicators,
    Processes and Modelling
    Edited by: Lucila Candela, Iñaki Vadillo and Francisco Javier Elorza
    2008, ISBN Hb: 978-0-415-47690-4
    Vogwill SERIES.tex 14/1/2016 13: 51 Page 2

15. Groundwater Governance in the Indo-Gangetic and Yellow River Basins –
    Realities and Challenges
    Edited by: Aditi Mukherji, Karen G. Villholth, Bharat R. Sharma
    and Jinxia Wang
    2009, ISBN Hb: 978-0-415-46580-9

16. Groundwater Response to Changing Climate
    Edited by: Makoto Taniguchi and Ian P. Holman
    2010, ISBN Hb: 978-0-415-54493-1

17. Groundwater Quality Sustainability
    Edited by: Piotr Maloszewski, Stanisław Witczak and Grzegorz Malina
    2013, ISBN Hb: 978-0-415-69841-2

18. Groundwater and Ecosystems
    Edited by: Luís Ribeiro, Tibor Y. Stigter, António Chambel,
    M. Teresa Condesso de Melo, José Paulo Monteiro and Albino Medeiros
    2013, ISBN Hb: 978-1-138-00033-9

19. Assesing and Managing Groundwater in Different Evironments
    Edited by: Jude Cobbing, Shafick Adams, Ingrid Dennis & Kornelius Rieman
    2013, ISBN Hb: 978-1-138-00100-8

20. Fractured Rock Hydrogeology
    Edited by: John M. Sharp
    2014, ISBN Hb: 978-1-138-00159-6

21. Calcium and Magnesium in Groundwater: Occurrence and
    Significance for Human Health
    Edited by: Lidia Razowska-Jaworek
    2014, ISBN Hb: 978-1-138-00032-2

22. Solving the Groundwater Challenges of the 21st Century
    Edited by: Ryan Vogwill
    2016, ISBN Hb: 978-1-138-02747-3